Modern Information Optics with MATLAB

Modern Information Optics with MATLAB is an easy-to-understand course book and is based on the authentic lectures and detailed research, conducted by the authors themselves, on information optics, holography, and MATLAB. This book is the first to highlight the incoherent optical system, provide up-to-date, novel digital holography techniques, and demonstrate MATLAB codes to accomplish tasks such as optical image processing and pattern recognition. This title is a comprehensive introduction to the basics of Fourier optics as well as optical image processing and digital holography. This is a step-by-step guide that details the vast majority of the derivations, without omitting essential steps, to facilitate a clear mathematical understanding. This book also features exercises at the end of each chapter, providing hands-on experience and consolidating the understanding. This book is an ideal companion for graduates and researchers involved in engineering and applied physics, as well as those interested in the growing field of information optics.

Yaping Zhang is Professor and Director of Yunnan Provincial Key Laboratory of Modern Information Optics, Kunming University of Science and Technology. Professor Zhang is an academic leader in optics in Yunnan Province and a member of the Steering Committee on Opto-Electronic Information Science and Engineering, Ministry of Education, China.

Ting-Chung Poon is Professor of Electrical and Computer Engineering at Virginia Tech, USA. Professor Poon is Fellow of the Institute of Electrical and Electronics Engineers (IEEE), the Institute of Physics (IOP), Optica, and the International Society for Optics and Photonics (SPIE). He also received the 2016 SPIE Dennis Gabor Award.

Modern Information Optics with MATLAB

YAPING ZHANG
Kunming University of Science and Technology

TING-CHUNG POON
Virginia Polytechnic Institute and State University

CAMBRIDGE
UNIVERSITY PRESS

CAMBRIDGE
UNIVERSITY PRESS

University Printing House, Cambridge CB2 8BS, United Kingdom

One Liberty Plaza, 20th Floor, New York, NY 10006, USA

477 Williamstown Road, Port Melbourne, VIC 3207, Australia

314–321, 3rd Floor, Plot 3, Splendor Forum, Jasola District Centre, New Delhi – 110025, India

103 Penang Road, #05–06/07, Visioncrest Commercial, Singapore 238467

Cambridge University Press is part of the University of Cambridge.

It furthers the University's mission by disseminating knowledge in the pursuit of education, learning, and research at the highest international levels of excellence.

www.cambridge.org
Information on this title: www.cambridge.org/9781316511596
DOI: 10.1017/9781009053204

@ Higher Education Press Limited Company 2023

First published 2023

A catalogue record for this publication is available from the British Library.

Library of Congress Cataloging-in-Publication Data
Names: Zhang, Yaping, 1978– author. | Poon, Ting-Chung, author.
Title: Modern information optics with MATLAB / Yaping Zhang, Kunming
University of Science and Technology, Ting-Chung Poon, Virginia
Polytechnic Institute and State University.
Description: Cambridge, United Kingdom ; New York, NY, USA : Cambridge
University Press, 2023. | Includes bibliographical references and index.
Identifiers: LCCN 2022024946 | ISBN 9781316511596 (hardback) |
ISBN 9781009053204 (ebook)
Subjects: LCSH: Information display systems. | Optical communications. |
Optics. | MATLAB.
Classification: LCC TK7882.I6 Z4364 2023 |
DDC 621.3815/422–dc23/eng/20220805
LC record available at https://lccn.loc.gov/2022024946

ISBN 978-1-316-51159-6 Hardback

To my parents, my husband, and my daughter

Yaping Zhang

To my grandchildren, Gussie, Sofia, Camden, and Aiden

Ting-Chung Poon

Contents

Preface

This book covers the basic principles used in information optics including some of its modern topics such as incoherent image processing, incoherent digital holography, modern approaches to computer-generated holography, and devices for optical information processing in information optics. These modern topics continue to find a niche in information optics.

This book will be useful for engineering or applied physics students, scientists, and engineers working in the field of information optics. The writing style of the book is geared toward juniors, seniors, and first-year graduate-level students in engineering and applied physics. We include details on most of the derivations without omitting essential steps to facilitate a clear mathematical development as we hope to build a strong mathematical foundation for undergraduate students. We also include exercises, challenging enough for graduate students, at the end of each chapter.

In the first three chapters of the book, we provide a background on basic optics including ray optics, wave optics, and important mathematical preliminaries for information optics. The book then extensively covers topics of incoherent image processing systems (Chapter 4), digital holography (Chapter 5), including important modern development on incoherent digital holography, and computer-generated holography (Chapter 6). In addition, the book covers in-depth principles of optical devices such as acousto-optic and electro-optic modulators for optical information processing (Chapter 7).

The material covered is enough for a one-semester course (Chapters 1–5) with course titles such as Fourier optics, holography, and modern information optics or a two-course sequence with the second course covering topics from Chapters 6 to 7 (with a brief review of Chapters 3 through 5). Example of a course title would be optical information processing. An important and special feature of this book is to provide the reader with experience in modeling the theory and applications using a commonly used software tool MATLAB®. The use of MATLAB allows the reader to visualize some of the important optical effects such as diffraction, optical image processing, and holographic reconstructions.

Our vision of the book is that there is an English and Chinese version of this book that are printed together as a single textbook. It is the first of its kind in textbooks and a pioneering project. Information optics is a growing field, and there is an enormous need for pioneering books of this kind. A textbook like this will allow students and scholars to appreciate the much-needed Chinese translation of English technical terms, and vice versa. The textbook also provides them with professional and technical translation in the area of information optics.

We would like to thank Jung-Ping Liu for his help on writing some of the MATLAB codes. Also thanks are extended to Yongwei Yao, Jingyuan Zhang, and Houxin Fan for drafting some initial figures used in the book and, last but not least, Christina Poon for reading parts of the manuscript and providing suggestions for improvements.

Yaping Zhang would like to thank her parents, her husband, and her daughter (Xinyi Xu) whose encouragement and support have enabled her to fulfill her dreams. In particular, she wishes to express her appreciation to her collaborator, Professor Poon, for his professional knowledge and language polishing that made the book more readable for users, and resulted in the publication of *Modern Information Optics*. Working with Professor Poon on this project was a great pleasure and resulted in further growth in her professional experience.

Ting-Chung Poon is greatly indebted to his parents, whose moral encouragement and sacrifice have enabled him to fulfill his dreams and further his achievements. They shall be remembered forever.

1 Gaussian Optics and Uncertainty Principle

This chapter contains *Gaussian optics* and employs a matrix formalism to describe optical image formation through light rays. In optics, a ray is an idealized model of light. However, in a subsequent chapter (Chapter 3, Section 3.5), we will also see that a matrix formalism can also be used to describe, for example, a Gaussian laser beam under diffraction through the wave optics approach. The advantage of the matrix formalism is that any ray can be tracked during its propagation through the optical system by successive matrix multiplications, which can be easily programmed on a computer. This is a powerful technique and is widely used in the design of optical elements. In the chapter, some of the important concepts in resolution, depth of focus, and depth of field are also considered based on the ray approach.

1.1 Gaussian Optics

Gaussian optics, named after Carl Friedrich Gauss, is a technique in geometrical optics that describes the behavior of light rays in optical systems under the paraxial approximation. We take the *optical axis* to be along the z-axis, which is the general direction in which the rays travel, and our discussion is confined to those rays that lie in the x–z plane and that are close to the optical axis. In other words, only rays whose angular deviation from the optical axis is small are considered. These rays are called *paraxial rays*. Hence, the sine and tangent of the angles may be approximated by the angles themselves, that is, $\sin\theta \approx \tan\theta \approx \theta$. Indeed, the mathematical treatment is simplified greatly because of the linearization process involved. For example, a linearized form of *Snell's law of refraction*, $n_1\sin\phi_i = n_2\sin\phi_t$, is $n_1\phi_i = n_2\phi_t$. Figure 1.1-1 shows ray refraction for Snell's law. ϕ_i and ϕ_t are the angles of incidence and refraction, respectively, which are measured from the normal, ON, to the interface POQ between Media 1 and 2. Media 1 and 2 are characterized by the constant refractive indices, n_1 and n_2, respectively. In the figure, we also illustrate the *law of reflection*, that is, $\phi_i = \phi_r$, where ϕ_r is the *angle of reflection*. Note that the incident ray, the refracted ray, and the reflected ray all lie in the same *plane of incidence*.

Consider the propagation of a paraxial ray through an optical system as shown in Figure 1.1-2. A ray at a given z-plane may be specified by its height x from the optical axis and by its launching angle θ. The convention for the angle is that θ is measured in radians and is anticlockwise positive from the z-axis. The quantities (x, θ) represent the

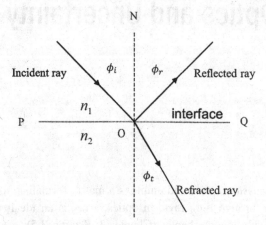

Figure 1.1-1 Geometry for Snell's law and law of reflection

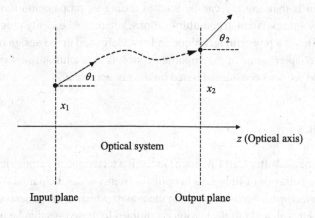

Figure 1.1-2 Ray propagating in an optical system

coordinates of the ray for a given z-plane. However, instead of specifying the angle the ray makes with the z-axis, another convention is used. We replace the angle θ by the corresponding $v = n\theta$, where n is the refractive index of the medium in which the ray is traveling. As we will see later, the use of this convention ensures that all the matrices involved are positive unimodular. A *unimodular matrix* is a real square matrix with determinant $+1$ or -1, and a positive unimodular matrix has determinant $+1$.

To clarify the discussion, we let a ray pass through the input plane with the *input ray coordinates* $(x_1, v_1 = n_1\theta_1)$. After the ray passes through the optical system, we denote the *output ray coordinates* $(x_2, v_2 = n_2\theta_2)$ on the output plane. In the paraxial approximation, the corresponding output quantities are linearly dependent on the input quantities. In other words, the output quantities can be expressed in terms of the weighted sum of the input quantities (known as the *principle of superposition*) as follows:

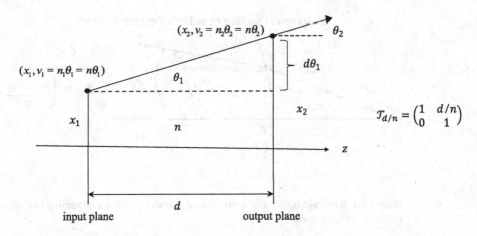

Figure 1.1-3 Ray propagating in a homogeneous medium with the input and output coordinates $(x_1, v_1 = n_1\theta_1)$ and $(x_2, v_2 = n_2\theta_2)$, respectively

$$x_2 = Ax_1 + Bv_1 \text{ and } v_2 = Cx_1 + Dv_1,$$

where $A, B, C,$ and D are the weight factors. We can cast the above equations into a matrix form as

$$\begin{pmatrix} x_2 \\ v_2 \end{pmatrix} = \begin{pmatrix} A & B \\ C & D \end{pmatrix} \begin{pmatrix} x_1 \\ v_1 \end{pmatrix}. \tag{1.1-1}$$

The *ABCD* matrix in Eq. (1.1-1) is called the *ray transfer matrix*, or the *system matrix* S, if it is represented by the multiplication of ray transfer matrices. In what follows, we shall derive several important ray transfer matrices.

1.1.1   Ray Transfer Matrices

Translation Matrix

A ray travels in a homogenous medium of refractive index n in a straight line (see Figure 1.1-3). Let us denote the input and output planes with the ray's coordinates, and then we try to relate the input and output coordinates with a matrix after the traveling of the distance d. Since $n_1 = n_2 = n$ and $\theta_1 = \theta_2, v_2 = n_2\theta_2 = n_1\theta_1 = v_1$. From the geometry, we also find $x_2 = x_1 + d\tan\theta_1 \approx x_1 + d\theta_1 = x_1 + dv_1/n$. Therefore, we can relate the output coordinates of the ray with its input coordinates as follows:

$$\begin{pmatrix} x_2 \\ v_2 \end{pmatrix} = \begin{pmatrix} 1 & d/n \\ 0 & 1 \end{pmatrix} \begin{pmatrix} x_1 \\ v_1 \end{pmatrix} = T_{d/n} \begin{pmatrix} x_1 \\ v_1 \end{pmatrix}, \tag{1.1-2}$$

where

$$T_{d/n} = \begin{pmatrix} 1 & d/n \\ 0 & 1 \end{pmatrix}, \tag{1.1-3a}$$

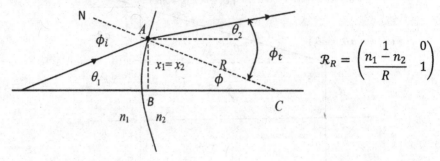

spherical surface of radius of curvature R

$$\mathcal{R}_R = \begin{pmatrix} 1 & 0 \\ \dfrac{n_1 - n_2}{R} & 1 \end{pmatrix}$$

Figure 1.1-4 Ray trajectory during refraction at a spherical surface separating two regions of refractive indices n_1 and n_2

which is called the *translation matrix*. The matrix describes the translation of a ray for a distance d along the optical axis in a homogenous medium of n. The determinant of $\mathcal{T}_{d/n}$ is

$$|\mathcal{T}_{d/n}| = \begin{vmatrix} 1 & d/n \\ 0 & 1 \end{vmatrix} = 1,$$

and hence $\mathcal{T}_{d/n}$ is a positive unimodular matrix. For a translation of the ray in air, we have $n = 1$, and the translation can be represented simply by

$$\mathcal{T}_d = \begin{pmatrix} 1 & d \\ 0 & 1 \end{pmatrix}. \tag{1.1-3b}$$

Refraction Matrix

We consider a spherical surface separating two regions of refractive indices n_1 and n_2 as shown in Figure 1.1-4. The center of the spherical surface is at C, and its radius of curvature is R. The convention for the *radius of curvature* is as follows. The radius of curvature of the surface is taken to be positive (negative) if the center C of curvature lies to the right (left) of the surface. The ray strikes the surface at the point A and gets refracted into media n_2. Note that the input and output planes are the same. Hence, the height of the ray at A, before and after refraction, is the same, that is, $x_2 = x_1$. ϕ_i and ϕ_t are the angles of incidence and refraction, respectively, which are measured from the normal NAC to the curved surface. Applying Snell's law and using the paraxial approximation, we have

$$n_1 \phi_i = n_2 \phi_t. \tag{1.1-4}$$

Now, from geometry, we know that $\phi_i = \theta_1 + \phi$ and $\phi_t = \theta_2 + \phi$ (Figure 1.1-4). Hence, the left side of Eq. (1.1-4) becomes

$$n_1 \phi_i = n_1 (\theta_1 + \phi) = v_1 + n_1 x_1 / R, \tag{1.1-5}$$

where we have used $\sin \psi = x_1 / R \approx \phi$. Now, the right side of Eq. (1.1-4) is

$$n_2 \phi_t = n_2 (\theta_2 + \phi) = v_2 + n_2 x_2 / R, \tag{1.1-6}$$

where $x_1 = x_2$ as the input and output planes are the same.

Finally, putting Eqs. (1.1-5) and (1.1-6) into Eq. (1.1-4), we have

$$v_1 + n_1 x_1 / R = v_2 + n_2 x_2 / R$$

or

$$v_2 = v_1 + (n_1 - n_2) x_1 / R, \quad \text{as} \quad x_1 = x_2. \tag{1.1-7}$$

We can formulate the above equation in terms of a matrix equation as

$$\begin{pmatrix} x_2 \\ v_2 \end{pmatrix} = \begin{pmatrix} 1 & 0 \\ \dfrac{n_1 - n_2}{R} & 1 \end{pmatrix} \begin{pmatrix} x_1 \\ v_1 \end{pmatrix} = \begin{pmatrix} 1 & 0 \\ -p & 1 \end{pmatrix} \begin{pmatrix} x_1 \\ v_1 \end{pmatrix} = \mathcal{R}_R \begin{pmatrix} x_1 \\ v_1 \end{pmatrix}, \tag{1.1-8}$$

where

$$\mathcal{R}_R = \begin{pmatrix} 1 & 0 \\ -p & 1 \end{pmatrix} = \begin{pmatrix} 1 & 0 \\ \dfrac{n_1 - n_2}{R} & 1 \end{pmatrix}.$$

The determinant of $\mathcal{R}_R$ is

$$|\mathcal{R}_R| = \begin{vmatrix} 1 & 0 \\ -p & 1 \end{vmatrix} = 1.$$

The 2×2 ray transfer matrix $\mathcal{R}_R$ is a positive unimodular matrix and is called the *refraction matrix*. The matrix describes refraction for the spherical surface. The quantity p given by

$$p = \frac{n_2 - n_1}{R}$$

is called the *refracting power* of the spherical surface. When R is measured in meters, the unit of p is called *diopters*. If an incident ray is made to converge on (diverge from) by a surface, the power is positive (negative) in sign.

Thick- and Thin-Lens Matrices

A thick lens consists of two spherical surfaces as shown in Figure 1.1-5. We shall find the system matrix that relates the system's input coordinates (x_1, v_1) to system's output ray coordinates (x_2, v_2). Let us first relate (x_1, v_1) to (x_1', v_1') through the spherical surface defined by R_1. (x_1', v_1') are the output coordinates due to the surface R_1. According to Eq. (1.1-8), we have

$$\begin{pmatrix} x_1' \\ v_1' \end{pmatrix} = \begin{pmatrix} 1 & 0 \\ \dfrac{n_1 - n_2}{R_1} & 1 \end{pmatrix} \begin{pmatrix} x_1 \\ v_1 \end{pmatrix} = \mathcal{R}_{R_1} \begin{pmatrix} x_1 \\ v_1 \end{pmatrix}. \tag{1.1-9}$$

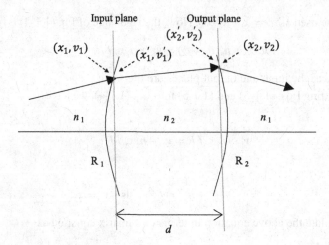

Figure 1.1-5 Thick lens

Now, (x_1', v_1') and (x_2', v_2') are related through a translation matrix as follows:

$$\begin{pmatrix} x_2' \\ v_2' \end{pmatrix} = \begin{pmatrix} 1 & d/n_2 \\ 0 & 1 \end{pmatrix} \begin{pmatrix} x_1' \\ v_1' \end{pmatrix} = \mathcal{T}_{d/n_2} \begin{pmatrix} x_1' \\ v_1' \end{pmatrix}, \tag{1.1-10}$$

where (x_2', v_2') are the output coordinates after translation, which are also the coordinates of the input coordinates for the surface R_2. Finally, we relate (x_2', v_2') to the system's output coordinates (x_2, v_2) through

$$\begin{pmatrix} x_2 \\ v_2 \end{pmatrix} = \begin{pmatrix} 1 & 0 \\ \dfrac{n_2 - n_1}{R_2} & 1 \end{pmatrix} \begin{pmatrix} x_2' \\ v_2' \end{pmatrix} = \mathcal{R}_{-R_2} \begin{pmatrix} x_2' \\ v_2' \end{pmatrix}. \tag{1.1-11}$$

If we now substitute Eq. (1.1-10) into Eq. (1.1-11), then we have

$$\begin{pmatrix} x_2 \\ v_2 \end{pmatrix} = \begin{pmatrix} 1 & 0 \\ \dfrac{n_2 - n_1}{R_2} & 1 \end{pmatrix} \begin{pmatrix} 1 & d/n_2 \\ 0 & 1 \end{pmatrix} \begin{pmatrix} x_1' \\ v_1' \end{pmatrix}.$$

Subsequently, substituting Eq. (1.1-9) into the above equation, we have the *system matrix equation* of the entire system as follows:

$$\begin{pmatrix} x_2 \\ v_2 \end{pmatrix} = \begin{pmatrix} 1 & 0 \\ \dfrac{n_2 - n_1}{R_2} & 1 \end{pmatrix} \begin{pmatrix} 1 & \dfrac{d}{n_2} \\ 0 & 1 \end{pmatrix} \begin{pmatrix} 1 & 0 \\ \dfrac{n_1 - n_2}{R_1} & 1 \end{pmatrix} \begin{pmatrix} x_1 \\ v_1 \end{pmatrix}$$

$$= \mathcal{R}_{-R_2} \mathcal{T}_{d/n_2} \mathcal{R}_{R_1} \begin{pmatrix} x_1 \\ v_1 \end{pmatrix} = \mathcal{S} \begin{pmatrix} x_1 \\ v_1 \end{pmatrix}. \tag{1.1-12}$$

We have now finally related the system's input coordinates to output coordinates. Note that the system matrix, $\mathcal{S} = \mathcal{R}_{-R_2} \mathcal{T}_{d/n_2} \mathcal{R}_{R_1}$, is a product of three ray transfer matrices. In general, the system matrix is made up of a collection of ray transfer matrices to account for the effects of a ray passing through the optical system. As the ray goes from left to right in the positive direction of the z-axis, we obtain the system matrix by writing the ray transfer matrices from right to left. This is precisely the advantage of the matrix formalism in that any ray, during its propagation through the optical system, can be tracked by successive matrix multiplications. Let A and B be the 2×2 matrices as follows:

$$A = \begin{pmatrix} a & b \\ c & d \end{pmatrix} \text{ and } B = \begin{pmatrix} e & f \\ g & h \end{pmatrix}.$$

Then the rule of matrix multiplication is

$$AB = \begin{pmatrix} a & b \\ c & d \end{pmatrix}\begin{pmatrix} e & f \\ g & h \end{pmatrix} = \begin{pmatrix} ae+bg & af+bh \\ ce+dg & cf+dh \end{pmatrix}.$$

Now, returning to the system matrix in Eq. (1.1-12), the determinant of the system matrix, $S = \mathcal{R}_{-R_2} \mathcal{T}_{d/n_2} \mathcal{R}_{R_1}$, is

$$|S| = \left|\mathcal{R}_{-R_2} \mathcal{T}_{d/n_2} \mathcal{R}_{R_1}\right| = \left|\mathcal{R}_{-R_2}\right| \times \left|\mathcal{T}_{d/n_2}\right| \times \left|\mathcal{R}_{R_1}\right| = 1.$$

Note that even the system matrix is also positive unimodular. The condition of a unit determinant is a necessary but not a sufficient condition on the system matrix.

We now derive a matrix of an idea thin lens of focal length f, called the *thin-lens matrix*, $\mathcal{L}_f$. For a thin lens in air, we let $d \to 0$ and $n_1 = 1$ in the configuration of Figure 1.1-5. We also write $n_2 = n$ for notational convenience. Then the system matrix in Eq. (1.1-12) becomes

$$\begin{aligned} S &= \begin{pmatrix} 1 & 0 \\ \dfrac{n-1}{R_2} & 1 \end{pmatrix}\begin{pmatrix} 1 & 0 \\ 0 & 1 \end{pmatrix}\begin{pmatrix} 1 & 0 \\ \dfrac{1-n}{R_1} & 1 \end{pmatrix} \\[2mm] &= \begin{pmatrix} 1 & 0 \\ \dfrac{n-1}{R_2} & 1 \end{pmatrix}\begin{pmatrix} 1 & 0 \\ \dfrac{1-n}{R_1} & 1 \end{pmatrix} \\[2mm] &= \begin{pmatrix} 1 & 0 \\ -\dfrac{1}{f} & 1 \end{pmatrix} = \mathcal{L}_f, \end{aligned} \qquad (1.1\text{-}13)$$

where f is the *focal length* of the thin lens and is given by

$$\frac{1}{f} = (n-1)\left(\frac{1}{R_1} - \frac{1}{R_2}\right).$$

For $f > 0$, we have a converging (convex) lens. On the other hand, we have a diverging (concave) lens when $f < 0$. Figure 1.1-6 summarizes the result for the ideal thin lens.

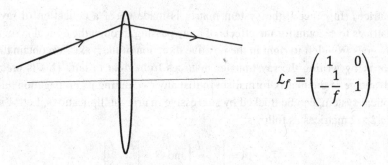

$$\mathcal{L}_f = \begin{pmatrix} 1 & 0 \\ -\dfrac{1}{f} & 1 \end{pmatrix}$$

Convex : $f > 0$; Concave : $f < 0$

Figure 1.1-6 Ideal thin lens of focal length f and its associated ray transfer matrix

Note that the determinant of $\mathcal{L}_f$ is

$$|\mathcal{L}_f| = \begin{vmatrix} 1 & 0 \\ -\dfrac{1}{f} & 1 \end{vmatrix} = 1.$$

1.1.2 Ray Tracing through a Thin Lens

As we have seen from Section 1.1.1, when a thin lens of focal length f is involved, then the matrix equation, from Eq. (1.1-13), is

$$\begin{pmatrix} x_2 \\ v_2 \end{pmatrix} = \mathcal{L}_f \begin{pmatrix} x_1 \\ v_1 \end{pmatrix} = \begin{pmatrix} 1 & 0 \\ -\dfrac{1}{f} & 1 \end{pmatrix} \begin{pmatrix} x_1 \\ v_1 \end{pmatrix}. \tag{1.1-14}$$

Input Rays Traveling Parallel to the Optical Axis

From Figure 1.1-7a, we recognize that $x_1 = x_2$ as the heights of the input and output rays are the same for the thin lens. Now, according to Eq. (1.1-14), $v_2 = -x_1/f + v_1$. For $v_1 = 0$, that is, the input rays are parallel to the optical axis, $v_2 = -x_1/f$. For positive $x_1, v_2 < 0$ as $f > 0$ for a converging lens. For negative $x_1, v_2 > 0$. All input rays parallel to the optical axis converge behind the lens to the back focal point (a distance of f away from the lens) of the lens as shown in Figure 1.1-7a. Note that for a thin lens, the front focal point is also a distance of f away from the lens.

Input Rays Traveling through the Center of the Lens

For input rays traveling through the center of the lens, their input ray coordinates are $(x_1, v_1) = (0, v_1)$. The output ray coordinates, according to Eq. (1.1.14), are $(x_2, v_2) = (0, v_1)$. Hence, we see as $v_2 = v_1$, all rays traveling through the center of the lens will pass undeviated as shown in Figure 1.1-7b.

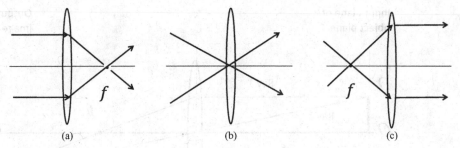

Figure 1.1-7 Ray tracing through a thin convex lens

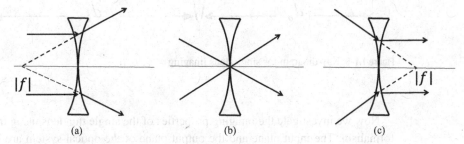

Figure 1.1-8 Ray tracing through a thin concave lens

Input Rays Passing through the Front Focal Point of the Lens

For this case, the input ray coordinates are $(x_1, v_1 = x_1 / f)$, and, according to Eq. (1.1-14), the output ray coordinates are $(x_2 = x_1, v_2 = 0)$, which means that all output rays will be parallel to the optical axis $(v_2 = 0)$, as shown in Figure 1.1-7c.

Similarly, in the case of a diverging lens, we can draw conclusions as follows. The ray after refraction diverges away from the axis as if it were coming from a point on the axis a distance $|f|$ in front of the lens, as shown in Figure 1.1-8a. The ray traveling through the center of the lens will pass undeviated, as shown in Figure 1.1-8b. Finally, for an input ray appearing to travel toward the back focus point of a diverging lens, the output ray will be parallel to the optical axis, as shown in Figure 1.1-8c.

Example: Imaging by a Convex Lens

We consider a single-lens imaging as shown in Figure 1.1-9, where we assume the lens is immersed in air. We first construct a *ray diagram* for the imaging system. An object OO′ is located a distance d_0 in front of a thin lens of focal length f. We send two rays from a point O′ towards the lens. Ray 1 from O′ is incident parallel to the optical axis, and from Figure 1.1-7a, the input ray parallel to the optical axis converges behind the lens to the back focal point. A second ray, that is, ray 2 also from O′, is now drawn through the center of the lens without bending and that is the result from Figure 1.1-7b. The interception of the two rays on the other side of the lens forms an image point of O′. The image point of O′ is labeled at I′ in the diagram. The final image is real, inverted, and is called a *real image*.

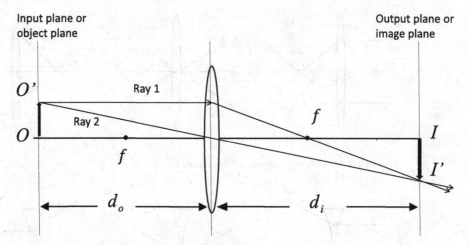

Figure 1.1-9 Ray diagram for single-lens imaging

Now we investigate the imaging properties of the single thin lens using the matrix formalism. The input plane and the output plane of the optical system are shown in Figure 1.1-9. We let (x_0, v_0) and (x_i, v_i) represent the coordinates of the ray at O' and I', respectively. We see there are three matrices involved in the problem. The overall system matrix equation becomes

$$\begin{pmatrix} x_i \\ v_i \end{pmatrix} = \mathcal{T}_{d_i} \mathcal{L}_f \mathcal{T}_{d_0} \begin{pmatrix} x_0 \\ v_0 \end{pmatrix} = \mathcal{S} \begin{pmatrix} x_0 \\ v_0 \end{pmatrix}. \tag{1.1-15}$$

The overall system matrix,

$$\mathcal{S} = \mathcal{T}_{d_i} \mathcal{L}_f \mathcal{T}_{d_0} = \begin{pmatrix} 1 & d_i \\ 0 & 1 \end{pmatrix} \begin{pmatrix} 1 & 0 \\ -\dfrac{1}{f} & 1 \end{pmatrix} \begin{pmatrix} 1 & d_0 \\ 0 & 1 \end{pmatrix},$$

is expressed in terms of the product of three matrices written in order from right to left as the ray goes from left to right along the optical axis, as explained earlier. According to the rule of matrix multiplication, Eq. (1.1-15) can be simplified to

$$\begin{pmatrix} x_i \\ v_i \end{pmatrix} = \begin{pmatrix} 1 - d_i / f & d_0 + d_i - d_0 d_i / f \\ -\dfrac{1}{f} & 1 - d_0 / f \end{pmatrix} \begin{pmatrix} x_0 \\ v_0 \end{pmatrix} = \begin{pmatrix} A & B \\ C & D \end{pmatrix} \begin{pmatrix} x_0 \\ v_0 \end{pmatrix}. \tag{1.1-16}$$

To investigate the conditions for imaging, let us concentrate on the *ABCD* matrix of $\mathcal{S}$ in Eq. (1.1-16). In general, we see that $x_i = Ax_0 + Bv_0 = Ax_0$ if $B = 0$, which means that all rays passing through the input plane at the same object point x_0 will pass through the same image point x_i in the output plane. This is the condition of *imaging*. In addition, for $B = 0$, $A = x_i / x_0$ is defined as the *lateral magnification* of

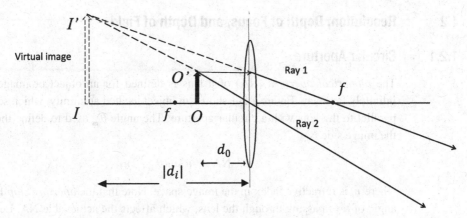

Figure 1.1-10 Imaging of a converging lens with object inside the focal length

the imaging system. Now in our case of thin-lens imaging, $B = 0$ in Eq. (1.1-16) leads to $d_0 + d_i - d_0 d_i / f = 0$, which gives the well-known *thin-lens formula*:

$$\frac{1}{d_0} + \frac{1}{d_i} = \frac{1}{f}. \tag{1.1-17}$$

The sign convention for Eq. (1.1-17) is that the object distance d_0 is positive (negative) if the object is to the left (right) of the lens. If the image distance d_i is positive (negative), the image is to the right (left) of the lens and it is real (virtual). In Figure 1.1-9, we have $d_0 > 0, d_i > 0$, and the image is therefore real, which means physically that light rays actually converge to the formed image I'. Hence, for imaging, Eq. (1.1-16) becomes

$$\begin{pmatrix} x_i \\ v_i \end{pmatrix} = \begin{pmatrix} 1 - d_i / f & 0 \\ -\dfrac{1}{f} & 1 - d_0 / f \end{pmatrix} \begin{pmatrix} x_0 \\ v_0 \end{pmatrix}, \tag{1.1-18}$$

which relates the input ray and output ray coordinates in the imaging system. Using Eq. (1.1-17), the *lateral magnification M* of the imaging system is

$$M = A = \frac{x_i}{x_0} = 1 - \frac{d_i}{f} = -\frac{d_i}{d_0}. \tag{1.1-19}$$

The sign convention is that if $M > 0$, the image is erect, and if $M < 0$, the image is inverted. As shown in Figure 1.1-9, we have an inverted image as both d_i and d_0 are positive.

If the object lies within the focal length, as shown in Figure 1.1-10, we follow the rules as given in Figure 1.1-8 to construct a ray diagram. However, now rays 1 and 2, after refraction by the lens, are divergent and do not intercept on the right side of the lens. They seem to come from a point I'. In this case, since $d_0 > 0$ and $d_i < 0$, $M > 0$. The final image is virtual, erect, and is called a *virtual image*.

1.2   Resolution, Depth of Focus, and Depth of Field

1.2.1   Circular Aperture

The *numerical aperture* (NA) of a lens is defined for an object or image located infinitely far away. Figure 1.2-1 shows an object located at infinity, which sends rays parallel to the lens with a circular aperture. The angle θ_{im} used to define the NA on the image side is

$$NA_i = n_i \sin(\theta_{im}/2), \tag{1.2-1}$$

where n_i is refractive index in the image space. Note that the *aperture stop* limits the angle of rays passing through the lens, which affects the achievable NA. Let us now find the lateral resolution, Δr.

Since we treat light as particles in geometrical optics, each particle then can be characterized by its momentum, p_0. According to the *uncertainty principle* in quantum mechanics, we relate the minimum uncertainty in a position of quantum, Δr, to the uncertainty of its momentum, Δp_r, according to the relationship

$$\Delta r \Delta p_r \geq h, \tag{1.2-2}$$

where h is *Planck's constant*, and Δp_r is the momentum spread in the r-component of the photons. The momentum of the FB ray (chief ray, a ray passes through the center of the lens) along the r-axis is zero, while the momentum of the FA ray (marginal ray, a ray passes through the edge of the lens) along the r-axis is $\Delta p_{AB} = p_0 \sin(\theta_{im}/2)$, where $p_0 = h/\lambda_0$ with λ_0 being the wavelength in the medium, that is, in the image space. Hence, Δp_r is $\Delta p_r = 2\Delta p_{AB} = 2p_0 \sin(\theta_{im}/2)$ to accommodate a maximum variation (or spread) of the momentum direction by an angle θ_{im}. By substituting this into Eq. (1.2-2), we have *lateral resolution*

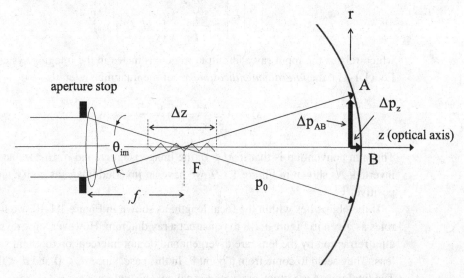

Figure 1.2-1 Uncertainty principle used in finding lateral resolution and depth resolution

$$\Delta r \geq \frac{h}{\Delta p_r} = \frac{h}{2p_0 \sin(\theta_{im}/2)} = \frac{\lambda_0}{2\sin(\theta_{im}/2)}. \tag{1.2-3}$$

Note that this equation reveals that in order to achieve high resolution, we can increase $\theta_{im}/2$. For example, when $\theta_{im}/2 = 90°$, Δr is half the wavelength, that is, $\lambda_0/2$, which is the theoretical maximum lateral resolution.

Since the wavelength in the image space, λ_0, is equal to λ_v/n_i, where λ_v is the wavelength in air or in vacuum, Eq. (1.2-3) becomes, using Eq. (1.2-1),

$$\Delta r \geq \frac{\lambda_v}{2n_i \sin(\theta_{im}/2)} = \frac{\lambda_v}{2\text{NA}_i}. \tag{1.2-4}$$

Similarly, we can calculate the *depth of focus*, Δz. Depth of focus, also called *longitudinal resolution* in the image space, is the axial distance over which the image can be moved without loss of sharpness in the object. To find the depth of focus, we use

$$\Delta z \Delta p_z \geq h, \tag{1.2-5}$$

where Δp_z is the momentum difference between rays FB and FA along the z-direction, as shown in Figure 1.2-1, which is given by

$$\Delta p_z = p_0 - p_0 \cos(\theta_{im}/2). \tag{1.2-6}$$

By substituting this expression into Eq. (1.2-5), we have

$$\Delta z \geq \frac{h}{\Delta p_z} = \frac{h}{p_0[1 - \cos(\theta_{im}/2)]} = \frac{\lambda_0}{[1 - \cos(\theta_{im}/2)]}. \tag{1.2-7}$$

This equation reveals that in order to achieve small depth of focus, we can increase $\theta_{im}/2$. For example, when $\theta_{im}/2 = 90°$, Δz is one wavelength, that is, λ_0, which is the theoretical maximum longitudinal resolution. Equation (1.2-7) can be written in terms of the numerical aperture and is given by

$$\Delta z \geq \frac{\lambda_0}{1 - \sqrt{1 - \sin^2(\theta_{im}/2)}} = \frac{\lambda_v}{n_i - \sqrt{n_i^2 - \text{NA}_i^2}}. \tag{1.2-8}$$

For small angles, that is, $\theta_{im} \ll 1$, one can use the approximation $\sqrt{1 - \sin^2 \beta} \approx 1 - (\sin^2 \beta)/2$ to get

$$\Delta z \geq \frac{2n_i \lambda_v}{\text{NA}_i^2}. \tag{1.2-9}$$

Taking the equality in Eq. (1.2-9) and combining with Eq. (1.2-4), we have

$$\frac{(\Delta r)^2}{\Delta z} \approx \frac{\lambda_0}{8}. \tag{1.2-10}$$

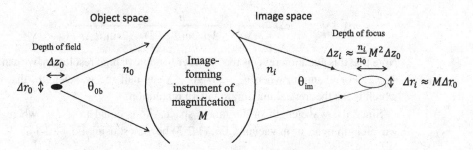

Figure 1.2-2 Image-forming instrument illustrating resolutions in the object space and image space

under the small-NA approximation. This equality is derived from uncertainty relationship, which says that during imaging the higher the lateral resolution is, the shorter the depth of focus. For example, by increasing the lateral resolution by a factor of 2 will result in the depth of focus decreased by a factor of 4.

Figure 1.2-2 shows an image-forming instrument with θ_{ob} and θ_{im} denoting the ray of maximum divergent angle and maximum convergent angle from the object side and the image side, respectively.

If the resolutions in the object space are given by $\Delta r_0 \approx \lambda_v / 2\mathrm{NA}_0$ and $\Delta z_0 \approx 2n_0\lambda_v / \mathrm{NA}_0^2$, where $\mathrm{NA}_0 = n_0 \sin(\theta_{ob}/2)$ and n_0 is the refractive index in the object space, and if the lateral magnification of the instrument is M, the resolution on the image space is then $\Delta r_i \approx M\Delta r_0$. Let us establish the relationship between the *depth of field* Δz_0 and the *depth of focus* Δz_i, where $\Delta z_i \approx 2n_i\lambda_v / \mathrm{NA}_i^2$ with $\mathrm{NA}_i = n_i \sin(\theta_{im}/2)$. Since $\Delta r_i \approx \lambda_v / 2\mathrm{NA}_i = M\Delta r_0 = M\lambda_v / 2\mathrm{NA}_0$, we have $\mathrm{NA}_0 = M \times \mathrm{NA}_i$. Hence,

$$\Delta z_i \approx 2n_i\lambda_v / \mathrm{NA}_i^2 = 2n_i\lambda_v / \left(\mathrm{NA}_0 / M\right)^2$$
$$= M^2(2n_i\lambda_v / \mathrm{NA}_0^2) \qquad (1.2\text{-}11)$$
$$= \frac{n_i}{n_0}M^2\Delta z_0.$$

This result indicates that the longitudinal resolutions in the object space and image space are related by a factor of M^2. Take a 40×, $\mathrm{NA}_0 \approx 0.6$ microscope objective as an example, we have $\Delta r_0 \approx \lambda_v / 2\mathrm{NA}_0 \approx 0.5\mu\mathrm{m}$ for red light with wavelength of 632 nm and the depth of field, $\Delta z_0 \approx 2n_0\lambda_v / \mathrm{NA}_0^2 \approx 3.5\mu\mathrm{m}$ for $n_0 = 1$ in air. In the image space, the lateral resolution is $\Delta r_i \approx M\Delta r_0 = 40 \times 0.5\mu\mathrm{m} = 20\mu\mathrm{m}$ and the depth of focus is $\Delta z_i \approx M^2\Delta z_0 = 40^2 \times 3.5\mu\mathrm{m} = 0.56\mathrm{mm}$ for $n_i = 1$.

1.2.2   Annular Aperture

Three-dimensional imaging in microscopy aims to develop techniques that can provide high lateral resolution, and at the same time maintain a large depth of focus in order to observe a thick specimen. However, calculations have shown that the depth of focus may be increased by reducing the numerical aperture of the lens, but this is achieved at the expense of a decrease in lateral resolution. In what follows, we

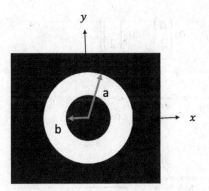

Figure 1.2-3 Annular aperture

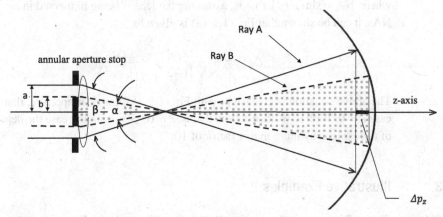

Figure 1.2-4 Uncertainty principle used in finding depth of focus for a lens with annular aperture stop

consider an annular aperture which has the property of increasing the depth of focus and at the same time maintaining the lateral resolution. An *annular aperture* is defined as a clear circular aperture with a central obstruction, as shown in Figure 1.2-3. If the aperture is an annulus with outer radius a and inner radius b, we define a *central obscuration ratio* $\varepsilon = b / a$. For $\varepsilon = 0$, we have a clear circular aperture.

Since $\beta = \theta_{im}$ as shown in Figure 1.2-1, the possible spread in the r-component of momentum Δp_r is the same in Figure 1.2-4 as it is in Figure 1.2-1. Hence, the lateral resolution remains the same as that in the case of the clear aperture, that is, $\varepsilon = 0$, given by Eq. (1.2-4). However, the spread or uncertainty in the momentum in the z-direction is different due to the central obscuration of the annulus. Δp_r in this case is the momentum difference between the ray passing the upper part of the annulus (Ray A) and the ray passing the lower part of the annulus (Ray B) and is given by

$$\Delta p_z = \Delta p_{\text{Ray A}} - \Delta p_{\text{Ray B}}, \tag{1.2-12}$$

where $\Delta p_{\text{Ray A}}$ and $\Delta p_{\text{Ray B}}$ are given by Eq. (1.2-6) with β and α substituted into the argument of cosine, respectively. Hence Eq. (1.2-12) becomes

$$\Delta p_z = p_0 \left[\cos\left(\frac{\alpha}{2}\right) - \cos\left(\frac{\beta}{2}\right) \right],$$

and the depth of focus is

$$\Delta z_{\text{ann}} = \frac{h}{\Delta p_z} = \frac{\lambda_0}{\cos\left(\dfrac{\alpha}{2}\right) - \cos\left(\dfrac{\beta}{2}\right)},$$
(1.2-13)

which can be shown to have the form

$$\Delta z_{\text{ann}} = \frac{\lambda_v}{\sqrt{\dfrac{1-\text{NA}^2}{1+\text{NA}^2\left(\varepsilon^2-1\right)}} - \sqrt{1-\text{NA}^2}},$$
(1.2-14)

where $\text{NA} = \sin\left(\beta/2\right)$, that is, assuming the lens is being immersed in air. For small NAs, it can be shown that Eq. (1.2-14) is given by

$$\Delta z_{\text{ann}} = \frac{2\lambda_v}{\text{NA}^2\left(1-\varepsilon^2\right)} = \frac{\Delta z}{\left(1-\varepsilon^2\right)}.$$
(1.2-15)

The above equation is consistent with Eq. (1.2-9) for a clear aperture, that is, for the case $\varepsilon = 0$. With 95% obstruction, that is, $\varepsilon = 0.95$, we can increase the depth of focus of a clear lens by more than a factor of 10.

1.3   Illustrative Examples

1.3.1   Three-Dimensional Imaging through a Single-Lens Example

Figure 1.3-1a shows the imaging of two objects in front of the lens, where both of the object lie beyond the focal length of the lens. We note that magnification is different, depending on the object distance to the lens. Let us consider longitudinal magnification, M_z, in addition to lateral magnification, M, considered earlier. Longitudinal magnification M_z is the ratio of an image displacement along the axial direction, δd_i, to the corresponding object displacement, δd_0, that is, $M_z = \delta d_i / \delta d_0$.

Using Eq. (1.1-17) and treating d_i and d_0 as variables, we take the derivative of d_i with respect to d_0 to obtain

$$M_z = \delta d_i / \delta d_0 = -M^2.$$
(1.3-1)

This equation is consistent with Eq. (1.2-11) and states that the longitudinal magnification is equal to the square of the lateral magnification. The minus sign in front of the equation signifies that a decrease in the distance of the object from the lens, $|d_0|$, will result in an increase in the image distance, $|d_i|$, and vice versa. The situation of a magnified volume is shown in Figure 1.3-1b, where a cube volume (abcd plus the dimension into the paper) is imaged into a truncated pyramid with a–b imaged into a′–b′ and c–d imaged into c′–d′.

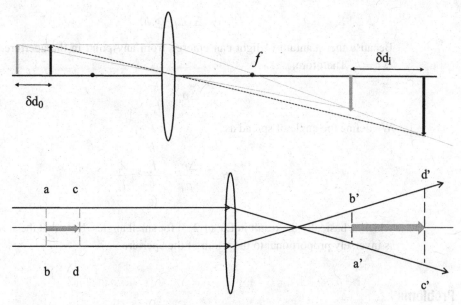

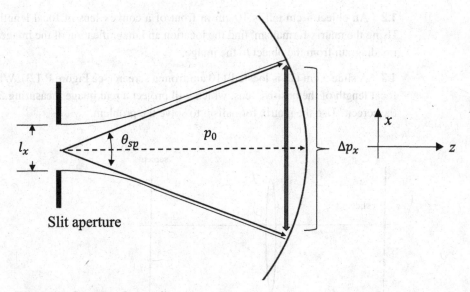

Figure 1.3-1 (a) Illustration of different magnifications and (b) volumetric imaging

Figure 1.3-2 Spreading from a slit

1.3.2   Angle of Spread from a Slit Example

Consider light emanating from a slit aperture of width l_x, as shown in Figure 1.3-2.

We relate the minimum uncertainty in position Δx of a quantum to the uncertainty in its momentum Δp_x according to

$$\Delta x \Delta p_x \geq h.$$

Because the quantum of light can emerge from any point in the aperture, we have $\Delta x = l_x$. Therefore,

$$\Delta p_x \sim \frac{h}{l_x}.$$

We define the angle of spread as

$$\theta_{sp} \sim \frac{\Delta p_x}{p_0} = \frac{\dfrac{h}{l_x}}{\dfrac{h}{\lambda_0}} = \frac{\lambda_0}{l_x}, \tag{1.3-2}$$

which is basically the result in Eq. (1.2-3) for small angles. Note that the spread angle is inversely proportional to the width of the aperture.

Problems

1.1   Find the transfer matrix for Snell's law.

1.2   An object 4 cm tall is 10 cm in front of a convex lens of focal length 20 cm. Using the matrix formalism, find the location and magnification of the image. Draw a ray diagram from the object to the image.

1.3   A slide 5 cm tall is located 110 cm from a screen (see Figure P.1.3). What is the focal length of the positive lens, which will project a real image measuring 50 cm on the screen? Use the matrix formalism to solve the problem.

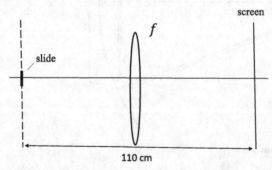

Figure P.1.3 Single-lens system

1.4   If an object 3 cm tall is located on the optical axial 24 cm to the left of the convex lens as shown in Figure P.1.4, find the position and size of the image using the matrix formalism. Also, draw the ray diagram from the object to the image.

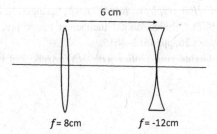

6 cm

f = 8cm f = -12cm

Figure P.1.4 Two-lens system

1.5 In 3-D imaging through a single lens example, it is stated that the longitudinal magnification is equal to the square of the lateral magnification, that is, $M_z = \delta d_i / \delta d_0 = -M^2$. Verify this statement.

1.6 Fill in the blanks below for several microscope objective lenses (assuming the lenses are immersed in air with wavelength of operation using red light of $\lambda_v = 632$ nm).

Magnification, NA	Resolution in image space	Depth of field (μm)	Depth of focus (mm)
10×, 0.1			
20×, 0.4			
40×, 0.6			
60×, 0.8			
100×, 0.95			

1.7 Starting from Eq. (1.2-13) and with reference to Figure 1.2-4, show that the depth of focus of an annular aperture with a thin lens is given by

$$\Delta z_{\text{ann}} = \frac{\lambda_v}{\sqrt{\dfrac{1 - NA^2}{1 + NA^2 \left(\varepsilon^2 - 1\right)}} - \sqrt{1 - NA^2}},$$

where $NA = \sin(\beta / 2)$ and for small NAs, it is approximately given by

$$\Delta z_{\text{ann}} = \frac{2\lambda_v}{NA^2 \left(1 - \varepsilon^2\right)} = \frac{\Delta z}{\left(1 - \varepsilon^2\right)},$$

where Δz is the depth of focus for a clear lens.

Bibliography

1. Banerjee, P. P. and T.-C. Poon (1991). *Principles of Applied Optics*. Irwin, Illinois.
2. Gerard, A. and J. M. Burch (1975). *Introduction to Matrix Methods in Optics*. Wiley, New York.
3. Hecht, E. (2002). *Optics*, 4th ed., Addison Wesley, California.
4. Korpel, A. (1970). *United State Patent (#3,614,310) Electrooptical Apparatus Employing a Hollow Beam for Translating an Image of an Object*.

5. Poon, T.-C. (2007). *Optical Scanning Holography with MATLAB®*, Springer, New York.

6. Poon, T.-C. and M. Motamedi (1987). "Optical/digital incoherent image processing for extended depth of field," *Applied Optics* 26, pp. 4612–4615.

7. Poon, T.-C. and T. Kim (2018). *Engineering Optics with MATLAB®*, 2nd ed., World Scientific, New Jersey.

2 Linear Invariant Systems and Fourier Analysis

The purpose of this chapter is twofold. We will first discuss basic aspect of signals and linear systems in the first part. We will see in subsequent chapters that diffraction as well as optical imaging systems can be modelled as linear systems. In the second part, we introduce the basic properties of Fourier series, Fourier transform as well as the concept of convolution and correlation. Indeed, many modern optical imaging and processing systems can be modelled with the Fourier methods, and Fourier analysis is the main tool to analyze such optical systems. We shall study time signals in one dimension and spatial signals in two dimensions will then be covered. Many of the concepts developed for one-dimensional.(1-D) signals and systems apply to two-dimensional (2-D) systems. This chapter also serves to provide important and basic mathematical tools to be used in subsequent chapters.

2.1 Signals and Systems

2.1.1 Signal Operations

A *signal* is a set of data. For electrical systems, a signal is a function of time, measured in terms of voltage or current. For optical systems, a signal is of the form of 1-D, 2-D, or even 3-D images. In this section, we introduce some signal operations and useful signal models. We first restrict ourselves to 1-D time signals for brevity. Two-dimensional signals will be covered subsequently. We denote a continuous-time signal by $x(t)$. There are three useful signal operations: shifting, scaling, and reversal.

Shifting
$x(t-t_0)$ is a *time-shifted version* of $x(t)$. If t_0 is positive, the time shift is to the right and it represents a time delay of t_0 seconds to the original signal $x(t)$. If t_0 is negative, the shift is to the left and the original signal $x(t)$ is advanced by t_0. The situation is illustrated in Figure 2.1-1.

Scaling
$x(at)$ is a *scaled version* of the original signal $x(t)$, where a is a scale factor and is a positive number. If $a > 1$, we have a compressed version of $x(t)$. On the other hand, if $a < 1$, we have an expanded version of $x(t)$. Figure 2.1-2 shows a compressed and expanded versions of $x(t)$ for $a = 2$ and 1/2, respectively.

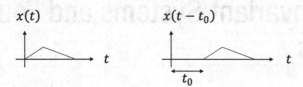

Figure 2.1-1 Time shifting a signal for t_0 positive

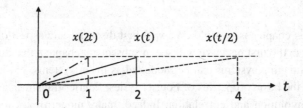

Figure 2.1-2 Time scaling of a signal

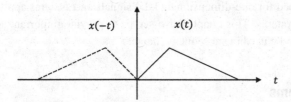

Figure 2.1-3 Time reversal a signal

Reversal

$x(-t)$ is a *time reversal* of $x(t)$. In other words, if we replace t by $-t$ in the original signal $x(t)$, we have a mirror image of $x(t)$ about the vertical axis $t = 0$. Figure 2.1-3 illustrates the situation.

Example: Signal Operations

Plot $x(1-t/2)$ if $x(t)$ is shown in Figure 2.1-4a.

Let us first rewrite $x(1-t/2)$ as $x\left(1-\dfrac{t}{2}\right) = x\left(\dfrac{2-t}{2}\right) = x\left(\dfrac{1}{2}(2-t)\right)$. We then recognize that there are three signal operations involved. The factors ½, 2, and −1 correspond to scaling, shifting, and reversal, respectively. Let us first perform reversal as $x(t)|_{t \to -t} = x(-t)$ and it is shown in Figure 2.1-4b. We then perform scaling according to $x(-t)|_{t \to t/2} = x(-t/2)$ and the result is shown in Figure 2.1-4c. Finally, we perform shifting according to $x\left(-\dfrac{t}{2}\right)\bigg|_{t \to t-2} = x\left(-\dfrac{t-2}{2}\right) = x\left(1-\dfrac{t}{2}\right)$, which is the desired signal we want to plot.

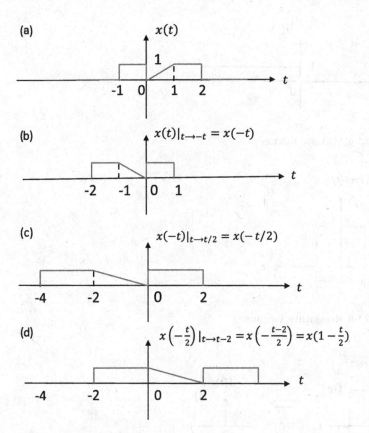

Figure 2.1-4 Signal operations example

2.1.2   Some Useful Signal Models

Unit Step Function, $u(t)$

The *unit step function*, $u(t)$, is shown in Figure 2.1-5. It is defined by

$$u(t) = \begin{cases} 0, & t < 0 \\ 1, & t > 0. \end{cases} \tag{2.1-1}$$

Because $u(t)$ does not have a unique value at $t = 0$ and its derivative is infinite at $t = 0$, it is a *singularity function*.

Rectangular Function, $\text{rect}(t / a)$

The *rectangular function* or simply *rect function*, $\text{rect}(t / a)$, is given by

$$\text{rect}\left(\frac{t}{a}\right) = \begin{cases} 1, & |t| < a/2 \\ 0, & \text{otherwise}, \end{cases} \tag{2.1-2}$$

where a is the width of the function. The rect function is typically called a *pulse* in electrical engineering. The function is shown in Figure 2.1-6. It is a singularity function as it is discontinuous at its two edges.

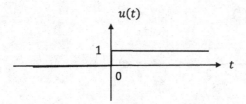

Figure 2.1-5 Unit step function

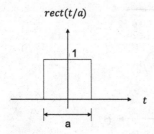

Figure 2.1-6 Rectangular function

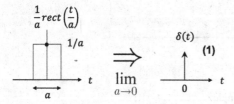

Figure 2.1-7 Dirac delta function defined using a rectangular function

Dirac Delta Function, $\delta(t)$

Another singularity function is the *Dirac delta function* $\delta(t)$, which is also known as the *unit impulse function* or just simply as the *delta function*. We can define the delta function as follows:

$$\delta(t) = \lim_{a \to 0} \frac{1}{a} \text{rect}\left(\frac{t}{a}\right).$$

The situation of the above equation is illustrated in Figure 2.1-7. Note that the area under the rectangular function is $a \times (1/a)$, which is always equal to 1 regardless the value of a. We denote this unity as "(1)" beside the arrow, which is the symbol of the delta function $\delta(t)$. This unity is sometimes called the "strength" of the delta function. For this reason, to plot $5\delta(t)$, we just plot the delta function as shown in Figure 2.1-7 but denote its strength as "(5)" beside the arrow.

The delta function is one of the most important functions in the study of system and signal analysis. We will discuss four important properties.

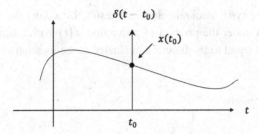

Figure 2.1-8 Illustration of the product property

Property#1: Unit Area

Since we have found that the area under the delta function is unity, we therefore write

$$\int_{-\infty}^{\infty} \delta(t)\,dt = 1. \tag{2.1-3}$$

Indeed, the following equality applies in general

$$\int_{a}^{b} \delta(t-t_0)\,dt = 1$$

if $a < t_0 < b$ since $\delta(t-t_0) = 0$ except at $t = t_0$.

Property#2: Product Property

$$x(t)\delta(t-t_0) = x(t_0)\delta(t-t_0). \tag{2.1-4}$$

The result of this property can be readily realized by the illustration shown in Figure 2.1-8, where an arbitrary function, $x(t)$, is shown to be overlapped with the time-shifted delta function, $\delta(t-t_0)$, located at $t = t_0$. The product of the two functions is clearly equal to $x(t_0)$ multiplied by $\delta(t-t_0)$. Therefore, the result has become a time-shifted delta function with its strength given by $x(t_0)$.

Property#3: Sampling Property

$$\int_{-\infty}^{\infty} x(t)\delta(t-t_0)\,dt = x(t_0). \tag{2.1-5}$$

The above result can be obtained readily by the use of Properties #1 and #2. Using Property #2, we write

$$\int_{-\infty}^{\infty} x(t)\delta(t-t_0)\,dt = \int_{-\infty}^{\infty} x(t_0)\delta(t-t_0)\,dt$$

$$= x(t_0)\int_{-\infty}^{\infty} \delta(t-t_0)\,dt = x(t_0),$$

where we have used Property#1 to obtain the final result. In words, the sampling property says that the area under the product of a function $x(t)$ with a time-shifted delta function, $\delta(t - t_0)$, is equal to the function evaluated at the location of the delta function, which is at t_0.

Property#4: Scaling Property

$$\delta(at) = \frac{1}{|a|}\delta(t) \text{ for } a \neq 0. \tag{2.1-6}$$

To determine how scaling affects the delta function, let us evaluate the area of $\delta(at)$ for $a > 0$:

$$\int_{-\infty}^{\infty} \delta(at)\,dt = \int_{-\infty}^{\infty} \delta(t')\frac{dt'}{a} = \frac{1}{a},$$

where we have made the substitution $t' = at$ to simplify the integration. So from the above equation, we have $\int_{-\infty}^{\infty} a\delta(at)\,dt = 1$. Comparing this with property #1, that is, $\int_{-\infty}^{\infty} \delta(t)\,dt = 1$, we deduce that $a\delta(at) = \delta(t)$, or

$$\delta(at) = \frac{1}{a}\delta(t). \tag{2.1-7a}$$

Similarly, for $a < 0$, we have

$$\int_{-\infty}^{\infty} \delta(at)\,dt = \frac{1}{-a},$$

giving

$$\delta(at) = \frac{1}{-a}\delta(t). \tag{2.1-7b}$$

For all values of a, Eqs. (2.1-7a) and (2.1-7b) give the result in Eq. (2.1-6).

Example: Relationship between $\delta(t)$ and $U(t)$
When we integrate the delta function from $-\infty$ to t, where $t > 0$, the result is always unity and hence, we have

$$\int_{-\infty}^{t} \delta(t)\,dt = u(t). \tag{2.1-8a}$$

After taking the derivative of Eq. (2.1-8a), we have

$$\frac{du(t)}{dt} = \delta(t). \tag{2.1-8b}$$

Figure 2.1-9 illustrates the above relationship.

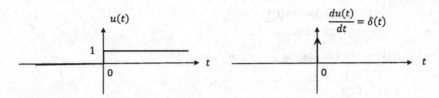

Figure 2.1-9 Illustration of the relationship between $\delta(t)$ and $u(t)$

2.1.3   Linear and Time-Invariant Systems

Systems are used to process signals. A system is a device that accepts an input $x(t)$ and gives out an output $y(t)$. Figure 2.1-10 shows a block diagram representation of a system, relating $x(t)$ and $y(t)$. We show this relation symbolically by

$$x(t) \rightarrow y(t).$$

We read this notation as "$x(t)$ produces $y(t)$." It has the same meaning as the block diagram shown in Figure 2.1-10.

We can represent the system by a mathematical operator as follows:

$$y(t) = \mathcal{O}\{x(t)\}, \tag{2.1-9}$$

where $\mathcal{O}\{.\}$ is an operator, mapping the input into the output. Equation (2.1-9) is the *system equation* as it relates the input $x(t)$ to the output $y(t)$.

A system is *linear* if superposition holds. Otherwise, it is *nonlinear*. If $x_1 \rightarrow y_1$ and $x_2 \rightarrow y_2$; then superposition means $x = ax_1 + bx_2 \rightarrow y = ay_1 + by_2$, where a and b are some constants. In words, the overall output y is a weighted sum of the outputs due to inputs x_1 and x_2.

Example: Linearity
Consider the system described by the system equation as follows: $y(t) = \mathcal{O}\{x(t)\} = tx(t)$ or $x(t) \rightarrow y(t) = tx(t)$. According to the system equation, we write $x_1 \rightarrow y_1 = tx_1$ and $x_2 \rightarrow y_2 = tx_2$. The system is linear, since

$$x = ax_1 + bx_2 \rightarrow y = tx = t(ax_1 + bx_2)$$
$$= tax_1 + tbx_2$$
$$= ay_1 + by_2.$$

Consider the system described by the system equation as follows: $y(t) = \mathcal{O}\{x(t)\} = x^2(t)$ or $x(t) \rightarrow y(t) = x^2(t)$. According to the system equation, we write $x_1 \rightarrow y_1 = x_1^2$ and $x_2 \rightarrow y_2 = x_2^2$. The system is nonlinear, since

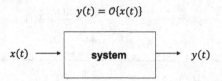

$$y(t) = O\{x(t)\}$$

$$x(t) \longrightarrow \boxed{\textbf{system}} \longrightarrow y(t)$$

Figure 2.1-10 Block diagram of a general system

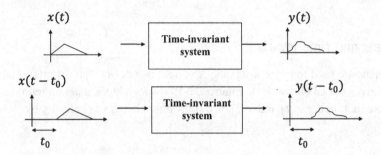

Figure 2.1-11 Time-invariant system

$$x = ax_1 + bx_2 \rightarrow y = x^2 = \left(ax_1 + bx_2\right)^2$$
$$= (ax_1)^2 + \left(bx_2\right)^2 + 2abx_1x_2$$
$$\neq ay_1 + by_2.$$

A system is *time invariant* if a time shift in the input results in the same time shift at the output. Otherwise, it is *time variant*. Figure 2.1-11 illustrates this property where t_0 is the time shift or the time delay.

Another way to illustrate the time-invariance property is shown in Figure 2.1-12. We can delay the output $y(t)$ of a system by applying the output to a t_0 delay, as shown in the top part of Figure 2.1-12. If the system is time invariant, then the delayed output $y(t-t_0)$ can also be obtained by first delaying the input by t_0 before applying it to the system, as shown in the bottom half of Figure 2.1-12. This illustration of time-invariance property is constructive as we can use this procedure to test if a system is time invariant or time variant.

Example: Time Invariance

Consider the system described by the system equation as follows: $y(t) = O\{x(t)\} = \dfrac{dx}{dt}$ or $x(t) \rightarrow y(t) = \dfrac{dx}{dt}$. The system is a *differentiator*. We determine time invariance according to the situations illustrated in Figure 2.1-12. From Figure 2.1-13, we see that the two final outputs are identical, and hence the system under investigation is time invariant.

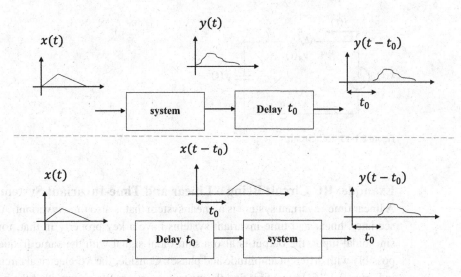

Figure 2.1-12 Illustration of time-invariance property

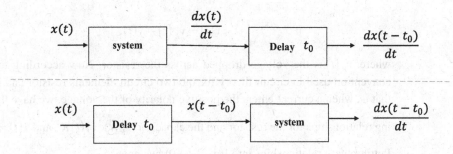

Figure 2.1-13 Time-invariance example: same output

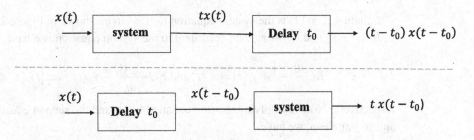

Figure 2.1-14 Time-variance example: different outputs

Consider the system described by the system equation as follows: $y(t) = \mathcal{O}\{x(t)\} = tx(t)$ or $x(t) \rightarrow y(t) = tx(t)$. From the results of Figure 2.1-14, we see that the final outputs are not the same, and hence the system is time variant.

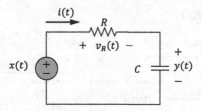

Figure 2.1-15 RC circuit

Example: RC Circuit Being a Linear and Time-Invariant System

A linear time-invariant system is a linear system that is also time invariant. As we will see later, linear and time-invariant systems have a key property in that, for a given sinusoidal input, the output is also a sinusoidal signal with the same frequency (but possibly with different amplitude and phase). Consider the RC electrical circuit shown in Figure 2.1-15. Let us first find the system equation. We assume that the input $x(t)$ is a voltage source and the output $y(t)$ is the voltage across the capacitor. Accordingly, using basic circuit theory, we get

$$x(t) = v_R(t) + y(t), \tag{2.1-10}$$

where $v_R(t)$ is the voltage dropped across the resistor. Now according to the sign convention used in circuit theory for the two circuit elements resistor and capacitor, that is, when a current enters the positive polarity of the voltage, we have the following relationships for the resistor and the capacitor: $v_R(t) = i(t)R$, and $i(t) = c\dfrac{dy(t)}{dt}$. Putting these relationships into Eq. (2.1-10), we have

$$RC\frac{dy(t)}{dt} + y(t) = x(t). \tag{2.1-11}$$

Equation (2.1-11) is the system equation for the circuit shown in Figure 2.1-15. If $x_1 \to y_1$ and $x_2 \to y_2$ then according to the system equation, we have

$$RC\frac{dy_1(t)}{dt} + y_1(t) = x_1(t) \text{ and } RC\frac{dy_2(t)}{dt} + y_2(t) = x_2(t),$$

respectively. Now multiplying the first equation by a and the second equation by b, and adding them, we have

$$RC\frac{d}{dt}[ay_1(t) + by_2(t)] + ay_1(t) + by_2(t) = ax_1(t) + bx_2(t)$$

This equation is the system equation [Eq. (2.1-11)] with

$$ax_1(t) + bx_2(t) \to ay_1(t) + by_2(t),$$

satisfying the principle of superposition. Hence, the system is linear. The system is also time invariant as the circuit elements are constant and not a function of time. As it turns out, any nth order linear differential equation with constant coefficients belongs to the class of linear time-invariant systems.

2.1.4　Impulse Responses

The impulse response of a system is the output owing to an impulse $\delta(t)$ input when all the initial conditions in the system have been set to zero. A practical illustrative example is that in our previously considered *RC* circuit in Figure 2.1-15, the capacitor has no charges stored when we apply an impulse function to obtain the system's output. Returning to Eq. (2.1-9), we have $y(t) = \mathcal{O}\{x(t)\}$. Now, for an impulse input of the form $\delta(t-t')$, the output $y(t) = \mathcal{O}\{x(t)\} = \mathcal{O}\{\delta(t-t')\} = h(t;t')$ is called the *impulse response* of the linear system. Using the sampling property of the delta function, we can write

$$x(t) = \int_{-\infty}^{\infty} x(t')\delta(t-t')\,dt',$$

that is, $x(t)$ is regarded as a linear combination of weighted and shifted delta functions. We can then write the output of the linear system as

$$
\begin{aligned}
y(t) = \mathcal{O}\{x(t)\} &= \mathcal{O}\left\{\int_{-\infty}^{\infty} x(t')\delta(t-t')\,dt'\right\} \\
&= \int_{-\infty}^{\infty} x(t')\mathcal{O}\{\delta(t-t')\}\,dt' = \int_{-\infty}^{\infty} x(t')h(t;t')\,dt'.
\end{aligned}
\tag{2.1-12}
$$

Note that in the above procedure, the operator has moved to pass $x(t')$ and to only operate on a time function. The above result relating the input and output is known as the *superposition integral*. This integral describes a *linear time-variant system*.

Now, a linear system is time invariant if the response to a time-shifted impulse function, $\delta(t-t')$, is given by $h(t-t')$, that is, $\mathcal{O}\{\delta(t-t')\} = h(t-t')$. Equation (2.1-12) then becomes

$$y(t) = \int_{-\infty}^{\infty} x(t')h(t-t')\,dt' = x(t) * h(t).
\tag{2.1-13}$$

This is an important result. The impulse $h(t;t')$ only depends on $t-t'$, that is, $h(t;t') = h(t-t')$ in linear and time-invariant systems. The integral can also be derived as illustrated in Figure 2.1-16.

The integral in Eq. (2.1-13) is known as the *convolution integral*, and $*$ is a symbol denoting the convolution of $x(t)$ and $h(t)$. $x(t) * h(t)$ reads as x convolves with h. For a linear and time-invariant system with all the initial conditions being zero, the system output to any input $x(t)$ is given by the convolution between the input and impulse response.

$$\delta(t) \rightarrow h(t)$$

$$\delta(t - t') \rightarrow h(t - t') \text{ , delaying } \delta(t) \text{ by } t' \text{ will delay the output by } t'$$

$$x(t')\delta(t - t') \rightarrow x(t')h(t - t') \text{ , when impulse response of amplitude } x(t') \text{ applied at } t'$$

$$\int_{-\infty}^{\infty} x(t')\delta(t - t')dt' \rightarrow \int_{-\infty}^{\infty} x(t')h(t - t')\,dt' \text{ , summing up all the responses at different } t'$$

$$x(t) \rightarrow y(t) = \int_{-\infty}^{\infty} x(t')h(t - t')\,dt'$$

Figure 2.1-16 Derivation of the superposition integral for a linear and time-invariant system

Convolution operation is actually *commutative*, which means the order in which we perform convolution of the two functions does not change the result. By changing variables in Eq. (2.1-13), we let $t - t' = t''$ so that $t' = t - t''$ and $dt' = -dt''$, we have

$$y(t) = x(t) * h(t) = \int_{-\infty}^{\infty} x(t')h(t - t')dt'$$

$$= \int_{\infty}^{-\infty} x(t - t'')h(t'')(-dt'') \qquad (2.1\text{-}14)$$

$$= \int_{-\infty}^{\infty} x(t - t'')h(t'')dt'' = h(t) * x(t).$$

Example: Graphical Method to Find Convolution

Given $x(t) = ae^{-at}u(t)$, $a > 0$ and $h(t) = u(t)$, we evaluate their convolution by a graphical method. According to the definition of convolution in Eq. (2.1-13), we have

$$x(t) * h(t) = \int_{-\infty}^{\infty} x(t')h(t - t')dt'.$$

The above integral simply means that we find the areas under the product of two functions, $x(t')$ and $h(t - t')$, for different values of t. Hence, the final answer is a function of t. Let us first visualize how to create the product $x(t')h(t - t')$. From the original functions $x(t)$ and $h(t)$, we plot $x(t')$ and $h(t')$ as shown in the Figure 2.1-17a. Now we have $x(t')$ on a graph. To create the graph of $h(t - t')$, we first take the reversal of $h(t')$ to get $h(-t')$, as shown in Figure 2.1-17b. Next, we shift $h(-t')$ by t to get $h(t - t')$, as shown in Figure 2.1-17c. We then overlap $x(t')$ and $h(t - t')$ on a single graph, which is shown on the left panel of Figure 2.1-17c. This graph guides us to set up the integrand and the limits for the integral of convolution as follows:

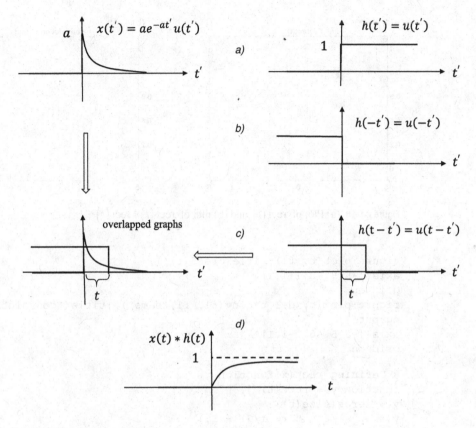

Figure 2.1-17 Graphical understanding of convolution

$$x(t) * h(t) = \int_{-\infty}^{\infty} x(t')h(t-t')dt'$$

$$= \int_{0}^{t} ae^{-at'}dt' = \begin{cases} 1-e^{-at}, & t>0 \\ 0, & t<0. \end{cases}$$

The result of the above integral is plotted in Figure 2.1-17d.

Example: Convolution MATLAB

We perform the following convolution: $\text{rect}(t) * \text{rect}(t)$. Figure 2.1-18a plots $\text{rect}(t)$ and the resulting convolution is plotted in Figure 2.1-18b.

```
% Convolution of rect(t) by itself; convolution_example.m
clear all; close all;
del_t=0.01;
t=-5.0:del_t:5.0;
f1 = rect(t);
```

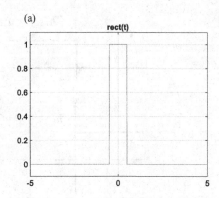
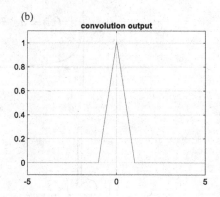

Figure 2.1-18 (a) Plot of $\mathrm{rect}(t)$ and (b) plot of $\mathrm{rect}(t) * \mathrm{rect}(t)$

```
figure;plot(t, f1);title('rect(t)');
axis([-5 5 -0.1 1.1])
grid on
figure;plot(t, del_t*conv(f1, f1,'same'));title('convolution
output');
axis([-5 5 -0.1 1.1])
grid on

% Defining rect(x)function
function y = rect(t)
y = zeros(size(t));
y(t >= -1/2 & t <= 1/2) = 1;
end
```

2.1.5   Frequency Response Functions

While we have analyzed responses in the time domain so far, much insight about the behavior of a linear and time-invariant system could be obtained through the frequency domain. Let us examine the response to a complex exponential $x(t) = e^{j\omega t}$, where $j = \sqrt{-1}$ and ω is *radian frequency* measured in radians per second if t is measured in seconds. According to Eq. (2.1-14), we have

$$y(t) = x(t) * h(t) = h(t) * x(t)$$

$$= \int_{-\infty}^{\infty} h(t') e^{j\omega(t-t')} dt' = e^{j\omega t} \int_{-\infty}^{\infty} h(t') e^{-j\omega t'} dt' = H(\omega) e^{j\omega t},$$

where we have defined the frequency response function, also called transfer function $H(\omega)$ as

$$H(\omega) = \int_{-\infty}^{\infty} h(t') e^{-j\omega t'} dt'. \tag{2.1-15}$$

Hence, we have established that

$$x(t) = e^{j\omega t} \rightarrow y(t) = H(\omega)e^{j\omega t}.$$

A frequency response function expresses the frequency domain relationship between an input $x(t)$ and output $y(t)$ of a linear time-invariant system and it is useful for finding the frequency response of a system. Note that the integral in Eq. (2.1-15) is called the *Fourier transform* of the impulse response $h(t)$, which we will discuss further in the next subsection. In words, the transfer function $H(\omega)$ is the Fourier transform of the impulse response $h(t)$ in linear and time-invariant systems.

Example: Frequency Response of an *RC* Circuit

With reference to the *RC* circuit shown in Figure 2.1-15, its system equation has been derived and given by

$$RC\frac{dy(t)}{dt} + y(t) = x(t).$$

By substituting $y(t) = H(\omega)e^{j\omega t}$ and $x(t) = e^{j\omega t}$ into the above equation, we obtain

$$RCH(\omega)(j\omega)e^{j\omega t} + H(\omega)e^{j\omega t} = e^{j\omega t}.$$

After canceling the $e^{j\omega t}$ term and solving for $H(\omega)$, we have

$$H(\omega) = \frac{1}{1 + j\omega RC}. \tag{2.1-16a}$$

The frequency response function is a complex function. Using the complex number identity $a + bj = \sqrt{a^2 + b^2}\, e^{j\tan^{-1}(b/a)}$, we write the transfer function of the circuit as

$$H(\omega) = |H(\omega)|e^{j\theta(\omega)} = \frac{1}{\sqrt{1 + (\omega RC)^2}}e^{-j\tan^{-1}(\omega RC)}, \tag{2.1-16b}$$

where $|H(\omega)| = \dfrac{1}{\sqrt{1 + (\omega RC)^2}}$ and $\theta(\omega) = -\tan^{-1}(\omega RC)$ are called the *magnitude frequency spectrum* and *phase frequency spectrum* of the system, respectively. In Figure 2.1-19, we show the plot of the magnitude frequency spectrum for $RC = 0.005$ s, where ω is measured in units of radian/second. Note that the *RC* circuit is a simple *signal processing* circuit. Indeed, the circuit performs low-pass filtering to the input $x(t)$ as it allows low-frequency signals to pass while attenuating high-frequency signals. In a subsequent chapter, for a given input image presented in an optical system, we will show how *image processing* can be performed and we can analyze an optical system by finding its transfer function.

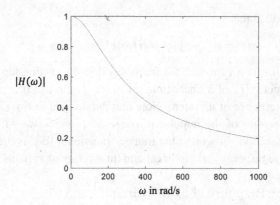

Figure 2.1-19 Magnitude frequency spectrum of an RC filter

Figure 2.1-19 is generated by RC_filter.m shown below.

```
% Plotting of the magnitude frequency response function; RC_
filter.m
clear all;
w=0:0.2:1000;
b=1;
H=1./(b+j*w*0.005);
plot(w,abs(H));
xlabel('w in rad/s')
ylabel('|H|')
```

Example: Response Due to a Sinusoidal Input

In the last example, for a linear and time-invariant system, we can find its transfer function $H(\omega)$. For a complex exponential input $x(t) = e^{j\omega t}$, we have the output $y(t) = H(\omega)e^{j\omega t}$. We now extend this result for sinusoidal inputs, which are real signals. Since $x(t) = e^{j\omega t} \to y(t) = H(\omega)e^{j\omega t}$, then $x(t) = e^{-j\omega t} \to y(t) = H(-\omega)e^{-j\omega t}$. So for a sinusoidal input, we have $x(t) = \cos\omega t = \frac{1}{2}e^{j\omega t} + \frac{1}{2}e^{-j\omega t} \to y(t) = \frac{1}{2}H(\omega)e^{j\omega t} + \frac{1}{2}H(-\omega)e^{-j\omega t}$ according to superposition. Now, from Eq. (2.1-15), we have $H(-\omega) = \int\limits_{-\infty}^{\infty} h(t')e^{j\omega t'}dt'$ and $H^*(\omega) = \int\limits_{-\infty}^{\infty} h^*(t')e^{j\omega t'}dt'$. So we can deduce that for $h(t)$ real, which is the case for physical systems with real inputs, we have $H(-\omega) = H^*(\omega)$. Therefore,

$$x(t) = \cos\omega t = \frac{1}{2}e^{j\omega t} + \frac{1}{2}e^{-j\omega t} \to y(t) = \frac{1}{2}H(\omega)e^{j\omega t} + \frac{1}{2}H(-\omega)e^{-j\omega t}$$
$$= \frac{1}{2}H(\omega)e^{j\omega t} + \frac{1}{2}H^*(\omega)e^{-j\omega t} = \mathrm{Re}\big[H(\omega)e^{j\omega t}\big].$$

Since $H(\omega) = |H(\omega)| e^{j\theta(\omega)}$,   we   write   the   final   output

$y(t) = \operatorname{Re}\left[H(\omega) e^{j\omega t}\right] = \operatorname{Re}\left[|H(\omega)| e^{j\theta(\omega)} e^{j\omega t}\right] = |H(\omega)| \cos(\omega t + \theta(\omega))$.   Summarizing

the result, we write

$$x(t) = \cos\omega t \rightarrow y(t) = |H(\omega)| \cos(\omega t + \theta(\omega)). \qquad (2.1\text{-}17)$$

This is an important result for sinusoidal inputs. As an example for the *RC* circuit, using Eq. (2.1-16), we have

$$x(t) = \cos\omega t \rightarrow y(t) = \frac{1}{\sqrt{1 + (\omega RC)^2}} \cos(\omega t - \tan^{-1}(\omega RC)).$$

While we have managed to find the output of linear and time-invariant systems for a single frequency input, how do we find the output for aperiodic signals, such as a pulse or a segment of audio signal, which as it turns out consists of many frequencies? A short answer to the question is through the use of Fourier analysis, to be discussed in the next section.

2.2   Fourier Analysis

2.2.1   Fourier Series

Any periodic function in time that appears in physical systems can be expanded into a *Fourier series*. A periodic signal $x(t)$ with period T_0 satisfies $x(t) = x(t + T_0)$ for all t. The smallest such T_0 is called the *fundamental period*. Its *fundamental frequency* is $f_0 = \dfrac{\omega_0}{2\pi} = 1/T_0$ and ω_0 is the *fundamental radian frequency*. For the units of T_0 in seconds [s], the units of f_0 are hertz [Hz] and of ω_0 are radians per second [rad/s]. Any function that is not periodic is called *aperiodic*. One of the simplest periodic functions is the function, for example, $\cos(\omega_0 t)$ as we can see that $\cos\omega_0(t + T_0) = \cos(\omega_0 t + \omega_0 T_0) = \cos(\omega_0 t + 2\pi) = \cos\omega_0 t$. Other examples of periodic functions are shown in Figure 2.2-1. Figure 2.2-1a is a *pulse train*, which is an interesting time signal. Its optical analog is a *diffraction grating*, as we will see it in a subsequent chapter. Figure 2.2-1b shows a sawtooth function.

Any periodic signal with T_0 can be represented by

$$x(t) = \sum_{n=-\infty}^{\infty} X_n e^{jn\omega_0 t}, \qquad (2.2\text{-}1)$$

where X_n are called the *Fourier coefficients*, ω_0 is the fundamental radian frequency or the *first harmonic* of $x(t)$, and the frequency $n\omega_0$ is called the nth *harmonic* with n taking on all integer values from $-\infty$ to ∞. The summation in Eq. (2.2-1) is called the *Fourier series* of $x(t)$. For a given $x(t)$, our objective is to find X_n. To derive

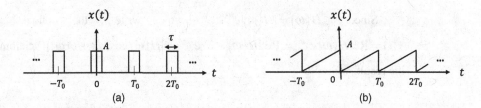

Figure 2.2-1 Examples of periodic functions with period T_0. (a) Pulse train with amplitude A and with pulse width τ and (b) sawtooth with amplitude A

X_n, we multiply both sides of Eq. (2.2-1) by $e^{-jm\omega_0 t}$, where m is an integer, and then integrate over one period $T_0 = 2\pi / \omega_0$. We, therefore, have

$$\int_{t_0}^{t_0+T_0} x(t) e^{-jm\omega_0 t} dt = \sum_{n=-\infty}^{\infty} X_n \int_{t_0}^{t_0+T_0} e^{j(n-m)\omega_0 t} dt \tag{2.2-2}$$

for any real number t_0. To evaluate the right side of the equation, we employ the *orthogonality of exponents* stated as follows:

$$\int_{t_0}^{t_0+T_0} e^{j(n-m)\omega_0 t} dt = \begin{cases} 0, & m \neq n \\ T_0, & m = n . \end{cases} \tag{2.2-3}$$

Hence, out of the infinite series, only one term will survive and that is when $n = m$. Equation (2.2-2) becomes

$$\int_{t_0}^{t_0+T_0} x(t) e^{-jm\omega_0 t} dt = X_m T_0$$

or

$$X_m = \frac{1}{T_0} \int_{t_0}^{t_0+T_0} x(t) e^{-jm\omega_0 t} dt.$$

To summarize, the Fourier series of $x(t)$ with a period T_0 is expressed as

$$x(t) = \sum_{n=-\infty}^{\infty} X_n e^{jn\omega_0 t}, \tag{2.2-4a}$$

where $\omega_0 = 2\pi / T_0$ and

$$X_n = \frac{1}{T_0} \int_{t_0}^{t_0+T_0} x(t) e^{-jn\omega_0 t} dt. \tag{2.2-4b}$$

Note that, for $n = 0$ in Eq. (2.2-4b), we have

$$X_0 = \frac{1}{T_0} \int_{t_0}^{t_0+T_0} x(t) dt.$$

X_0 is the *average value* of the signal $x(t)$. The average value is also called the (Direct Current) *DC value*, a term that is originated from circuit analysis. For some signals, the DC value can be found easily by inspection. Since the Fourier coefficients X_n are

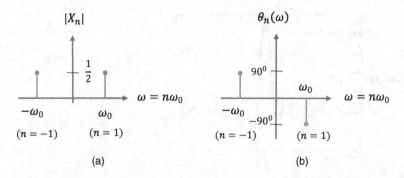

Figure 2.2-2 (a) Magnitude spectrum and (b) phase spectrum of a sine signal

functions of $\omega = n\omega_0$ and are complex in general, we can write $X_n = |X_n| e^{j\theta_n(\omega)}$, where $|X_n|$ and $\theta_n(\omega)$ are called the *magnitude spectrum* and *phase spectrum* of the signal $x(t)$, respectively.

Example: Fourier Series of a Sine Function

Consider $x(t) = \sin \omega_0 t$. Let us find the Fourier coefficients of $\sin \omega_0 t$, and also plot its magnitude and phase spectra. In a straightforward way, we simply put $x(t) = \sin \omega_0 t$ into Eq. (2.2-4b) to evaluate X_n. However, a simpler approach is to recognize that we can obtain an expression for the sine function in terms of exponential functions through the identity $\sin \omega_0 t = \left(e^{j\omega_0 t} - e^{-j\omega_0 t} \right)/2j$. So we write

$$x(t) = \sin \omega_0 t = \frac{e^{j\omega_0 t} - e^{-j\omega_0 t}}{2j} = \sum_{n=-\infty}^{\infty} X_n e^{jn\omega_0 t}.$$

We can now identify that n only exists for $+1$ and -1. Therefore, the Fourier coefficients are $X_1 = |X_1| e^{j\theta_1(\omega)} = \dfrac{1}{2j} = \dfrac{1}{2} e^{-j90^\circ}$ and $X_{-1} = |X_{-1}| e^{j\theta_{-1}(\omega)} = \dfrac{-1}{2j} = \dfrac{1}{2} e^{j90^\circ}$. Figure 2.2-2a and b shows the plots for the magnitude spectrum and phase spectrum, respectively. These plots are called *line spectra* because vertical lines indicate the magnitudes and phases. It is easy to understand what a positive value of the angular frequency ω_0 means in both of the plots. How do we interpret a negative frequency? It does not have a physical meaning, as defining ω along the negative axis is purely a mathematical convenience. For example, we represent $\sin \omega_0 t$ by two complex exponents $\left(e^{j\omega_0 t} - e^{-j\omega_0 t} \right)/2j$.

Example: Fourier Series of a Pulse Train

Let us find the Fourier series of a pulse train of an amplitude A and a pulse width of τ shown in Figure 2.2-1a. According to Eq. (2.2-4b), we have

$$X_n = \frac{1}{T_0} \int_{t_0}^{t_0+T_0} x(t) e^{-jn\omega_0 t} dt = \frac{1}{T_0} \int_{-\frac{T_0}{2}}^{\frac{T_0}{2}} x(t) e^{-jn\omega_0 t} dt = \frac{1}{T_0} \int_{-\frac{\tau}{2}}^{\frac{\tau}{2}} A e^{-jn\omega_0 t} dt$$

$$= \frac{A}{T_0} \frac{e^{-jn\omega_0 t}}{-jn\omega_0} \bigg|_{-\tau/2}^{\tau/2} = \frac{A}{T_0(-jn\omega_0)} \left[e^{-jn\omega_0 \tau/2} - e^{+jn\omega_0 \tau/2} \right].$$

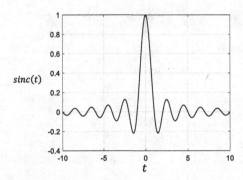

Figure 2.2-3 Plot of $\text{sinc}(t)$ function

```
% Plotting of the sinc function, sinc_function.m
t = -10:0.01:10;
plot(t,sinc(t)), grid on
xlabel('t')
ylabel('sinc(t)')
```

Now using $2j\sin\theta = e^{j\theta} - e^{-j\theta}$, and recognizing $\omega_0 T_0 = 2\pi$, the above equation is manipulated to become

$$X_n = \frac{A}{n\pi}\sin\left(\frac{n\omega_0\tau}{2}\right) = \frac{A}{n\pi}\sin\left(\frac{n\pi\tau}{T_0}\right). \tag{2.2-5}$$

Note that all the information of the pulse train, that is, A, τ, and T_0, contain in the Fourier coefficients X_n. X_n can be written in terms of the *sinc function*, which plays an important role in signal processing as well as in the theory of Fourier analysis.

Let us introduce the sinc function [pronounced "sink"]: $\text{sinc}(t) = \dfrac{\sin\pi t}{\pi t}$. The plot of the sinc function is shown in Figure 2.2-3 along with its MATLAB code. Note that the $\text{sinc}(t)$ function has a value of unity at $t = 0$. $\text{sinc}(t)$ has zeros at the points at which $\sin(\pi t)$ is zero, which is, at $\pm 1, \pm 2, \ldots$

With the definition of the sinc function, we can rewrite X_n as

$$X_n = \frac{A}{n\pi}\sin\left(\frac{n\omega_0\tau}{2}\right) = \frac{A\tau}{T_0}\text{sinc}\left(\frac{n\omega_0\tau}{2\pi}\right).$$

In Figure 2.2-4, we plot X_n for $\tau = 2\pi$, $T_0 = 6\pi$, and $A = 3$. Note that since X_n is real, we can present the whole spectrum of the pulse train with a single graph, and such a graph is called the *amplitude spectrum*.

2.2.2   Fourier Transform

In the last section, we have seen that we can represent a given periodic signal by the Fourier series and the Fourier coefficients are displayed as discrete line spectra. In this

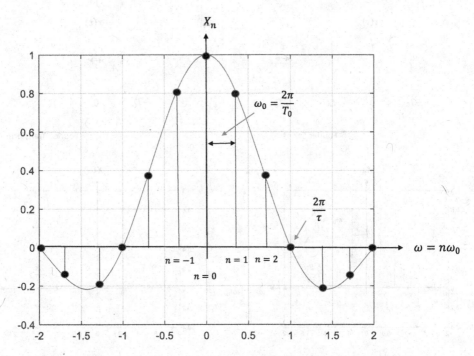

Figure 2.2-4 Amplitude spectrum of a pulse train for $\tau = 2\pi$, $T_0 = 6\pi$, and $A = 3$

section, we investigate aperiodic signals and introduce the *Fourier transform*. Let us take the pulse train discussed from the last section as an example and see what happens when we let $T_0 \to \infty$. Figure 2.2-5 illustrates the situation. From Figure 2.2-5a, we can see that as T_0 increases, the line spectrum becomes denser. At $T_0 \to \infty$, $\omega = n\omega_0$ becomes a continuum as it is illustrated in Figure 2.2-5b. Hence, the discrete sinc function in Figure 2.2-5a has become a continuous function in ω. Note that, however, the first zero of the sinc function remains at $2\pi / \tau$.

Mathematically, we can derive the Fourier transform of a signal from X_n. From Eq. (2.2-4b) and re-stating it below:

$$X_n = \frac{1}{T_0} \int_{t_0}^{t_0+T_0} x(t) e^{-jn\omega_0 t} dt.$$

In order not to have a zero when $T_0 \to \infty$ on the right-hand side of the equation, we re-write the equation and take the limit $T_0 \to \infty$ as follows:

$$\lim_{T_0 \to \infty} T_0 X_n = \lim_{T_0 \to \infty} \int_{-\frac{T_0}{2}}^{\frac{T_0}{2}} x(t) e^{-jn\omega_0 t} dt = \int_{-\infty}^{\infty} x(t) e^{-j\omega t} dt, \tag{2.2-6}$$

where we have replaced $n\omega_0$ by ω as $T_0 \to \infty$ in the last step of Eq. (2.2-6). The above integral is known as the *forward Fourier transform* $X(\omega)$ of the signal $x(t)$ and is written as

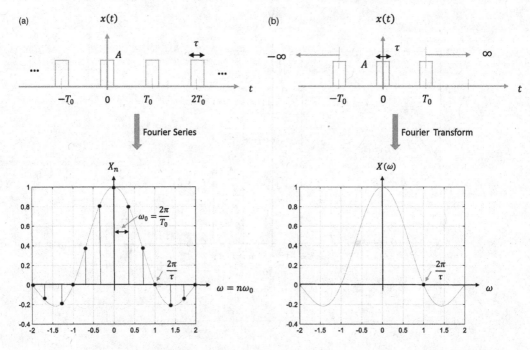

Figure 2.2-5 (a) Periodic pulse train and its amplitude spectrum, X_n and (b) aperiodic pulse and its spectrum, $X(\omega)$ when $T_0 \to \infty$ in the pulse train in (a)

$$\mathcal{F}\{x(t)\} = \int_{-\infty}^{\infty} x(t)e^{-j\omega t}dt = X(\omega), \qquad (2.2\text{-}7a)$$

where $\mathcal{F}\{x(t)\}$ is an operator notation for the Fourier transform of $x(t)$. $X(\omega)$ is called the *spectrum* of the signal $x(t)$ in signal processing as well as in circuit analysis. Since $X(\omega)$ is complex in general, we can write

$$X(\omega) = |X(\omega)|e^{j\theta(\omega)},$$

where $|X(\omega)|$ is the *magnitude spectrum* and $\theta(\omega)$ is the *phase spectrum* of the signal $x(t)$.

For completeness, we state the *inverse Fourier transform* as $\mathcal{F}^{-1}\{X(\omega)\}$ and defined as

$$\mathcal{F}^{-1}\{X(\omega)\} = \frac{1}{2\pi}\int_{-\infty}^{\infty} X(\omega)e^{j\omega t}d\omega = x(t). \qquad (2.2\text{-}7b)$$

Equations (2.2-7a) and (2.2-7b) are referred to as the *Fourier transform pair,* and their relationship is often represented symbolically as

$$x(t) \Leftrightarrow X(\omega).$$

The Fourier transform is *invertible,* that is, it is possible to recover a function from its Fourier transform. Indeed, the *Fourier integral theorem* states that

$$x(t) = \mathcal{F}^{-1}\{\mathcal{F}\{x(t)\}\}.$$ (2.2-8a)

We can also check easily that

$$X(\omega) = \mathcal{F}\{\mathcal{F}^{-1}\{X(\omega)\}\}.$$ (2.2-8b)

Note that square integrability, that is,

$$\int_{-\infty}^{\infty} |x(t)|^2\, dt < \infty,$$ (2.2-9)

is a *sufficient condition* for the existence of the Fourier transform. In other words, $x(t)$ is Fourier transformable if Eq. (2.2-9) is satisfied. Square integrability, however, is not a *necessary condition* for the function to be Fourier transformable. Some functions such as a step function or a sinusoidal function do not satisfy Eq. (2.2-9) and yet have Fourier transforms.

Example: Fourier Transform of a Pulse
We simply evaluate the integral given by Eq. (2.2-7a) by recognizing that $x(t) = \text{rect}(t)$. Therefore, we write

$$\mathcal{F}\{\text{rect}(t)\} = \int_{-\infty}^{\infty} \text{rect}(t) e^{-j\omega t}\, dt$$

$$= \int_{-1/2}^{1/2} 1\, e^{-j\omega t}\, dt \;=\; \frac{e^{-j\omega t}}{-j\omega}\bigg|_{-1/2}^{1/2} = \frac{e^{-j\omega/2} - e^{j\omega/2}}{-j\omega}$$

$$= 2\frac{e^{j\omega/2} - e^{-j\omega/2}}{2j\omega} = \frac{\sin(\omega/2)}{\omega/2} = \text{sinc}\left(\frac{\omega}{2\pi}\right).$$

Example: Use of Fourier Transform for RC-Circuit Analysis
We have found the frequency response function $H(\omega)$ of an *RC* circuit earlier and shown that the magnitude of $H(\omega)$ allows us to investigate the filtering characteristic of the circuit. Let us start with the system equation of the circuit, stated below for convenience:

$$RC\frac{dy(t)}{dt} + y(t) = x(t).$$

Instead of investigating the behavior of the system in the time domain, we could analyze it in the *frequency domain* with the Fourier transform. Taking the Fourier transform of both sides of the above equation, we obtain

$$RC\mathcal{F}\left\{\frac{dy(t)}{dt}\right\}+\mathcal{F}\left\{y(t)\right\}=\mathcal{F}\left\{x(t)\right\}. \tag{2.2-10}$$

Let $\mathcal{F}\left\{x(t)\right\}=X(\omega)$ and $\mathcal{F}\left\{y(t)\right\}=Y(\omega)$. We now need to find $\mathcal{F}\left\{\frac{dy(t)}{dt}\right\}$. Using Eq. (2.2-7b), we write

$$y(t)=\mathcal{F}^{-1}\left\{Y(\omega)\right\}=\frac{1}{2\pi}\int_{-\infty}^{\infty}Y(\omega)e^{j\omega t}d\omega.$$

After differentiating the above equation, we get

$$\frac{dy(t)}{dt}=\frac{1}{2\pi}\int_{-\infty}^{\infty}j\omega Y(\omega)e^{j\omega t}d\omega=\mathcal{F}^{-1}\left\{j\omega Y(\omega)\right\}.$$

Now, taking the Fourier transform on both sides of the above equation, we have

$$\mathcal{F}\left\{\frac{dy(t)}{dt}\right\}=\mathcal{F}\left\{\mathcal{F}^{-1}\left\{j\omega Y(\omega)\right\}\right\}=j\omega Y(\omega). \tag{2.2-11}$$

Using this result, Eq. (2.2-10) becomes

$$RCj\omega Y(\omega)+Y(\omega)=X(\omega),$$

which gives

$$\frac{Y(\omega)}{X(\omega)}=\frac{1}{1+j\omega RC}=H(\omega). \tag{2.2-12}$$

This result is identical to that shown in Eq. (2.1-16a). However, from the above result, we see that the frequency response function $H(\omega)$ is simply the ratio of the output spectrum $Y(\omega)$ to the input spectrum $X(\omega)$. Hence, we can see that the Fourier transform is very useful in determining the frequency response (or transfer function) of the system. In a subsequent chapter, we will also take advantage of the Fourier transform to introduce the *coherent transfer function* and *optical transfer function* for optical systems. As a final note to this example, and with reference to Eq. (2.1-15) re-stated below

$$H(\omega)=\int_{-\infty}^{\infty}h(t')e^{-j\omega t'}dt',$$

we see that from the definition of the Fourier transform, we have $H(\omega)=\mathcal{F}\left\{h(t)\right\}$. In words, it says the frequency response function of the system is the Fourier transform of the impulse response $h(t)$.

Example: Response Due to an Aperiodic Signal
We have discussed a response in a linear time-invariant system due to a sinusoidal input. Now, we can find the response due to any aperiodic signal. Using Eq. (2.2-12), we write

$$Y(\omega) = X(\omega)H(\omega). \tag{2.2-13a}$$

In the time domain, we have

$$y(t) = \mathcal{F}^{-1}\{Y(\omega)\} = \mathcal{F}^{-1}\{X(\omega)H(\omega)\}. \tag{2.2-13b}$$

This is an important result for linear and time-invariant systems. It says that the input spectrum $X(\omega)$ is modified or filtered by the system's transfer function $H(\omega)$ as the input signal goes through the system to give a filtered version of the input as an output. Indeed, we have seen a simple RC circuit [see Figure 2.1-15] as a low-pass filter. Filtering is characterized by the transfer function of the system.

2.3 Fourier Analysis in Two Dimensions

The study of two-dimensional (2-D) Fourier transform follows closely the study of the one-dimensional (1-D) Fourier transform discussed in the last section. Hence, our devotion to the 1-D Fourier transform will prove to be useful. Many engineering problems such as in digital image processing and optical system analysis, the inputs and outputs are typically 2-D and, in some cases, higher dimensional. For the case of planar images, $f(x, y)$, we deal with two spatial dimensions, which we label x and y. In this section, we devote our time to 2-D signals.

2.3.1 The Two-Dimensional Fourier Transform

A function, $f(x, y)$, of two spatial variables x and y is called a 2-D signal or a 2-D spatial function. A common example is a photographic image or a computer-generated image, wherein the variables x and y are the spatial coordinates of the image, and $f(x, y)$ is the amplitude of gray values. However, in general, the function $f(x, y)$ takes on complex values. In that case, depiction of a 2-D complex function can be represented in terms of the real and imaginary parts, or the magnitude and phase parts since a complex number $z = a + bj = \sqrt{a^2 + b^2}\, e^{j\tan^{-1}(b/a)}$, where a and b are the real and imaginary parts of z, and $\sqrt{a^2 + b^2}$ and $\tan^{-1}(b/a)$ are the magnitude and phase parts of z. When dealing with 1-D time signals, we have two independent transform variables $(t; \omega)$, where t is measured in seconds, and radian temporal frequency ω is measured in rad/s. With $f(x, y)$, we have four independent transform variables $(x, y; k_x, k_y)$, where x and y are measured in meters, and k_x and k_y are called *radian spatial frequencies* in rad/m.

The *two-dimensional (2-D) forward Fourier transform* of a spatial signal $f(x, y)$ is defined as

$$\mathcal{F}\{f(x, y)\} = F(k_x, k_y) = \iint\limits_{-\infty}^{\infty} f(x, y)e^{jk_x x + jk_y y}\, dx dy, \tag{2.3-1a}$$

and the *two-dimensional (2-D) inverse Fourier transform* is

$$\mathcal{F}^{-1}\{F(k_x,k_y)\}=f(x,y)=\frac{1}{4\pi^2}\iint\limits_{-\infty}^{\infty}F(k_x,k_y)e^{-jk_xx-jk_yy}dk_xdk_y. \qquad (2.3\text{-}1b)$$

$f(x,y)$ and $F(k_x,k_y)$ is a Fourier transform pair, and symbolically we write

$$f(x,y)\Leftrightarrow F(k_x,k_y).$$

We remark that the definitions of 2D "spatial" Fourier and inverse transforms are different from the definitions of "temporal" Fourier and inverse transforms in that, for the former, we use jk_xx+jk_yy and $-jk_xx-jk_yy$ as the exponents for the forward and inverse transforms, respectively, whereas $-j\omega t$ and $j\omega t$ serve as the exponents in the latter. This is done to be consistent with the engineering convention for a travelling wave. In the convention, $\text{Re}\left[e^{j\omega t-jkz}\right]$ denotes a wave travelling in the $+z$ direction with temporal frequency ω and a propagation constant k, where $\text{Re}[\cdot]$ denotes taking the real part of the bracketed complex quantity.

2.3.2 Calculation Examples of Some 2-D Fourier Transforms

2-D Delta Function and Its Alternative Definition

The 2-D delta function is defined as

$$\delta(x,y)=\delta(x)\delta(y).$$

The function is used to model an ideal point source or a pin hole aperture in optical systems. From the definition of the Fourier transform in Eq. (2.3-1a), we have

$$\mathcal{F}\{\delta(x,y)\}=\iint\limits_{-\infty}^{\infty}\delta(x,y)e^{jk_xx+jk_yy}dxdy.$$

Since $\delta(x,y)$ is a function of two independent variables and is a *separable function*, we re-write the above equation as a product of two functions, each of them depending on only one variable as follows:

$$\mathcal{F}\{\delta(x,y)\}=\mathcal{F}\{\delta(x)\delta(y)\}=\int\limits_{-\infty}^{\infty}\delta(x)e^{jk_xx}dx\times\int\limits_{-\infty}^{\infty}\delta(y)e^{jk_yy}dy$$

$$=e^{jk_xx}\Big|_{x=0}\times e^{jk_yy}\Big|_{y=0}=1,$$

where we have used the sampling property of a delta function from Eq. (2.1-5), that is, $\int\limits_{-\infty}^{\infty}x(t)\delta(t-t_0)dt=x(t_0)$ to obtain the final result. We can state the result symbolically as

$$\delta(x,y)\Leftrightarrow 1. \qquad (2.3\text{-}2)$$

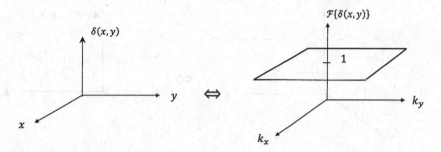

Figure 2.3-1 A 2-D delta function and its Fourier transform

The Fourier transform pair is illustrated in Figure 2.3-1.

We have seen the definition of the delta function with a rectangular function [see Figure 2.1-7]. In what follows, we obtain an integral definition of the delta function, which is useful for simplifying some integrals. Let us find the 2-D Fourier transform of 1. According to the definition of the Fourier transform, we write

$$\mathcal{F}\{1\} = \iint\limits_{-\infty}^{\infty} 1 \times e^{jk_x x + jk_y y} dx dy = \int\limits_{-\infty}^{\infty} 1 \times e^{jk_x x} dx \times \int\limits_{-\infty}^{\infty} 1 \times e^{jk_y y} dy.$$

Now, employing standard techniques of calculus,

$$\int\limits_{-\infty}^{\infty} 1 \times e^{jk_x x} dx = \frac{e^{jk_x x}}{jk_x}\Bigg|_{-\infty}^{\infty},$$

which is indeterminate when we evaluate the limits. As we can see $e^{j\theta}\big|_{\theta \to \infty} = (\cos\theta + j\sin\theta)\big|_{\theta \to \infty} \to$ indeterminate. To bypass such an issue, let us take the inverse transform of the delta function as follows:

$$\mathcal{F}^{-1}\{\delta(k_x, k_y)\} = \frac{1}{4\pi^2} \iint\limits_{-\infty}^{\infty} \delta(k_x, k_y) e^{-jk_x x - jk_y y} dk_x dk_y$$

$$= \frac{1}{4\pi^2} e^{-jk_x x - jk_y y}\Bigg|_{k_x = k_y = 0} = \frac{1}{4\pi^2},$$

where we have used the sampling property of the delta function to evaluate the integral. Now, let us take the transform on both sides of the above equation, we then have

$$\mathcal{F}^{-1}\{4\pi^2 \delta(k_x, k_y)\} = 1.$$

Finally, taking the transform of the above and using the Fourier integral theorem,

$$\mathcal{F}\{\mathcal{F}^{-1}\{4\pi^2 \delta(k_x, k_y)\}\} = 4\pi^2 \delta(k_x, k_y) = \mathcal{F}\{1\}, \qquad (2.3\text{-}3a)$$

or

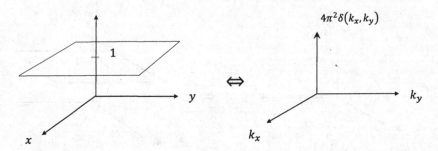

Figure 2.3-2 A constant and its Fourier transform

$$1 \Leftrightarrow 4\pi^2 \delta\left(k_x, k_y\right). \tag{2.3-3b}$$

The Fourier transform pair is illustrated in Figure 2.3-2.

From the result of Eq. (2.3-3a), we write

$$\mathcal{F}\{1\} = 4\pi^2 \delta\left(k_x, k_y\right),$$

or using the definition of the Fourier transform, we obtain

$$\mathcal{F}\{1\} = \int\!\!\!\int_{-\infty}^{\infty} 1 \times e^{jk_x x + jk_y y} \, dx \, dy = 4\pi^2 \delta\left(k_x, k_y\right).$$

From the above equality, we write

$$\delta\left(k_x, k_y\right) = \frac{1}{4\pi^2} \int\!\!\!\int_{-\infty}^{\infty} e^{jk_x x + jk_y y} \, dx \, dy.$$

If we make the following substitutions in the integral: $\left(k_x, k_y\right)$ by (x, y), and (x, y) by (x', y'), the integral becomes

$$\delta(x, y) = \frac{1}{4\pi^2} \int\!\!\!\int_{-\infty}^{\infty} e^{jxx' + jyy'} \, dx' \, dy'.$$

Since the delta function is an even function, which follows clearly from the definition according to Figure (2.1-7), we have

$$\delta(-x, -y) = \frac{1}{4\pi^2} \int\!\!\!\int_{-\infty}^{\infty} e^{-jxx' - jyy'} \, dx' \, dy' = \delta(x, y).$$

So we can write a general integral definition of the delta function as

$$\delta(x, y) = \frac{1}{4\pi^2} \int\!\!\!\int_{-\infty}^{\infty} e^{\pm jxx' \pm jyy'} \, dx' \, dy'. \tag{2.3-4}$$

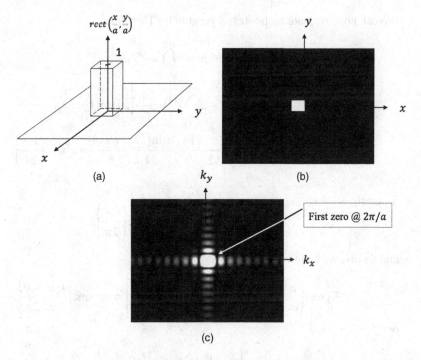

Figure 2.3-3 (a) 3-D plot, (b) gray scale plot, and (c) magnitude of the Fourier transform, of the 2-D rectangular function for $a = b$

2-D Rectangular Function

The 2-D rectangular function is a separable function and is defined as

$$\text{rect}\left(\frac{x}{a}\right)\text{rect}\left(\frac{y}{b}\right) = \text{rect}\left(\frac{x}{a}, \frac{y}{b}\right). \tag{2.3-5}$$

In Figure 2.3-3a and b, we have shown the 3-D plot and the gray scale plot of the function for $a = b$, respectively. In the gray scale plot, we have assumed that an amplitude of 1 translates to "white" and an amplitude of zero to "black." The function could also be effective in representing a rectangular aperture within an optical system, where an amplitude of "1" means the opening or clear area of the aperture, and an amplitude of zero means the area of the aperture is opaque. We will return to this aspect of the representation of an aperture in a subsequent chapter when we deal with diffraction.

Let us find the Fourier transform of the rectangular function. By definition,

$$\mathcal{F}\left\{\text{rect}\left(\frac{x}{a}, \frac{y}{b}\right)\right\} = \iint\limits_{-\infty}^{\infty}\text{rect}\left(\frac{x}{a}, \frac{y}{b}\right)e^{jk_x x + jk_y y}\,dxdy$$

$$= \int\limits_{-\infty}^{\infty}\text{rect}\left(\frac{x}{a}\right)e^{jk_x x}\,dx \times \int\limits_{-\infty}^{\infty}\text{rect}\left(\frac{y}{b}\right)e^{jk_y y}\,dy.$$

We can now evaluate each integral separately. Therefore,

$$\int_{-\infty}^{\infty} \text{rect}\left(\frac{x}{a}\right) e^{jk_x x} dx = \int_{-a/2}^{a/2} 1 \times e^{jk_x x} dx$$

$$= \frac{e^{jk_x x}}{jk_x}\bigg|_{-a/2}^{a/2} = \frac{1}{jk_x}\left[e^{jk_x a/2} - e^{-jk_x a/2}\right] = \frac{1}{jk_x}\left[2j \sin\left(k_x a/2\right)\right]$$

$$= \frac{\sin\left(k_x a/2\right)}{k_x/2} = \frac{a\sin\left(k_x a/2\right)}{k_x a/2} = \frac{a\sin\left(\pi k_x a/2\pi\right)}{\pi k_x a/2\pi} = a \text{ sinc}\left(\frac{k_x a}{2\pi}\right).$$

Similarly,

$$\int_{-\infty}^{\infty} \text{rect}\left(\frac{y}{b}\right) e^{jk_y y} dy = b \text{ sinc}\left(\frac{k_y b}{2\pi}\right),$$

and finally, we have

$$\mathcal{F}\left\{\text{rect}\left(\frac{x}{a},\frac{y}{b}\right)\right\} = ab \text{ sinc}\left(\frac{k_x a}{2\pi}\right) \text{sinc}\left(\frac{k_y b}{2\pi}\right) = ab \text{ sinc}\left(\frac{k_x a}{2\pi},\frac{k_y b}{2\pi}\right),$$

or

$$\text{rect}\left(\frac{x}{a},\frac{y}{b}\right) \Leftrightarrow ab \text{ sinc}\left(\frac{k_x a}{2\pi},\frac{k_y b}{2\pi}\right). \tag{2.3-6}$$

Note that the 2-D sinc function is also a separable function. Figure 2.3-3c shows the magnitude of the 2-D Fourier transform of a rectangular function for $a = b$. Figure 2.3-3b and 2.3-3c is generated using the m-file shown below.

```
%Fourier transform of a square function; FT2D_rect.m
Clear

L=1;%length of display area
N=256;% number of sampling points
dx=L/(N-1);% dx : step size

% Create square image, M by M square, rect(x/a), M=odd number
M=17;
a=M/256
R=zeros(256);%assign a matrix (256x256) of zeros
r=ones(M);% assign a matrix (MxM) of ones
n=(M-1)/2;
R(128-n:128+n,128-n:128+n)=r;
%End of creating square input image

%Axis Scaling
for k=1:256
X(k)=1/255*(k-1)-L/2;
Y(k)=1/255*(k-1)-L/2;

%Kx=(2*pi*k)/((N-1)*dx)
%in our case, N=256, dx=1/255
```

```
Kx(k)=(2*pi*(k-1))/((N-1)*dx)-((2*pi*(256-1))/((N-1)*dx))/2;
Ky(k)=(2*pi*(k-1))/((N-1)*dx)-((2*pi*(256-1))/((N-1)*dx))/2;
end

%Image of the rectangular function
figure(1)
image(X+dx/2,Y+dx/2,255*R);
colormap(gray(256));
axis off

%Computing Fourier transform
FR=(1/256)^2*fft2(R);
FR=fftshift(FR);

%Magnitude spectrum of the rectangular function
figure(2);
gain=10000;
image(Kx,Ky,gain*(abs(FR)).^2/max(max(abs(FR))).^2)
axis off
colormap(gray(256))
```

Circular Function

The circular function is defined as

$$\text{circ}\left(\frac{r}{r_0}\right) = \begin{cases} 1, & r < r_0 \\ 0, & \text{otherwise}, \end{cases} \tag{2.3-7}$$

where $r = \sqrt{x^2 + y^2}$. This function is particularly useful in optics where lenses and aperture stops are often circular in shape. Since the problem of interest has circular symmetry, let us first express the Fourier transform in polar coordinates. We make a transformation from rectangular coordinates to polar coordinates in both the (x, y) and the (k_x, k_y) planes as follows:

$$x = r\cos\theta, \ y = r\sin\theta, \ k_x = k_r\cos\phi, \ k_y = k_r\sin\phi. \tag{2.3-8}$$

Figure 2.3-4 shows the relations between the two coordinates in the spatial domain as well as frequency domain.

Hence, the 2-D Fourier transform in rectangular coordinates of $f(x, y)$, that is,

$$\mathcal{F}\{f(x,y)\} = \iint_{-\infty}^{\infty} f(x,y) e^{jk_x x + jk_y y} dx dy = F(k_x, k_y),$$

transforms to

$$\mathcal{F}\{\bar{f}(r,\theta)\} = \int_0^\infty \int_0^{2\pi} \bar{f}(r,\theta) e^{jk_r r(\cos\theta\cos\phi + \sin\theta\sin\phi)} r d\theta dr$$

$$= \int_0^\infty \int_0^{2\pi} \bar{f}(r,\theta) e^{jk_r r\cos(\theta - \phi)} r d\theta dr = \bar{F}(k_r, \phi), \tag{2.3-9}$$

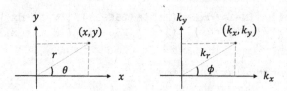

Figure 2.3-4 Relations between rectangular and polar coordinates

where we have used Eq. (2.3-8) to transform the integral and recognized that the differential area element $dxdy = rd\theta dr$. Note that $f(x,y)$ in polar coordinates is $\overline{f}(r,\theta)$ and $F(k_x,k_y)$ in polar coordinates is $\overline{F}(k_r,\phi)$.

A function $\overline{f}(r,\theta)$ is separable in polar coordinates if $\overline{f}(r,\theta) = \overline{f}_r(r)\overline{f}_\theta(\theta)$. When $\overline{f}_\theta(\theta) = 1$, $\overline{f}(r,\theta)$ is *circularly symmetric*, that is, $\overline{f}(r,\theta) = \overline{f}_r(r)$, Eq. (2.3-9) becomes

$$\mathcal{F}\{\overline{f}_r(r)\} = \int_0^\infty r\overline{f}_r(r)\left[\int_0^{2\pi} e^{jk_r r\cos(\theta-\phi)}d\theta\right]dr. \tag{2.3-10}$$

Since

$$J_0(\beta) = \frac{1}{2\pi}\int_0^{2\pi} e^{j\beta\cos(\theta-\phi)}d\theta, \tag{2.3-11}$$

where $J_0(\beta)$ is the *zero-order Bessel function of the first kind*, Eq. (2.3-10) can be written as

$$\mathcal{F}\{\overline{f}_r(r)\} = 2\pi\int_0^\infty r\overline{f}_r(r)J_0(k_r r)dr = \mathcal{B}\{\overline{f}_r(r)\}. \tag{2.3-12}$$

Equation (2.3-12) defines the *Fourier–Bessel transform* and arises in circularly symmetric problems. After developing the Fourier–Bessel transform, we can now find the Fourier transform of the circular function. According to Eq. (2.3-12), for $\overline{f}_r(r) = \text{circ}\left(\dfrac{r}{r_0}\right)$, we have

$$\mathcal{B}\left\{\text{circ}\left(\frac{r}{r_0}\right)\right\} = 2\pi\int_0^\infty r\,\text{circ}\left(\frac{r}{r_0}\right)J_0(k_r r)dr = 2\pi\int_0^{r_0} rJ_0(k_r r)dr. \tag{2.3-13}$$

Now, using the following identity of the Bessel functions:

$$J_1(\alpha) = \frac{1}{\alpha}\int_0^\alpha \beta J_0(\beta)d\beta, \tag{2.3-14}$$

where $J_1(\alpha)$ is the *first-order Bessel function of the first kind*, Eq. (2.3-13) can be simplified to give

$$\mathcal{B}\left\{\text{circ}\left(\frac{r}{r_0}\right)\right\} = \frac{2\pi r_0}{k_r}J_1(r_0 k_r). \tag{2.3-15}$$

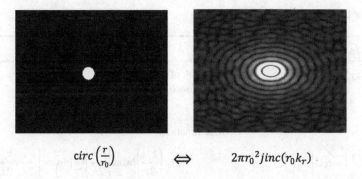

$$circ\left(\frac{r}{r_0}\right) \qquad \Longleftrightarrow \qquad 2\pi r_0^2 jinc(r_0 k_r)$$

Figure 2.3-5 A circular function and the magnitude of its Fourier transform

Defining the *jinc function* as

$$jinc(r) = \frac{J_1(r)}{r}, \qquad (2.3\text{-}16)$$

Eq. (2.3-15) becomes

$$\mathcal{B}\left\{circ\left(\frac{r}{r_0}\right)\right\} = 2\pi r_0^2 jinc\left(r_0 k_r\right). \qquad (2.3\text{-}17)$$

Figure 2.3-5 illustrates the Fourier transform pair and it is generated using the m-file shown below.

```
%fft_circular_function.m
%Simulation of Fourier transformation of a circular function
Clear

I=imread('smallcircle.bmp','bmp');%Input image of 256 by 256
I=I(:,:,1);
figure(1)%displaying input
colormap(gray(255));
image(I)
axis off

FI=fft2(I);
FI=fftshift(FI);
max1=max(FI);
max2=max(max1);
scale=1.0/max2;
FI=FI.*scale;

figure(2)%Gray scale image of the absolute value of the
transform
colormap(gray(255));
image(10*(abs(256*FI)));
axis off
```

Table 2.1 Select Fourier transform pairs

$f(x,y)$	$F(k_x,k_y)$
1. Delta function: $\delta(x,y)$	Constant of unity: 1
2. Constant of unity: $\qquad 1$	Delta function: $\qquad 4\pi^2\delta(k_x,k_y)$
3. Rectangular function : $\text{rect}\left(\dfrac{x}{a}\right)\text{rect}\left(\dfrac{y}{b}\right)=\text{rect}\left(\dfrac{x}{a},\dfrac{y}{b}\right)$	Sinc function: $a\,\text{sinc}\left(\dfrac{k_x a}{2\pi}\right)b\,\text{sinc}\left(\dfrac{k_y b}{2\pi}\right)=ab\,\text{sinc}\left(\dfrac{k_x a}{2\pi},\dfrac{k_y b}{2\pi}\right)$
4. Circular function: $\text{circ}\left(\dfrac{r}{r_0}\right)$	Jinc function: $2\pi r_0^2\,\text{jinc}(r_0 k_r)$
5. Gaussian function: $e^{-\alpha(x^2+y^2)}$	Gaussian function: $\dfrac{\pi}{\alpha}e^{-\frac{1}{4\alpha}(k_x^2+k_y^2)}$
6. Complex Fresnel zone plate (CFZP) $e^{-j\alpha(x^2+y^2)}$	Complex Fresnel zone plate (CFZP) $\dfrac{-j\pi}{\alpha}e^{\frac{j}{4\alpha}(k_x^2+k_y^2)}$

Table 2.1 summarizes our results in the section and lists some of the useful Fourier transform pairs.

2.3.3   Properties of the Fourier Transform

In this section, we will discuss some of the useful properties of the Fourier transform.

1. Linearity Property

Since the Fourier transform is a linear operation, superposition applies.
    If

$$f_1(x,y) \Leftrightarrow F_1(k_x,k_y) \text{ and } f_2(x,y) \Leftrightarrow F_2(k_x,k_y),$$

then

$$af_1(x,y)+bf_2(x,y) \Leftrightarrow aF_1(k_x,k_y)+bF_2(k_x,k_y), \qquad (2.3\text{-}18)$$

where a and b are any arbitrary complex constants in general. The derivation of this property follows from the integral definition of the Fourier transform. As an example, let us find the Fourier transform of a "fringe" pattern given by

$$f(x,y)=1+\cos(ax+by). \qquad (2.3\text{-}19)$$

Taking the Fourier transform, we have

$$
\mathcal{F}\left\{f\left(x,y\right)\right\} = \mathcal{F}\left\{1+\cos\left(ax+by\right)\right\}
$$

$$
= \mathcal{F}\left\{1+\frac{1}{2}\left[e^{j(ax+by)}+e^{-j(ax+by)}\right]\right\}
$$

$$
= \iint_{-\infty}^{\infty} 1 e^{jk_x x + jk_y y}\,dxdy \;+\; \iint_{-\infty}^{\infty}\frac{1}{2}e^{j(ax+by)}e^{jk_x x + jk_y y}\,dxdy \;+\; \iint_{-\infty}^{\infty}\frac{1}{2}e^{-j(ax+by)}e^{jk_x x + jk_y y}\,dxdy.
$$

By grouping similar terms in the exponents of the second and the third terms on the right side of the equation, we have

$$
\mathcal{F}\left\{1+\cos\left(ax+by\right)\right\}
$$

$$
= \iint_{-\infty}^{\infty} 1 e^{jk_x x + jk_y y}\,dxdy \;+\; \iint_{-\infty}^{\infty}\frac{1}{2}e^{j(k_x+a)x+j(k_y+b)y}\,dxdy \;+\; \iint_{-\infty}^{\infty}\frac{1}{2}e^{j(k_x-a)x+j(k_y-b)y}\,dxdy.
$$

Now, by recognizing the integral definition of the delta function from Eq. (2.3-4), we finally obtain

$$
\mathcal{F}\left\{1+\cos\left(ax+by\right)\right\} = 4\pi^2\delta\left(k_x,k_y\right)+2\pi^2\delta\left(k_x+a,k_y+b\right)+
$$
$$
2\pi^2\delta\left(k_x-a,k_y-b\right). \tag{2.3-20}
$$

In Figure 2.3-6a, we illustrate the grayscale image of $f\left(x,y\right)=1+\cos\left(ax+by\right)$ with $a=b$. Its corresponding Fourier transform is illustrated with a 3-D plot with three delta functions in Figure 2.3-6b. Figure 2.3-6 is generated using the m-file shown below.

```
% plot2d.m
clear
close all
x=-15:0.15:15;
y=-x;
[x,y]=meshgrid(x,y);
z=1 + cos(2*x+2*y);
figure(1)
imshow(z);
Fz=fftshift(fft2(z));
Fzmax=max(max(Fz));
Fz1=Fz./Fzmax;
figure(2)
imshow(abs(Fz1));
figure(3)
plot3(x,y,abs(Fz1));
```

2.   Shifting Property

If

$$
f\left(x,y\right)\Leftrightarrow F\left(k_x,k_y\right),
$$

then

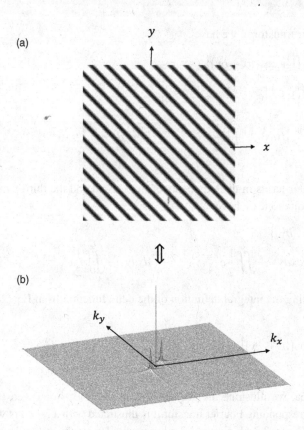

Figure 2.3-6 (a) The fringe pattern and (b) its corresponding Fourier transform

$$f(x-x_0, \ y-y_0) \Leftrightarrow e^{jk_x x_0 + jk_y y_0} F(k_x, k_y).$$ (2.3-21)

By definition,

$$\mathcal{F}\{f(x-x_0, \ y-y_0)\} = \int\!\!\!\int\limits_{-\infty}^{\infty} f(x-x_0, \ y-y_0) e^{jk_x x + jk_y y} dxdy.$$

Letting $x-x_0 = x'$ and $y-y_0 = y'$, we have

$$\mathcal{F}\{f(x-x_0, \ y-y_0)\} = \int\!\!\!\int\limits_{-\infty}^{\infty} f(x', \ y') e^{jk_x(x'+x_0)+jk_y(y'+y_0)} dx'dy'$$

$$= e^{jk_x x_0 + jk_y y_0} \int\!\!\!\int\limits_{-\infty}^{\infty} f(x', \ y') e^{jk_x x' + jk_y y'} dx'dy'$$

$$= e^{jk_x x_0 + jk_y y_0} F(k_x, k_y).$$

The result shows that shifting a signal by (x_0, y_0) does not change its magnitude spectrum. The phase spectrum, however, is changed by $(k_x x_0, k_y y_0)$. A simple example is

$$\mathcal{F}\left\{\text{rect}\left(x-x_0,y-x_0\right)\right\} = e^{jk_x x_0 + jk_y y_0} \, \text{sinc}\left(\frac{k_x}{2\pi},\frac{k_y}{2\pi}\right).$$

3.   Reciprocal Shifting Property

If

$$f(x,y) \Leftrightarrow F(k_x,k_y),$$

then

$$e^{j\alpha x + j\beta y} f(x,y) \Leftrightarrow F(k_x + \alpha, k_y + \beta). \qquad (2.3\text{-}22)$$

By definition,

$$\mathcal{F}\left\{e^{j\alpha x + j\beta y} f(x,y)\right\} = \iint\limits_{-\infty}^{\infty} e^{j\alpha x + j\beta y} f(x,y) e^{jk_x x + jk_y y} dxdy$$

$$= \iint\limits_{-\infty}^{\infty} f(x,y) e^{j(k_x+\alpha)x + j(k_y+\beta)y} dxdy = F(k_x + \alpha, k_y + \beta).$$

The result indicates that the multiplication of a linear phase factor $e^{j\alpha x + j\beta y}$ shifts the spectrum of that signal by $k_x = -\alpha$ and $k_y = -\beta$. A simple example is

$$\mathcal{F}\left\{e^{j\alpha x + j\beta y} \text{rect}(x,y)\right\} = \text{sinc}\left(\frac{k_x + \alpha}{2\pi}, \frac{k_y + \beta}{2\pi}\right).$$

Figure 2.3-7 shows the original magnitude spectrum and its shifted magnitude spectrum when the original function $\text{rect}(x,y)$ is multiplied by the linear phase factor $e^{j\alpha x + j\beta y}$.

4.   Scaling Property

If

$$f(x,y) \Leftrightarrow F(k_x,k_y),$$

then

$$f(ax,by) \Leftrightarrow \frac{1}{|ab|} F\left(\frac{k_x}{a},\frac{k_y}{b}\right), \qquad (2.3\text{-}23)$$

where a and b are any arbitrary complex constants. For brevity, we consider positive real constants $a > 0$ and $b > 0$. By definition, we have

$$\mathcal{F}\left\{f(ax,by)\right\} = \iint\limits_{-\infty}^{\infty} f(ax,by) e^{jk_x x + jk_y y} dxdy.$$

Letting $ax = x'$ and $by = y'$, we have

$$\mathcal{F}\left\{f(ax,by)\right\} = \iint\limits_{-\infty}^{\infty} f(x',y') e^{\frac{jk_x}{a}x' + \frac{jk_y}{b}y'} (dx'/a)(dy'/b)$$

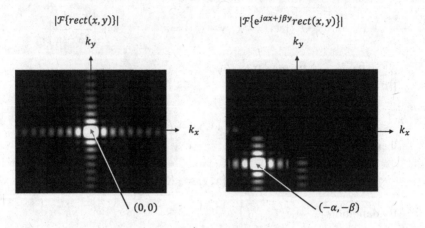

$|\mathcal{F}\{rect(x,y)\}|$

$|\mathcal{F}\{e^{j\alpha x+j\beta y}rect(x,y)\}|$

Figure 2.3-7 Illustration of the reciprocal shifting property

$$= \frac{1}{ab} \iint\limits_{-\infty}^{\infty} f(x',y') e^{j\frac{k_x}{a}x'+j\frac{k_y}{b}y'} dx'\, dy' = \frac{1}{ab}\, F\left(\frac{k_x}{a},\frac{k_y}{b}\right).$$

When a is negative, the limits on the x' integral are reversed as the variable of integration is changed so that

$$\mathcal{F}\{f(ax,by)\} = -\frac{1}{ab} F\left(\frac{k_x}{a},\frac{k_y}{b}\right)$$

for $a < 0$. These two results can be combined into the more compact form as follows:

$$\mathcal{F}\{f(ax,by)\} = \frac{1}{|ab|} F\left(\frac{k_x}{a},\frac{k_y}{b}\right).$$

The scaling property is illustrated graphically by Figure 2.3-8. Figure 2.3-8a displays a square function with "white" area of $x_0 \times x_0$, that is, $rect\left(\frac{x}{x_0},\frac{y}{x_0}\right)$. According to Eq. (2.3-6), its corresponding transform, $x_0^2 \, sinc\left(\frac{x_0 k_x}{2\pi},\frac{x_0 k_y}{2\pi}\right)$, is also shown. Note that the first zero of the sinc function is at $2\pi / x_0$. A similar square function is shown in Figure 2.3-8b, except that its area is now $2x_0 \times 2x_0$. Hence, the square function is a stretched-out version of the first one with a scaling factor $a = 1/2$. Hence, for $a < 1$, we have an expansion of the original spatial function. However, stretching the spatial function by a factor of 2 leads to compression of the spectrum by the same factor as we can see that the spectrum of $rect\left(\frac{x}{2x_0},\frac{y}{2x_0}\right)$ is $4x_0^2 \, sinc\left(\frac{2x_0 k_x}{2\pi},\frac{2x_0 k_y}{2\pi}\right)$ with its first zero at π / x_0.

These conclusions lead us to state that the function $f(ax,by)$ represents the function $f(x,y)$ scaled by factors a and b along the x- and y-directions, respectively. Similarly, the function $F\left(\frac{k_x}{a},\frac{k_y}{b}\right)$ represents the function $F(k_x,k_y)$ scaled in frequencies

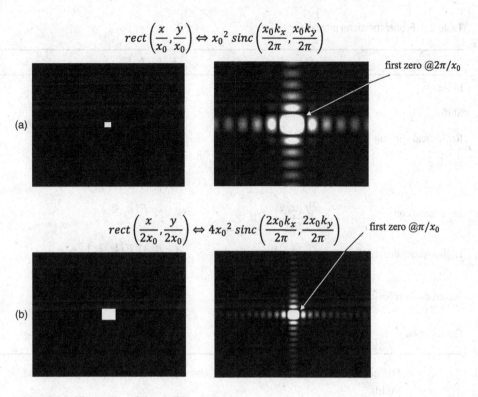

$$rect\left(\frac{x}{x_0},\frac{y}{x_0}\right) \Leftrightarrow x_0^2\, sinc\left(\frac{x_0 k_x}{2\pi},\frac{x_0 k_y}{2\pi}\right)$$

first zero @$2\pi/x_0$

(a)

$$rect\left(\frac{x}{2x_0},\frac{y}{2x_0}\right) \Leftrightarrow 4x_0^2\, sinc\left(\frac{2x_0 k_x}{2\pi},\frac{2x_0 k_y}{2\pi}\right)$$

first zero @π/x_0

(b)

Figure 2.3-8 Illustration of the scaling property

by $1/a$ and $1/b$ along the k_x and k_y, respectively. For $a<1$ and $b<1$, we have spatial expansion of the original signal, which leads to spectral compression by the same factor. On the other hand, for $a>1$ and $b>1$, we have spatial compression of the original signal, which leads to spectral expansion by the same factor.

5.   Differentiation Property

If

$$f(x,y) \Leftrightarrow F(k_x,k_y),$$

then

$$\frac{\partial f(x,y)}{\partial x} \Leftrightarrow -jk_x\, F(k_x,k_y). \tag{2.3-24}$$

The derivation of the above property is similar to that in the 1-D Fourier transform in the earlier example on the use of the Fourier Transform for an *RC* circuit. Starting from Eq. (2.3-1b), that is, the definition of the 2-D inverse Fourier transform, we have

$$f(x,y) = \frac{1}{4\pi^2}\iint\limits_{-\infty}^{\infty} F(k_x,k_y)e^{-jk_x x - jk_y y}\, dk_x dk_y.$$

Table 2.2 Fourier transform properties

Property	$f(x,y)$	$F(k_x,k_y)$
Linearity	$af_1(x,y)+bf_2(x,y)$	$aF_1(k_x,k_y)+bF_2(k_x,k_y)$
Shifting	$f(x-x_0,\,y-y_0)$	$e^{jk_x x_0+jk_y y_0}F(k_x,k_y)$
Reciprocal shifting	$e^{j\alpha x+j\beta y}f(x,y)$	$F(k_x+\alpha,k_y+\beta)$
Scaling	$f(ax,by)$	$\dfrac{1}{\lvert ab\rvert}F\!\left(\dfrac{k_x}{a},\dfrac{k_y}{b}\right)$
Differentiation	$\dfrac{\partial f(x,y)}{\partial x}$	$-jk_x\,F(k_x,k_y)$
Higher-order differentiation	$\dfrac{\partial^n f(x,y)}{\partial x^n}$	$(-jk_x)^n\,F(k_x,k_y)$
Mixed differentiation	$\dfrac{\partial^2 f(x,y)}{\partial x\partial y}$	$(-jk_x)(-jk_y)F(k_x,k_y)$
Conjugation	$f^*(x,y)$	$F^*(-k_x,-k_y)$

Differentiation of both sides of the above equation with respect to the x variable yields

$$\frac{\partial f(x,y)}{\partial x}=\frac{1}{4\pi^2}\iint_{-\infty}^{\infty}-jk_x F(k_x,k_y)e^{-jk_x x-jk_y y}\,dk_x dk_y=\mathcal{F}^{-1}\left\{-jk_x F(k_x,k_y)\right\}.$$

Taking the Fourier transform on both sides, we have

$$\frac{\partial f(x,y)}{\partial x}\Leftrightarrow -jk_x\,F(k_x,k_y).$$

Repeated application of this property yields

$$\frac{\partial^n f(x,y)}{\partial x^n}\Leftrightarrow(-jk_x)^n\,F(k_x,k_y). \tag{2.3-25}$$

Table 2.2 summarizes some of the useful properties of the Fourier transform.

2.3.4   2-D Convolution, Correlation, and Matched Filtering

2-D Convolution

We have discussed that in a linear and time-invariant system, the convolution integral is involved. In this section, we will discuss the concept of 2-D convolution. Subsequently, we will discuss another important operation called *correlation*.

In optics, we can extend the concept of a linear and time-invariant system to the so-called *linear space-invariant* system. Hence, we can write the 2-D convolution integral as follows:

$$g(x,y) = \iint\limits_{-\infty}^{\infty} f(x',y')h(x-x',y-y')dx'dy' = f(x,y)*h(x,y), \quad (2.3\text{-}26)$$

where $f(x,y)$ is the 2-D input to a linear and space-invariant system. $h(x,y)$ and $g(x,y)$ are the corresponding impulse response and the output of the system, respectively. In optical systems, $h(x,y)$ is often called the *point spread function (PSF)*, which describes the response of an optical system to a point source. The concept of time invariance has been described in Figure 2.1-11. Delaying the input signal $x(t)$ by any constant t_0 gives the same output but delayed by exactly t_0. What does it mean by space invariance in optical systems? In Figure 2.3-9, we clarify this concept. From the bottom half of the figure, we see that as the input image $f(x,y)$ is shifted to a new origin (x_0,y_0), its output, remaining the same functional form as $g(x,y)$, is shifted accordingly on the output plane. The *isoplanatic patch* is the name given to the input plane over which the optical system obeys space invariance. We will encounter linear and space invariant optical systems later in the coming chapters.

As the convolution operation in two dimensions is a bit complicated concept to grasp, we try to provide some clarity in the process. In Figure 2.3-10, we illustrate the concept of convolution involving two 2-D functions, $f(x,y)$ and $h(x,y)$. The discussion on the illustration follows closely to that in Figure 2.1-17 for 1-D signals. According to the definition of convolution, the convolution of the two functions basically involves the calculation of the different areas under the product of two functions, $f(x',y')$ and $h(x-x',y-y')$, for different shifts, (x, y) So we need first to create the product $f(x',y')h(x-x',y-y')$ from the two original functions. The two original functions are shown in the first row in Figure 2.3-10. On the second row, we put them on the x',y'-axis. Now we have $f(x',y')$ on a graph. To create the graph of $h(x-x',y-y')$, we first take the reversal of $h(x',y')$ upon the x'-axis and then upon the y'-axis to get $h(-x',-y')$, as shown in the second row. Once we have $h(-x',-y')$ we shift it by x and y to get $h(x-x',y-y')$ on the x',y'-plane. We can now overlap $f(x',y')$ and $h(x-x',y-y')$ as shown in the left-hand side of the third row on the x',y'-plane. Finally, we calculate the areas of the product for different shifts (x,y) to obtain a 2-D plot of $g(x,y)$, shown as a gray-scale plot in the figure

Convolution Theorems

If

$$f_1(x,y) \Leftrightarrow F_1(k_x,k_y) \text{ and } f_2(x,y) \Leftrightarrow F_2(k_x,k_y),$$

then

$$f_1(x,y)*f_2(x,y) \Leftrightarrow F_1(k_x,k_y)F_2(k_x,k_y). \quad (2.3\text{-}27)$$

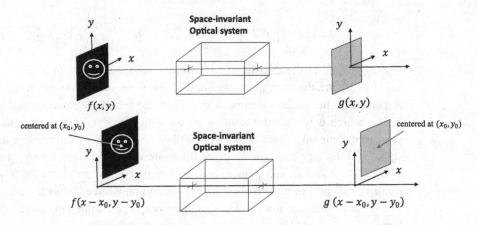

Figure 2.3-9 Space-invariance concept

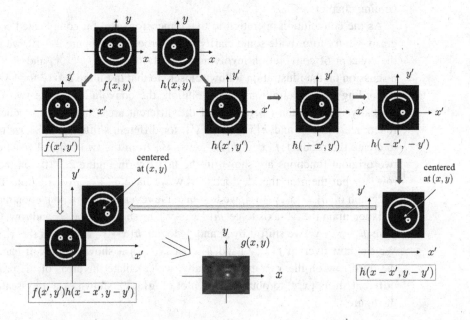

Figure 2.3-10 Concept of 2-D convolution

To prove Eq. (2.3-27), we use the definition of the Fourier transform:

$$\mathcal{F}\{f_1(x,y) * f_2(x,y)\} = \iint_{-\infty}^{\infty} f_1(x,y) * f_2(x,y) e^{jk_x x + jk_y y} dxdy$$

$$= \iint_{-\infty}^{\infty} \iint_{-\infty}^{\infty} f_1(x',y') f_2(x-x',y-y') dx'dy' e^{jk_x x + jk_y y} dxdy. \tag{2.3-28}$$

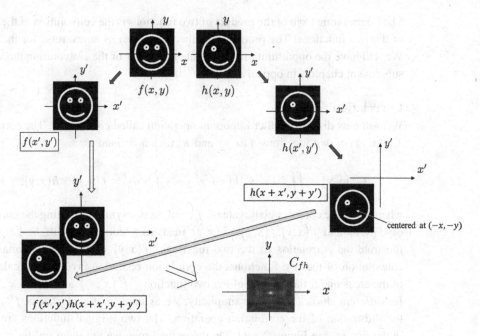

Figure 2.3-11 Concept of 2-D correlation

Let us integrate over x and y first, and the integral over x and y is

$$\iint\limits_{-\infty}^{\infty} f_2\left(x-x',y-y'\right)e^{jk_xx+jk_yy}dxdy = e^{jk_xx'+jk_yy'}F_2\left(k_x,k_y\right),$$

which is according to the shifting property of the Fourier transform shown in Table 2.2. Substituting the above result into Eq. (2.3-28), we have

$$\mathcal{F}\left\{f_1\left(x,y\right)*f_2\left(x,y\right)\right\} = \iint\limits_{-\infty}^{\infty} f_1\left(x',y'\right)e^{jk_xx'+jk_yy'}F_2\left(k_x,k_y\right)dx'dy'$$

$$= F_2\left(k_x,k_y\right)\iint\limits_{-\infty}^{\infty} f_1\left(x',y'\right)e^{jk_xx'+jk_yy'}dx'dy' = F_1\left(k_x,k_y\right)F_2\left(k_x,k_y\right).$$

The Fourier transform of the convolution of two functions is the product of the spectra of the two convolving functions. The converse of this theorem also holds.
If

$$f_1\left(x,y\right) \Leftrightarrow F_1\left(k_x,k_y\right) \text{ and } f_2\left(x,y\right) \Leftrightarrow F_2\left(k_x,k_y\right),$$

then

$$f_1\left(x,y\right)f_2\left(x,y\right) \Leftrightarrow F_1\left(k_x,k_y\right)*F_2\left(k_x,k_y\right). \tag{2.3-29}$$

The Fourier transform of the product of two functions is the convolution of the spectra of the two functions. The proof of this theorem is left as an exercise for the reader. We will have the opportunity to appreciate the power of the convolution theorems in subsequent chapters in optical systems.

Correlation

We will now discuss another important operation called *correlation*. The correlation, $C_{fh}(x,y)$, of two functions $f(x,y)$ and $h(x,y)$, is defined as

$$C_{fh}(x,y) = \iint\limits_{-\infty}^{\infty} f^*(x',y')h(x+x',y+y')dx'dy' = f(x,y) \otimes h(x,y), \quad (2.3\text{-}30)$$

where f^* is the complex conjugate of f, and $\otimes$ is a symbol denoting the correlation of $f(x,y)$ and $h(x,y)$. $f(x,y) \otimes h(x,y)$ reads as f correlates with h. Let us now illustrate the correlation of the two functions, $f(x,y)$ and $h(x,y)$. Similar to the convolution of the two functions, the correlation operation involves the calculation of the areas under the product of the two functions, $f^*(x',y')$ and $h(x+x',y+y')$, for different shifts (x,y). For simplicity, we assume f is real, that is, $f^* = f$, for the illustration of the correlation operation. The two original functions are shown in the first row in Figure 2.3-11. On the second row, we put them on the x',y'-axis. Now we have $f(x',y')$ on a graph. To create the graph of $h(x+x',y+y')$, we shift $h(x',y')$ by x and y to get $h(x+x',y+y')$ on the x',y'-plane. We can now overlap $f(x',y')$ and $h(x+x',y+y')$ as shown in the left-hand side of the third row on the x',y'-plane. Finally, we calculate the areas of the product for different shifts (x,y) to obtain a 2-D plot of $C_{fh}(x,y)$, shown as a gray-scale plot in the figure. $C_{fh}(x,y)$ is called the *cross-correlation* of $f(x,y)$ and $h(x,y)$, and $C_{ff}(x,y)$ is the *auto-correlation* of $f(x,y)$.

Example: Graphical Method to Find Correlation

Given $x(t) = ae^{-at}u(t)$, $a > 0$ and $h(t) = u(t)$, we evaluate their correlation by a graphical method. According to the definition of convolution in Eq. (2.3-30) for 1-D signals, we have

$$C_{xh}(t) = x(t) \otimes h(t) = \int_{-\infty}^{\infty} x^*(t')h(t+t')dt'.$$

We need to find the areas under the product of two functions, $x^*(t')$ and $h(t+t')$, for different values of t. Let us visualize how to create the product $x^*(t')h(t+t')$. From the original functions $x(t)$ and $h(t)$, we plot $x(t')$ and $h(t')$ as shown in Figure 2.3-12a. Now we have $x(t')$ on a graph. To create the graph of $h(t+t')$, we shift $h(t')$ by t to get $h(t+t')$, as shown in Figure 2.3-12b. We then overlap $x(t')$ and $h(t+t')$ on a single graph, which is shown in Figure 2.3-12c. This graph guides us to set up the integrand and the limits for the integral of correlation. There are two cases involved in terms of setting up the limits of the correlation integral. In case I, when $t > 0$, we use Figure 2.3-12d to help setting up the correlation integral:

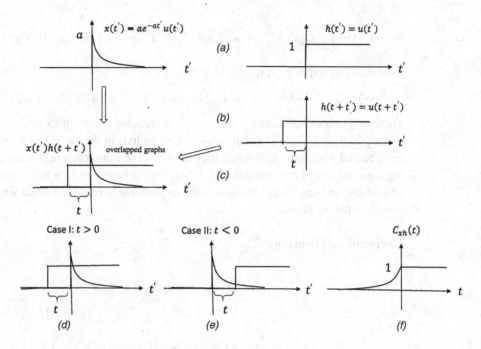

Figure 2.3-12 Graphical method for correlation

$$C_{xh}(t) = x(t) \otimes h(t) = \int_{-\infty}^{\infty} x(t')h(t+t')dt' = \int_{0}^{\infty} ae^{-at'}dt' = 1.$$

In case II, when $t < 0$, we use Figure 2.3-12e to set up the correlation integral:

$$C_{xh}(t) = x(t) \otimes h(t) = \int_{-\infty}^{\infty} x(t')h(t+t')dt' = \int_{|t|}^{\infty} ae^{-at'}dt' = e^{-a|t|}.$$

The final plot of $C_{xh}(t)$ is shown in Figure 2.3-12f.

Matched Filtering

In a linear and space-invariant (LSI) system, a *matched filter* is said to be matched to the particular signal (image) $s(x,y)$ if its impulse response $h(x,y)$ is given by $s^*(-x,-y)$. In LSI systems, we find the output $o(x,y)$ by convolving the input $s(x,y)$ with its impulse response:

$$o(x,y) = s(x,y) * h(x,y) = s(x,y) * s^*(-x,-y).$$

Let us rewrite the above as follows:

$$o(x,y) = s^*(-x,-y) * s(x,y) = \int\int_{-\infty}^{\infty} s^*(-x',-y')s(x-x',y-y')dx'dy'.$$

We let $-x' = x''$ and $-y' = y''$, the above integral becomes

$$o(x,y) = \int\limits_{-\infty}^{\infty}\!\!\int s^*(x'',y'')s(x+x'',y+y'')dx''dy'' = s(x,y) \otimes s(x,y) = C_{ss}(x,y).$$

Note that if the input is $g(x,y)$, we have

$$o(x,y) = g(x,y) * h(x,y) = g(x,y) * s^*(-x,-y) = s(x,y) \otimes g(x,y) = C_{sg}(x,y).$$

Therefore, in matched filtering, we perform correlation of two functions. As it turns out, $|C_{ss}(0,0)| \geq |C_{ss}(x,y)|$, which means the magnitude of the autocorrelation always has a central maximum, facilitating the use of autocorrelation as a means for comparing the similarity of two functions. *Matched filtering* is, therefore, a useful concept in *optical pattern recognition*. We will return to this topic in Chapter 5 when we discuss *complex filtering* in optics.

Correlation Theorems
If

$$f_1(x,y) \Leftrightarrow F_1(k_x,k_y) \text{ and } f_2(x,y) \Leftrightarrow F_2(k_x,k_y),$$

then

$$f_1(x,y) \otimes f_2(x,y) \Leftrightarrow F_1^*(k_x,k_y)F_2(k_x,k_y). \qquad (2.3\text{-}31)$$

The Fourier transform of the correlation of two functions is the product of the complex conjugate of the spectrum of the first function with the spectrum of the second function. The converse of this theorem also holds, that is,

$$f_1^*(x,y)f_2(x,y) \Leftrightarrow F_1(k_x,k_y) \otimes F_2(k_x,k_y). \qquad (2.3\text{-}32)$$

The proof of the above theorems follows similarly to that of the convolution theorem derived earlier.

Problems

2.1 For the signal $x(t)$ shown in Figure P. 2.1, sketch (a) $x(2t-4)$ and (b) $x(2-4t)$

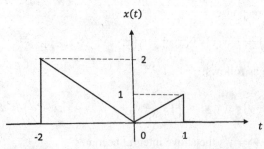

Figure P.2.1

2.2 Evaluate the following integrals:

(a) $\int\limits_{-\infty}^{\infty} \delta(t+2) e^{-2(t-1)} dt$

(b) $\int\limits_{-\infty}^{\infty} \delta(t+2) e^{-2(t-1)} u(t) dt$

(c) $\int\limits_{-\infty}^{\infty} \delta(at) dt$, where a is real and positive

2.3 Determine if the system described by $y(t) = t^2 x(t)$ is (i) linear, (ii) time-invariant.

2.4 Determine if the system described by

$$\left(\frac{dy(t)}{dt}\right)^2 + 2y(t) = x(t)$$

is linear or nonlinear.

2.5 For the circuit shown in Figure P.2.5,

(a) find the differential equation relating the output $y(t)$ to the input $x(t)$.
(b) Find the frequency response function of the circuit $H(\omega)$ using two methods and determine if the circuit performs low-pass filtering or high-pass filtering (Figure P.2.5).

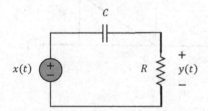

Figure P.2.5

2.6 Given a signal $x(t) = 10 \operatorname{rect}\left(\frac{t}{6}\right) e^{-t^2}$, a noise source $n(t) = \sin(10t) + \frac{1}{2}\sin(12t)$, an impulse response of system A being $h_A(t) = \operatorname{rect}(10t)$, and an impulse response of system B being $h_B(t) = \operatorname{rect}\left(\frac{t}{2}\right)$,

write MATLAB codes to perform the following tasks:

(a) Plot $x(t)$
(b) Plot noisy signal $x_n(t) = x(t) + n(t)$
(c) Plot $x_n(t) * h_A(t)$
(d) Plot $x_n(t) * h_B(t)$
(e) From the above results, which system does a better job of cleaning up the noise? Please explain.

2.7 Suppose a signal is described by $s(t) = e^{-t^2/10}\left[4e^{-t^2/10} + 2e^{-(t-1)^2}\right]$ and this signal is fed to a linear and time-invariant system with its impulse response given by $h(t)$.

(a) Design $h(t)$ such that the system is matched to the input $s(t)$.
(b) Write a MTALAB program to plot $s(t)$ and the system's output.
(c) Another signal $g(t)$ described by

$$g(t) = e^{-t^2/10}\left[4e^{-t^2/10} + 2e^{-(t-2)^2}\right]$$

is fed to the same system with $h(t)$ defined in part a. Plot $g(t)$ and the system's output.
(d) Make some concluding remarks from the results of (a), (b), and (c).

2.8 Find $\text{rect}\left(\dfrac{x}{x_0}\right) * \text{rect}\left(\dfrac{x}{x_0}\right)$

(a) using the graphical method of convolution and
(b) using the Fourier transform.

2.9 For $x_1(t)$ and $x_2(t)$ shown in Figure P.2.9, find $x_1(t) * x_2(t)$ for $-1 < t < 1$ using the graphical method.

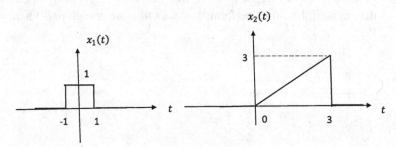

Figure P.2.9

2.10 Find the Fourier series and sketch the corresponding spectrum for the impulse train with period T_0. The impulse train is depicted in Figure P.2.10 and defined mathematically as follows:

$$x(t) = \sum_{n=-\infty}^{\infty} \delta(t - nT_0) = \delta_{T_0}(t).$$

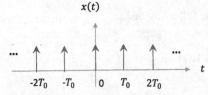

Figure P.2.10

2.11 Find the Fourier transform of the impulse train $\delta_{T_0}(t)$.

2.12 Verify the Fourier transform pair of item #5 in Table 2.1.

2.13 (a) Using the identity

$$J_1(\alpha) = \frac{1}{\alpha} \int_0^\alpha \beta J_0(\beta) d\beta,$$

show that

$$\mathcal{B}\left\{ \text{circ}\left(\frac{r}{r_0}\right) \right\} = \frac{2\pi r_0}{k_r} J_1(r_0 k_r) = 2\pi r_0^2 \text{jinc}(r_0 k_r).$$

(b) Using the result in part a, find $\text{jinc}(0)$.

2.14 Let us define a 1-D version of the Fourier transform of a spatial function as follows:

$$\mathcal{F}\{f(x)\} = F(k_x) = \int_{-\infty}^\infty f(x) e^{jk_x x} dx.$$

Find $\mathcal{F}\left\{ e^{j\pi x} f\left(x - \frac{1}{2}\right) \right\}$.

2.15 Verify the Fourier transform property of conjugation in Table 2.2.

2.16 If

$$f_1(x, y) \Leftrightarrow F_1(k_x, k_y) \text{ and } f_2(x, y) \Leftrightarrow F_2(k_x, k_y),$$

show that

(a) $f_1(x, y) \otimes f_2(x, y) \Leftrightarrow F_1^*(k_x, k_y) F_2(k_x, k_y),$

(b) $f_1^*(x, y) f_2(x, y) \Leftrightarrow F_1(k_x, k_y) \otimes F_2(k_x, k_y).$

Bibliography

Banerjee, P. P. and T.-C. Poon (1991). *Principles of Applied Optics*. Irwin, Illinois.

Blahut, R. (2004). *Theory of Remote Image Formation*. Cambridge University Press, Cambridge.

Lathi, B. P. and R. Green (2018). *Linear Systems and Signals*, 3rd ed., Oxford University Press, Oxford.

Poon, T.-C. (2007). *Optical Scanning Holography with MATLAB®*. Springer, New York.

Poon, T.-C. and T. Kim (2018). *Engineering Optics with MATLAB®*, 2nd ed., World Scientific, New Jersey.

Ulaby, F. T. and A. E. Yagle (2016). *Engineering Signals and Systems in Continuous and Discrete Time*, 2nd ed., National Technology & Science Press.

3   Wave Propagation and Wave Optics

We have introduced Gaussian optics and used a matrix formalism to describe light rays through optical systems in Chapter 1. Light rays are based on the particle nature of light. Since light has a dual nature, light is waves as well. In 1924, de Broglie formulated the *de Broglie hypothesis*, which relates wavelength and momentum as $\lambda_0 = h/p$, where P is the momentum [in kg-m/s] of the particle (called photon) when the particle moves, λ_0 is the wavelength [in m] of the wave, and $h = 6.626 \times 10^{-34}$ J s is Planck's constant. In this chapter, we explore the wave nature of light, which accounts for wave effects such as interference and diffraction.

3.1   Maxwell's Equations

In electromagnetics, we are concerned with four vector quantities called *electromagnetic (EM) fields*: the electric field E (V/m), the electric flux density D (C/m²), the magnetic field H (A/m), and the magnetic flux density B (Wb/m²). These EM fields are functions of position in space, that is, x, y, z, and time, t, and they are governed by four Maxwell's equations as follows.

Electric Gauss's Law

Electric Gauss's law describes the relationship between a static electric field and the electric charges that generate it. The law is stated as

$$\nabla \cdot D = \rho_v, \tag{3.1-1}$$

where ρ_v is the electric charge density (C/m³). The symbol ∇ is a vector partial-differentiation operator, often called the ∇ (del) operator and is given by

$$\nabla = \frac{\partial}{\partial x}\hat{x} + \frac{\partial}{\partial y}\hat{y} + \frac{\partial}{\partial z}\hat{z}, \tag{3.1-2}$$

where $\hat{x}$, $\hat{y}$, and $\hat{z}$ are the *unit vectors* in the x, y, and z-directions, respectively. For a vector A with components A_x, A_y, and A_z, we write $A = A_x\hat{x} + A_y\hat{y} + A_z\hat{z}$. $\nabla \cdot A$ is called the *divergence* of A or simply called "del dot A," where the dot denotes an operation equivalent to the *dot product* or *scalar product* of two vectors. The alternative name "scalar product" emphasizes that the result of the product is a scalar, rather than a vector. $\nabla \cdot A$ is the dot product of the vectors ∇ and A given by

$$\nabla \cdot A = \left(\frac{\partial}{\partial x} \hat{x} + \frac{\partial}{\partial y} \hat{y} + \frac{\partial}{\partial z} \hat{z} \right) \cdot \left(A_x \hat{x} + A_y \hat{y} + A_z \hat{z} \right) = \frac{\partial A_x}{\partial x} + \frac{\partial A_y}{\partial y} + \frac{\partial A_z}{\partial z}. \qquad (3.1\text{-}3)$$

Given two vectors A and B, the dot product is defined as

$$A \cdot B = |A||B| \cos(\theta_{AB}) = AB \cos(\theta_{AB}),$$

where $|A| = A$ and $|B| = B$ are the magnitudes of A and B, respectively. θ_{AB} is the smaller angle between A and B. Note that the dot product is commutative and distributive, that is,

$$A \cdot B = B \cdot A \text{ (commutative property)},$$

$$A \cdot (B + C) = A \cdot B + A \cdot C \text{ (distributive property)}.$$

Example : Component of a Vector in a Given Direction

We can use the dot product to find the component of a vector in a given direction. Let us denote B on the x–y plane as shown in Figure 3.1-1. The component of B along the unit vector $\hat{x}$ is

$$B_x = B \cdot \hat{x} = |B||\hat{x}| \cos\theta_{Bx} = B \cos\theta_{Bx}.$$

Similarly, the component of B along the unit vector $\hat{y}$ is

$$B_y = B \cdot \hat{y} = |B||\hat{y}| \cos\theta_{By} = B \cos\theta_{By} = B \cos(90° - \theta_{Bx}) = B \sin(\theta_{Bx}).$$

So we can write

$$B = B_x \hat{x} + B_y \hat{y} = (B \cdot \hat{x})\hat{x} + (B \cdot \hat{y})\hat{y}.$$

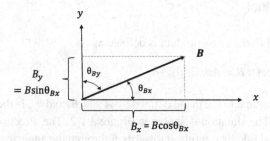

Figure 3.1-1 Components of a vector

Magnetic Gauss's Law

Magnetic Gauss's Law states that there are no "magnetic monopoles." In other words, a magnetic monopole, an analogue of an electric charge, does not exist. For example, the two ends of a bar magnet are referred to as North and South poles. The two poles exist at the same time and they are called a *magnetic dipole*. The law is stated as

$$\nabla \cdot \boldsymbol{B} = 0. \tag{3.1-4}$$

Faraday's Law

Faraday's law describes how a time-varying magnetic field generates an electric field. The law is stated as

$$\nabla \times \boldsymbol{E} = -\frac{\partial \boldsymbol{B}}{\partial t}. \tag{3.1-5}$$

With the vector operator ∇ defined in Eq. (3.1-2), $\nabla \times \boldsymbol{A}$ is called the *curl* of $\boldsymbol{A}$ or simply called "del cross $\boldsymbol{A}$," where the cross denotes an operation equivalent to the *cross product* or *vector product* of two vectors. The alternative name "vector product" emphasizes that the result of the product is a vector. $\nabla \times \boldsymbol{A}$ is the cross product of the vectors ∇ and $\boldsymbol{A}$ given by

$$\nabla \times \boldsymbol{A} = \begin{vmatrix} \hat{\boldsymbol{x}} & \hat{\boldsymbol{y}} & \hat{\boldsymbol{z}} \\ \dfrac{\partial}{\partial x} & \dfrac{\partial}{\partial y} & \dfrac{\partial}{\partial z} \\ A_x & A_y & A_z \end{vmatrix}$$

$$= \hat{\boldsymbol{x}} \begin{vmatrix} \dfrac{\partial}{\partial y} & \dfrac{\partial}{\partial z} \\ A_y & A_z \end{vmatrix} - \hat{\boldsymbol{y}} \begin{vmatrix} \dfrac{\partial}{\partial x} & \dfrac{\partial}{\partial z} \\ A_x & A_z \end{vmatrix} + \hat{\boldsymbol{z}} \begin{vmatrix} \dfrac{\partial}{\partial x} & \dfrac{\partial}{\partial y} \\ A_x & A_y \end{vmatrix}$$

$$= \hat{\boldsymbol{x}} \left(\frac{\partial A_z}{\partial y} - \frac{\partial A_y}{\partial z} \right) - \hat{\boldsymbol{y}} \left(\frac{\partial A_z}{\partial x} - \frac{\partial A_x}{\partial z} \right) + \hat{\boldsymbol{z}} \left(\frac{\partial A_y}{\partial x} - \frac{\partial A_x}{\partial y} \right). \tag{3.1-6}$$

Example : Cross Product

Given two vectors $\boldsymbol{A}$ and $\boldsymbol{B}$, the cross product is defined as

$$\boldsymbol{A} \times \boldsymbol{B} = AB\sin\left(\theta_{AB}\right)\hat{\boldsymbol{n}},$$

where $\hat{\boldsymbol{n}}$ is a unit vector normal to the plane defined by $\boldsymbol{A}$ and $\boldsymbol{B}$, and θ_{AB} is the smaller angle between $\boldsymbol{A}$ and $\boldsymbol{B}$. The situation is shown in Figure 3.1-2. The direction of $\hat{\boldsymbol{n}}$ is found using the right-hand rule by rotating $\boldsymbol{A}$ towards $\boldsymbol{B}$ through the angle θ_{AB} with the fingers, and the thumb points to the $\hat{\boldsymbol{n}}$ direction.

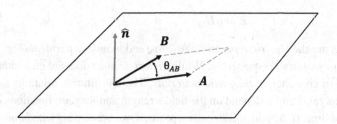

Figure 3.1-2 The cross product by the right-hand rule

The cross product is anti-commutative and distributive, that is,

$$A \times B = -B \times A \text{ (anti-commutative property)},$$

$$A \times (B + C) = A \times B + A \times C \text{ (distributive property)}.$$

Finally, there is the vector triple product known as the "bac-cab" rule given by

$$A \times (B \times C) = B(A \cdot C) - C(A \cdot B).$$

Ampere's Law

Ampere's law states that magnetic fields can be generated in two ways: by electric current (J_c) and by time-varying electric fields ($\partial D / \partial t$). The law is stated as

$$\nabla \times H = J = J_c + \frac{\partial D}{\partial t}. \tag{3.1-7}$$

The vector quantity J_c is the conduction current density (A/m^2) and it comes from the movement of electric charges such as electrons being present in conductors.

We summarize Maxwell's four equations as follows:

$$\nabla \cdot D = \rho_v \text{ (Electric Gauss's law)} \tag{3.1-8a}$$

$$\nabla \cdot B = 0 \text{ (Magnetic Gauss's law)} \tag{3.1-8b}$$

$$\nabla \times E = -\frac{\partial B}{\partial t} \text{ (Faraday's law)} \tag{3.1-8c}$$

$$\nabla \times H = J = J_c + \frac{\partial D}{\partial t} \text{ (Ampere's Law)}. \tag{3.1-8d}$$

ρ_v and J_c are the sources responsible for generating the electromagnetic fields. In order to completely determine the four field quantities in Maxwell's equations, we need also the *constitutive relations*:

$$D = \varepsilon E \tag{3.1-9}$$

and

$$B = \mu H, \tag{3.1-10}$$

where ε and μ are the *electrical permittivity* (F/m) and *magnetic permeability* (H/m) of the medium or material, respectively. While *dielectric materials* are characterized by ε, μ is used to characterize *magnetic materials*. The constitutive relations are rarely simple. In general, ε and μ depend on the field strength and they are functions of space (position) and time. If they do not depend on position, we are dealing with *homogeneous* materials. If they do not depend on the amplitude of the fields in the medium, the materials are *linear*. We have *isotropic materials* if ε and μ are the same in all directions from any given point. In the case for a linear, homogenous, and isotropic medium such as in vacuum or free space, $\varepsilon = \varepsilon_0 \approx \left(\dfrac{1}{36\pi}\right) \times 10^{-9}$ F/m, and $\mu = \mu_0 \approx 4\pi \times 10^{-7}$ H/m.

Throughout this book, we deal with nonmagnetic materials, and hence we take $\mu = \mu_0$. For light travelling though inhomogeneous media such as optical fibers, we deal with $\varepsilon(x, y, z)$. In Chapter 7, we deal with electro-optics in which light traveling through crystals is analyzed by ε of a 3×3 matrix for anisotropic media. Also in the same chapter, we discuss acousto-optics when ε is a function of time.

3.2   Vector Wave Equations

In the last section, we have enunciated Maxwell's equations and the constitutive relations. Indeed, we could solve for the electromagnetic fields for a given ρ_v and $\boldsymbol{J}_C$. We shall now derive the wave equation for one of the important special cases in that the medium is characterized by ε and μ both being constant, that is, the medium is linear, homogeneous and isotropic. A good example of such a medium is air. We will see the wave equations describe the propagation of the electromagnetic fields.

To derive the wave equation in $\boldsymbol{E}$, we take the curl of the curl equation for $\boldsymbol{E}$. The curl equation for $\boldsymbol{E}$ is Faraday's law given in Eq. (3.1-8c). We, therefore, have

$$\nabla \times (\nabla \times \boldsymbol{E}) = \nabla \times \left(-\frac{\partial \boldsymbol{B}}{\partial t}\right) = -\frac{\partial}{\partial t}(\nabla \times \boldsymbol{B}),$$

where we have interchanged ∇ and $\partial / \partial t$ operations. Using Eq. (3.1.8d), and the two constitutive equations in Eqs. (3.1-9) and (3.1-10) in the right-hand side of the above equation, we have

$$\nabla \times (\nabla \times \boldsymbol{E}) = -\mu\varepsilon\frac{\partial^2 \boldsymbol{E}}{\partial t^2} - \mu\frac{\partial \boldsymbol{J}_C}{\partial t}. \tag{3.2-1}$$

Using the vector identity involving the del operator:

$$\nabla \times (\nabla \times \boldsymbol{A}) = \nabla(\nabla \cdot \boldsymbol{A}) - \nabla^2 \boldsymbol{A},$$

where $\nabla^2 \boldsymbol{A}$ is the *Laplacian* of $\boldsymbol{A}$, which in Cartesian coordinates is given by

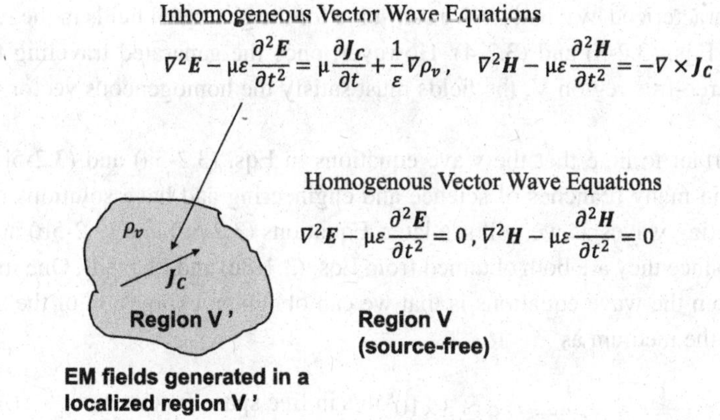

Inhomogeneous Vector Wave Equations

$$\nabla^2 E - \mu\varepsilon \frac{\partial^2 E}{\partial t^2} = \mu \frac{\partial J_c}{\partial t} + \frac{1}{\varepsilon}\nabla\rho_v, \quad \nabla^2 H - \mu\varepsilon \frac{\partial^2 H}{\partial t^2} = -\nabla \times J_c$$

Homogenous Vector Wave Equations

$$\nabla^2 E - \mu\varepsilon \frac{\partial^2 E}{\partial t^2} = 0, \nabla^2 H - \mu\varepsilon \frac{\partial^2 H}{\partial t^2} = 0$$

ρ_v

J_c

Region V'

Region V
(source-free)

EM fields generated in a
localized region V'

Figure 3.2-1 Region V' contains sources, whereas region V is source-free

$$\nabla^2 A = \frac{\partial^2 A}{\partial x^2} + \frac{\partial^2 A}{\partial y^2} + \frac{\partial^2 A}{\partial z^2},$$

Eq. (3.2-1) becomes

$$\nabla(\nabla \cdot E) - \nabla^2 E = -\mu\varepsilon \frac{\partial^2 E}{\partial t^2} - \mu \frac{\partial J_c}{\partial t}. \tag{3.2-2}$$

Using Eqs. (3.1-8a) and (3.1-9), we have $\nabla(\nabla \cdot E) = \frac{1}{\varepsilon}\nabla\rho_v$. Therefore, Eq. (3.2-2) finally becomes

$$\nabla^2 E - \mu\varepsilon \frac{\partial^2 E}{\partial t^2} = \mu \frac{\partial J_c}{\partial t} + \frac{1}{\varepsilon}\nabla\rho_v, \tag{3.2-3}$$

which is the *inhomogeneous vector wave equation* for E in a linear, homogenous, and isotropic medium in the presence of sources J_c and ρ_v. To derive the wave equation for H, we simply take the curl of the curl equation for H and proceed with similar procedures to obtain the vector wave equation for H:

$$\nabla^2 H - \mu\varepsilon \frac{\partial^2 H}{\partial t^2} = -\nabla \times J_c. \tag{3.2-4}$$

For a source-free medium, that is, $J_c = \rho_v = 0$, the two wave equations become

$$\nabla^2 E - \mu\varepsilon \frac{\partial^2 E}{\partial t^2} = 0, \tag{3.2-5a}$$

and

$$\nabla^2 H - \mu\varepsilon \frac{\partial^2 H}{\partial t^2} = 0 \tag{3.2-5b}$$

which are the *homogeneous vector wave equations*. Figure 3.2-1 summarizes the situation. In Figure 3.2-1, we assume that the sources J_c and ρ_v are localized in a limited region V' characterized by ε and μ. We can solve for the generated fields in the region according to Eqs. (3.2-3) and (3.2-4). However, once the generated traveling fields reach the source-free region V, the fields must satisfy the homogeneous vector wave equations.

It is important to note that the wave equations in Eqs. (3.2-5a) and (3.2-5b) are encountered in many branches of science and engineering and have solutions in the form of traveling waves as we will see later. Equations (3.2-5a) and (3.2-5b) are not independent since they are both obtained from Eqs. (3.1-8c) and (3.1-8d). One important result from the wave equations is that we can obtain an expression of the speed of light, v, in the medium as

$$v = \frac{1}{\sqrt{\varepsilon\mu}} = \frac{1}{\sqrt{\varepsilon_0\mu_0}} = c \approx 3\times10^8 \text{ m/s in free space (vacuum).} \qquad (3.2\text{-}6)$$

This theoretical value of the speed of light was first calculated by Maxwell and indeed this led Maxwell to conclude that light is an electromagnetic disturbance in the form of waves.

3.3   Traveling-Wave Solutions and the Poynting Vector

Let us find some of the simplest solutions in the wave equation. The wave equation in Eq. (3.2-5a) is equivalent to three scalar equations – one for each component of $\boldsymbol{E}$. We let $\boldsymbol{E} = E_x\hat{\boldsymbol{x}} + E_y\hat{\boldsymbol{y}} + E_z\hat{\boldsymbol{z}}$ and substitute this into Eq. (3.2-5a) to obtain

$$\nabla^2\boldsymbol{E} = \frac{\partial^2\boldsymbol{E}}{\partial x^2} + \frac{\partial^2\boldsymbol{E}}{\partial y^2} + \frac{\partial^2\boldsymbol{E}}{\partial z^2} = \left(\frac{\partial^2}{\partial x^2} + \frac{\partial^2}{\partial y^2} + \frac{\partial^2}{\partial z^2}\right)(E_x\hat{\boldsymbol{x}} + E_y\hat{\boldsymbol{y}} + E_z\hat{\boldsymbol{z}})$$

$$= \mu\varepsilon\frac{\partial^2\boldsymbol{E}}{\partial t^2} = \mu\varepsilon\frac{\partial^2}{\partial t^2}(E_x\hat{\boldsymbol{x}} + E_y\hat{\boldsymbol{y}} + E_z\hat{\boldsymbol{z}}).$$

By comparing the $\hat{\boldsymbol{x}}$-component on both sides of the equation, we obtain

$$\frac{\partial^2E_x}{\partial x^2} + \frac{\partial^2E_x}{\partial y^2} + \frac{\partial^2E_x}{\partial z^2} = \mu\varepsilon\frac{\partial^2E_x}{\partial t^2},$$

or

$$\nabla^2E_x = \mu\varepsilon\frac{\partial^2E_x}{\partial t^2}.$$

Similarly, by comparing other components, we have the same type of equation shown above for the E_y and E_z components. Therefore, we can write

$$\nabla^2\psi = \mu\varepsilon\frac{\partial^2\psi}{\partial t^2} = \frac{1}{v^2}\frac{\partial^2\psi}{\partial t^2}, \qquad (3.3\text{-}1)$$

where $\psi(x,y,z;t)$ may represent any component, E_x, E_y, or E_z, of the electric field $\boldsymbol{E}$. Equation (3.3-1) is called the *3-D scalar wave equation*. The starting point for

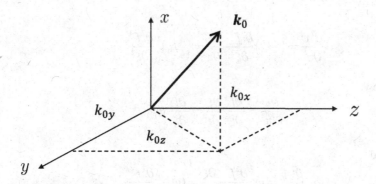

Figure 3.3-1 Propagation vector

wave optics is the scalar wave equation in which *scalar diffraction theory* is based, as we will see in subsequent subsections.

For monochromatic light fields oscillating at the frequency ω_0, the general solution to Eq. (3.3-1) is

$$\psi(x,y,z;t) = c_1 f^+ \left(\omega_0 t - \mathbf{k}_0 \cdot \mathbf{R} \right) + c_2 f^- \left(\omega_0 t + \mathbf{k}_0 \cdot \mathbf{R} \right), \tag{3.3-2}$$

with the condition that

$$\frac{\omega_0^2}{k_{0x}^2 + k_{0y}^2 + k_{0z}^2} = \frac{\omega_0^2}{k_0^2} = v^2,$$

where

$$\mathbf{R} = x\hat{\mathbf{x}} + y\hat{\mathbf{y}} + z\hat{\mathbf{z}}$$

is the *position vector*, and

$$\mathbf{k}_0 = k_{0x}\hat{\mathbf{x}} + k_{0y}\hat{\mathbf{y}} + k_{0z}\hat{\mathbf{z}}$$

is the *propagation vector* with $|\mathbf{k}_0| = k_0 = \sqrt{k_{0x}^2 + k_{0y}^2 + k_{0z}^2}$ being the *propagation constant* (rad/m). c_1 and c_2 are some constants. $f^+ (\cdot)$ and $f^- (\cdot)$ are arbitrary functions, where f^+ depicts a wave propagating along the $\mathbf{k}_0$ direction, and f^- means a wave is propagating along the $-\mathbf{k}_0$ direction. Figure 3.3-1 illustrates $\mathbf{k}_0$ with its components.

Example: Travelling-Wave Solution

With reference to Eq. (3.3-2), let us take $c_2 = 0$, $c_1 = 1$, and $f^+ (\cdot)$ of the variable $\xi = \omega_0 t - k_0 z$ as a simple example. In this example, we show that $\psi(z;t) = f^+ \left(\omega_0 t - k_0 z \right)$ is a solution of the scalar wave equation. To start with, we have

$$\frac{\partial \psi}{\partial z} = \frac{\partial f^+}{\partial \xi} \frac{\partial \xi}{\partial z} = \frac{\partial f^+}{\partial \xi} (-k_0),$$

and

$$\frac{\partial^2 \psi}{\partial z^2} = (-k_0) \frac{\partial^2 f^+}{\partial \xi^2} \frac{\partial \xi}{\partial z} = (-k_0)^2 \frac{\partial^2 f^+}{\partial \xi^2}. \tag{3.3-3}$$

Now,

$$\frac{\partial \psi}{\partial t} = \frac{\partial f^+}{\partial \xi} \frac{\partial \xi}{\partial t} = \frac{\partial f^+}{\partial \xi} \omega_0,$$

and

$$\frac{\partial^2 \psi}{\partial t^2} = \omega_0 \frac{\partial^2 f^+}{\partial \xi^2} \frac{\partial \xi}{\partial t} = (\omega_0)^2 \frac{\partial^2 f^+}{\partial \xi^2}. \tag{3.3-4}$$

Substituting Eq. (3.3-4) into Eq. (3.3-3), we obtain

$$\frac{\partial^2 \psi}{\partial z^2} = \frac{(k_0)^2}{(\omega_0)^2} \frac{\partial^2 \psi}{\partial t^2}.$$

Since $\frac{\omega_0^2}{k_0^2} = v^2$, the above equation becomes

$$\frac{\partial^2 \psi}{\partial z^2} = \frac{1}{v^2} \frac{\partial^2 \psi}{\partial t^2}, \tag{3.3-5}$$

which is a 1-D version of the 3-D wave equation shown in Eq. (3.3-1). Hence, $\psi(z;t) = f^+(\omega_0 t - k_0 z)$ is a solution of the scalar wave equation. Indeed, the general solution to Eq. (3.3-5) is

$$\psi(z;t) = c_1 f^+(\omega_0 t - k_0 z) + c_2 f^-(\omega_0 t + k_0 z). \tag{3.3-6}$$

Let us take $f^+(\cdot) = e^{j(\cdot)}$ as an example. We then have

$$\psi(z;t) = e^{j(\omega_0 t - k_0 z)} \tag{3.3-7}$$

as a simple solution. This solution is called a *plane-wave solution* to the wave equation. The wave is called a *plane wave* of unit amplitude, albeit written in a complex notation. Since ψ represents a component of the electromagnetic field, it must be a real function of space and time. To obtain a real quantity, we simply take the real part of the complex function. We, therefore, have $\mathrm{Re}\{\psi(z;t)\} = \mathrm{Re}\{e^{j(\omega_0 t - k_0 z)}\} = \cos(\omega_0 t - k_0 z)$ to represent a physical quantity, where $\mathrm{Re}\{\cdot\}$ means taking the real part of the quantity being bracketed.

Taking the plane wave solution as an example, let us now illustrate that $f(\omega_0 t - k_0 z) = \cos(\omega_0 t - k_0 z)$ is a travelling wave. For simplicity, we take $\omega_0 = k_0 = 1$, and hence $v = \omega_0 / k_0 = 1 \mathrm{m/s}$. In Figure 3.3-2a and b, we plot $\cos(t - z)$ for $t = 0$ and $t = 2$ s, respectively. Note that if we track a given point on the wave, such as the peak P, and follow it as time advances, we can measure the so-called *phase velocity* of the

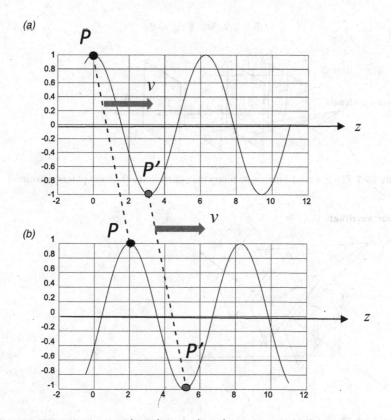

Figure 3.3-2 Plots of $f(t-z) = \cos(t-z)$ as a function of z at (a) $t = 0$ and (b) $t = 2$ s. z is measured in meters

wave, which is 1 m/s in this particular example. Similarly, we can track the trough P' of the wave travelling at the same phase velocity.

Plane-Wave Solution

We have briefly discussed the plane-wave solution in the above example. $\psi(z;t) = e^{j(\omega_0 t - k_0 z)}$ is the plane-wave solution for Eq. (3.3-5), where the plane wave is propagating along the z-direction. Let us write

$$\psi(z;t) = e^{j(\omega_0 t - k_0 z)} = e^{j\omega_0 t} e^{-j\theta(z)}, \tag{3.3-8}$$

where $\theta(z) = k_0 z = \dfrac{2\pi}{\lambda_0} z$ is called the phase of the wave with λ_0 being the wave-

length of the wave. Let us take the origin of the coordinates as a zero-phase reference position, that is, $\theta(z = 0) = 0$. What it means is that over the plane $z = 0$, we have zero phase. At $z = \lambda_0$, $\theta(z = \lambda_0) = \dfrac{2\pi}{\lambda_0}\lambda_0 = 2\pi$. So for every distance of propaga-

tion of a wavelength, the phase of the wave gains 2π. We, therefore, have what is

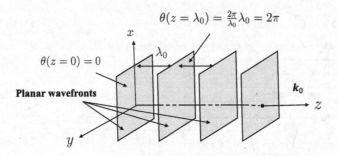

Figure 3.3-3 Plane wave propagating along the z-direction exhibiting planar wavefronts

Planar wavefronts

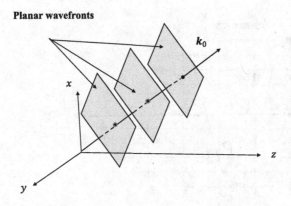

Figure 3.3-4 Plane wave propagating along the k_0-direction exhibiting planar wavefronts

known as the *planar wavefronts* along the z-direction. The situation is illustrated in Figure 3.3-3.

For a general direction of plane-wave propagation along k_0, according to Eq. (3.3-2), we write

$$\psi(x,y,z;t) = e^{j(\omega_0 t - k_0 \cdot R)} = e^{j(\omega_0 t - k_{0x}x - k_{0y}y - k_{0z}z)} \tag{3.3-9}$$

and the planar wavefronts propagate along the direction of k_0 as shown in Figure 3.3-4, where k_0 has been defined in Figure 3.3-1.

Spherical-Wave Solution

Another important basic solution to the scalar wave equation is the *spherical-wave solution*. We consider the spherical coordinates shown in Figure 3.3-5. For a spherical wave, ψ depends only on R and t, that is, $\psi(R,t)$, and we have *spherical symmetry*. $\nabla^2\psi$ in the wave equation of Eq. 3.3-1 is reduced to $\nabla^2\psi = \dfrac{\partial^2\psi}{\partial R^2} + \dfrac{2}{R}\dfrac{\partial\psi}{\partial R}$. Hence, the wave equation becomes

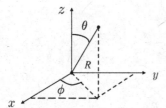

Figure 3.3-5 Spherical coordinates system

$$\nabla^2\psi = \frac{\partial^2\psi}{\partial R^2} + \frac{2}{R}\frac{\partial\psi}{\partial R} = \frac{1}{v^2}\frac{\partial^2\psi}{\partial t^2}. \tag{3.3-10}$$

Using the chain rule in calculus, we have

$$\frac{\partial^2(R\psi)}{\partial R^2} = R\left(\frac{\partial^2\psi}{\partial R^2} + \frac{2}{R}\frac{\partial\psi}{\partial R}\right),$$

and Eq. (3.3-10) becomes

$$\frac{\partial^2(R\psi)}{\partial R^2} = \frac{1}{v^2}\frac{\partial^2(R\psi)}{\partial t^2}. \tag{3.3-11}$$

This equation is of the same form as that of Eq. (3.3-5), and therefore, the general solution of Eq. (3.3-11) is, according to Eq. (3.3-6),

$$R\psi = c_1 f^+(\omega_0 t - k_0 R) + c_2 f^-(\omega_0 t + k_0 R).$$

Therefore, the final general solution to Eq. (3.3-10) is

$$\psi(R;t) = c_1\frac{f^+(\omega_0 t - k_0 R)}{R} + c_2\frac{f^-(\omega_0 t + k_0 R)}{R}. \tag{3.3-12}$$

Again, for $c_2 = 0$, $c_1 = 1$, and $f^+(\cdot) = e^{j(\cdot)}$, it has a simple solution given by

$$\psi(R;t) = \frac{1}{R}e^{j(\omega_0 t - k_0 R)}, \tag{3.3-13}$$

which is called a *spherical wave*. This is a wave propagating along R, that is, away from the origin of the coordinates and with the factor $1/R$ term implying that the amplitude of a spherical wave decreases inversely with R. We can write

$$\psi(R;t) = \frac{1}{R}e^{j\omega_0 t}e^{-j\theta(R)},$$

where $\theta(R) = k_0 R = \frac{2\pi}{\lambda_0}R$. Again we take the origin of the coordinates as a zero-phase reference position, that is, $\theta(R = 0) = 0$, and then $\theta(R = \lambda_0) = \frac{2\pi}{\lambda_0}\lambda_0 = 2\pi$.

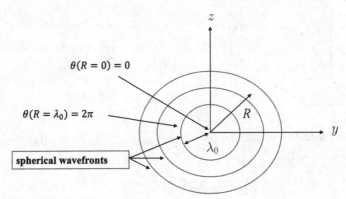

Cross section of wavefronts of a spherical wave

Figure 3.3-6 Spherical wavefronts

Therefore, for every distance of propagation of a wavelength, the phase of the wave gains 2π. Figure 3.3-6 illustrates the situation with a cross-section on the y–z plane.

The Poynting Vector

Let us now consider the relationship between existing electric and magnetic fields in free space. In this connection, we introduce the concept of power flow during electromagnetic propagation and define the *Poynting vector*.

We have seen from the last section that $\psi(x,y,z;t) = e^{j(\omega_0 t - k_0 z)}$ may represent any component, E_x, E_y, or E_z, of the electric field $\boldsymbol{E}$ for the wave equation given by Eq. (3.2-5a). Hence, we can write the general solution of the wave equations given by Eqs. (3.2-5a) and (3.2-5b) as

$$\boldsymbol{E} = E_x\hat{\boldsymbol{x}} + E_y\hat{\boldsymbol{y}} + E_z\hat{\boldsymbol{z}} = E_{0x}e^{j(\omega_0 t - k_0 z)}\hat{\boldsymbol{x}} + E_{0y}e^{j(\omega_0 t - k_0 z)}\hat{\boldsymbol{y}} + E_{0z}e^{j(\omega_0 t - k_0 z)}\hat{\boldsymbol{z}} \qquad (3.3\text{-}14)$$

and

$$\boldsymbol{H} = H_x\hat{\boldsymbol{x}} + H_y\hat{\boldsymbol{y}} + H_z\hat{\boldsymbol{z}} = H_{0x}e^{j(\omega_0 t - k_0 z)}\hat{\boldsymbol{x}} + H_{0y}e^{j(\omega_0 t - k_0 z)}\hat{\boldsymbol{y}} + H_{0z}e^{j(\omega_0 t - k_0 z)}\hat{\boldsymbol{z}} , \qquad (3.3\text{-}15)$$

respectively. E_{0x}, E_{0y}, E_{0z} and H_{0x}, H_{0y}, H_{0z} are complex constants in general. Again, to represent a physical quantity, we simply take the real part of $\boldsymbol{E}$ or $\boldsymbol{H}$. However, these general solutions must satisfy the Maxwell's equations. From electric Gauss's law [see Eq. (3.1-1)] along with a source-free situation, that is, $\rho_v = 0$, we have

$$\nabla \cdot \boldsymbol{E} = 0,$$

which gives

$$\left(\frac{\partial}{\partial x}\hat{\boldsymbol{x}} + \frac{\partial}{\partial y}\hat{\boldsymbol{y}} + \frac{\partial}{\partial z}\hat{\boldsymbol{z}}\right) \cdot \left(E_{0x}e^{j(\omega_0 t - k_0 z)}\hat{\boldsymbol{x}} + E_{0y}e^{j(\omega_0 t - k_0 z)}\hat{\boldsymbol{y}} + E_{0z}e^{j(\omega_0 t - k_0 z)}\hat{\boldsymbol{z}}\right) = 0$$

or

$$\frac{\partial}{\partial z}E_{0z}e^{j(\omega_0 t - k_0 z)} = -jk_0 E_{0z}e^{j(\omega_0 t - k_0 z)} = 0,$$

implying

$$E_{0z} = 0 \qquad\qquad (3.3\text{-}16)$$

as $-jk_0 e^{j(\omega_0 t - k_0 z)}$ in general is not zero. Similarly, by substituting H into magnetic Gauss's Law [see Eq. (3.1-4)], we get

$$H_{0z} = 0. \qquad\qquad (3.3\text{-}17)$$

Equations (3.3-16) and (3.3-17) mean that there is no component of the electric and magnetic fields in the direction of propagation in a linear, homogeneous, and isotropic medium (such as free space). Such electromagnetic (EM) waves are called *transverse electromagnetic (TEM) waves*.

Now, let us work with one of the curl equations from Maxwell's equations, which relates E to H. Substituting Eqs. (3.3-14) and (3.3-15) with $E_{0z} = H_{0z} = 0$ into Faraday's law [see Eq. (3.1-5)], we have

$$\nabla \times E = \begin{vmatrix} \hat{x} & \hat{y} & \hat{z} \\ \dfrac{\partial}{\partial x} & \dfrac{\partial}{\partial y} & \dfrac{\partial}{\partial z} \\ E_{0x}e^{j(\omega_0 t - k_0 z)} & E_{0y}e^{j(\omega_0 t - k_0 z)} & 0 \end{vmatrix}$$

$$= jk_0 E_{0y}e^{j(\omega_0 t - k_0 z)}\hat{x} - jk_0 E_{0x}e^{j(\omega_0 t - k_0 z)}\hat{y},$$

and

$$-\frac{\partial B}{\partial t} = -\frac{\partial \mu H}{\partial t} = -\mu\frac{\partial}{\partial t}[H_{0x}e^{j(\omega_0 t - k_0 z)}\hat{x} + H_{0y}e^{j(\omega_0 t - k_0 z)}\hat{y}]$$

$$= -j\mu\omega_0 H_{0x}e^{j(\omega_0 t - k_0 z)}\hat{x} - j\mu\omega_0 H_{0y}e^{j(\omega_0 t - k_0 z)}\hat{y}.$$

Hence,

$$\nabla \times E = -\frac{\partial B}{\partial t}$$

gives

$$jk_0 E_{0y}e^{j(\omega_0 t - k_0 z)}\hat{x} - jk_0 E_{0x}e^{j(\omega_0 t - k_0 z)}\hat{y} = -j\mu\omega_0 H_{0x}e^{j(\omega_0 t - k_0 z)}\hat{x} - j\mu\omega_0 H_{0y}e^{j(\omega_0 t - k_0 z)}\hat{y}.$$

By comparing the $\hat{x}$ and $\hat{y}$ components on both sides of the above equation, we have

$$H_{0x} = -\frac{1}{\eta}E_{0y}, \quad H_{0y} = \frac{1}{\eta}E_{0x}, \tag{3.3-17}$$

where η is called the *characteristic impedance* of the medium and given by

$$\eta = \frac{\omega_0}{k_0}\mu = v\mu = \sqrt{\frac{\mu}{\epsilon}}. \tag{3.3-18}$$

The characteristic impendence has the units of ohm and its value for free space is $\eta = \sqrt{\mu_0/\epsilon_0} = 377\,\Omega$.

Using Eq. (3.3-17), we can now write down the E and H fields as follows:

$$E = E_{0x}e^{j(\omega_0 t - k_0 z)}\hat{x} + E_{0y}e^{j(\omega_0 t - k_0 z)}\hat{y}, \tag{3.3-19a}$$

and

$$H = H_{0x}e^{j(\omega_0 t - k_0 z)}\hat{x} + H_{0y}e^{j(\omega_0 t - k_0 z)}\hat{y} = -\frac{1}{\eta}E_{0y}e^{j(\omega_0 t - k_0 z)}\hat{x} + \frac{1}{\eta}E_{0x}e^{j(\omega_0 t - k_0 z)}\hat{y}. \tag{3.3-19b}$$

Once we know the components of the E field, that is, E_{0x} and E_{0y}, we can find the H field directly as η is known for a given medium. Additionally, we note that

$$E \cdot H = -E_{0x}e^{j(\omega_0 t - k_0 z)}\frac{1}{\eta}E_{0y}e^{j(\omega_0 t - k_0 z)} + E_{0y}e^{j(\omega_0 t - k_0 z)}\frac{1}{\eta}E_{0x}e^{j(\omega_0 t - k_0 z)} = 0,$$

meaning that the electric and magnetic fields are orthogonal to each other.

The quantity

$$S = E \times H \tag{3.3-20}$$

is called the Poynting vector and is measured in watts per square meter [W/m²]. Evaluating S, we have

$$S = \begin{vmatrix} \hat{x} & \hat{y} & \hat{z} \\ E_{0x}e^{j(\omega_0 t - k_0 z)} & E_{0y}e^{j(\omega_0 t - k_0 z)} & 0 \\ -\frac{1}{\eta}E_{0y}e^{j(\omega_0 t - k_0 z)} & \frac{1}{\eta}E_{0x}e^{j(\omega_0 t - k_0 z)} & 0 \end{vmatrix} = \frac{1}{\eta}(E_{0x}^2 + E_{0y}^2)e^{2j(\omega_0 t - k_0 z)}\hat{z}, \tag{3.3-21}$$

where have assumed both E_{0x} and E_{0y} are real for simplicity. We notice that the EM waves carry energy and the energy flow occurs in the direction of propagation of the wave. Figure 3.3-7 illustrates the direction of the Poynting vector if the electrical field is in the $\hat{x}$-direction.

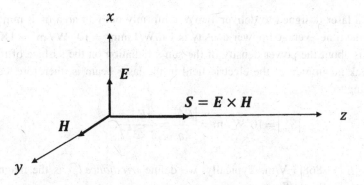

Figure 3.3-7 A transverse electromagnetic (TEM) wave propagating in the $\hat{z}$-direction

Example: Average Power Calculation and Intensity

In a linear, homogenous, and isotropic medium characterized by ϵ and μ, we consider a simple case in that the electric field is along the $\hat{x}$-direction and propagates along the $\hat{z}$-direction. The situation is shown in Figure 3.3-7. Hence, we write

$$E = E_0 e^{j(\omega_0 t - k_0 z)}\hat{x} = |E_0| e^{-j\phi} e^{j(\omega_0 t - k_0 z)}\hat{x},$$

taking into account that E_0 could be complex. Since the electrical field must be a real function of space and time, we simply take the real part of the above to have a real function expression: $\mathrm{Re}[E] = |E_0|\cos(\omega_0 t - k_0 z - \phi)\,\hat{x}$. The magnitude field, according to Eq. (3.3-17), is

$$H = \frac{1}{\eta}|E_0| e^{j(\omega_0 t - k_0 z - \phi)}\hat{y},$$

or $\mathrm{Re}\,[H] = \frac{1}{\eta}|E_0|\cos(\omega_0 t - k_0 z - \phi)\hat{y}$ as $S \propto \hat{x} \times \hat{y} = \hat{z}$. Note that S is a function of time and it is more convenient to define the time-averaged power density:

$$
\begin{aligned}
S_{\text{ave}} &= \frac{1}{T_0}\int_0^{T_0}\mathrm{Re}[S]\,dt = \frac{1}{T_0}\int_0^{T_0}\mathrm{Re}[E]\times\mathrm{Re}[H]\,dt \\
&= \frac{1}{T_0}\int_0^{T_0}|E_0|\cos(\omega_0 t - k_0 z - \phi)\hat{x}\times\frac{1}{\eta}|E_0|\cos(\omega_0 t - k_0 z - \phi)\hat{y}\,dt \\
&= \frac{1}{T_0}\int_0^{T_0}\frac{|E_0|^2}{\eta}\cos^2(\omega_0 t - k_0 z - \phi)\,dt\,\hat{z} = \frac{1}{2}\frac{|E_0|^2}{\eta}\hat{z},
\end{aligned}
$$

as

$$\frac{1}{T_0}\int_0^{T_0}\cos^2(\omega_0 t - \theta)\,dt = \frac{1}{2},$$

where θ is some constant, and $T_0 = 2\pi/\omega_0$ is the period of the sinusoidal function. Note that we are taking the real part of S before the integration because the average power is a real quantity.

For a laser designed to deliver 1 mW uniformly onto an area of 1 mm^2 in free space, the time-averaged power density is $1 \text{ mW/1 mm}^2 = 10^3 \text{ W/m}^2 = 1 \text{KW/m}^2$, which is about the power density of the sun's radiation on the surface of the Earth. The peak magnitude of the electric field in the laser beam is, therefore, calculated according to

$$\left|S_{ave}\right| = 10^3 \, W / m = \frac{\left|E_0\right|^2}{2\eta} = \frac{\left|E_0\right|^2}{2 \times 377},$$

giving $\left|E_0\right| = 868.3$ V/m. Typically, we define *irradiance* I_{ir} as the magnitude of S_{ave}, that is, $\left|S_{ave}\right| = S_{ave}$,

$$I_{ir} = S_{ave} = \frac{\left|E_0\right|^2}{2\eta} \propto I = \left|E_0\right|^2. \tag{3.3-22}$$

In information optics, we usually speak of the *intensity I* being taken to be the square magnitude of the field, that is, $I = \left|E_0\right|^2$.

3.4   Fourier Transform-Based Scalar Diffraction Theory

We have discussed wave propagation in a linear, homogeneous, and isotropic medium in which waves travel with no obstruction. *Diffraction* refers to phenomena that occur when a wave encounters an aperture. An *aperture* is an opening through which light travels. If we want to deal with the vectorial nature of diffraction, we employ *electromagnetic optics* in which vectorial Maxwell's equations [see Eq. (3.1-8)] are needed to be solved under boundary conditions. In that case, we have *vector diffraction theory* for electromagnetic waves. In this section, we deal with *scalar diffraction*. In scalar diffraction theory, we solve the 3-D scalar wave equation [see Eq. (3.3-1)] subject to boundary conditions. The scalar theory is only valid when the diffracting structures are large compared with the wavelength of light. As it turns out, the 3-D scalar wave equation satisfies the principle of superposition. Hence, the theory to be developed is valid for linear systems as discussed in Chapter 2. The principle of superposition provides the basis for our subsequent study of diffraction and *interference*.

Scalar diffraction theory typically is developed by starting with the Helmholtz equation and converting the equation using highly mathematical Green's theorem and as such, some of the physical insight is easily obscured. We will develop scalar diffraction theory through the use of Fourier transform and the 3D scalar wave equation. Indeed, as we shall see, the concept of diffraction can be derived in a much simpler way. We call such an approach as *Fourier Transform-based (FT-based) scalar diffraction theory*. As usual, the theory developed is for monochromatic waves, that is, waves with a single wavelength. The result, nevertheless, can be generalized to non-monochromatic waves simply by using Fourier analysis. The situation of such generalization is analogous to finding the output due to a pulse input to a circuit discussed in Chapter 2.

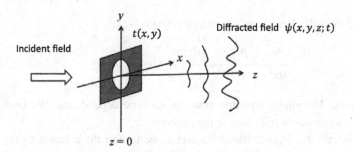

Figure 3.4-1 Diffraction geometry

Figure 3.4-1 illustrates the diffraction geometry. The aperture $t(x,y)$ is located at $z=0$. The incident wave of temporal frequency of ω_0 at $z=0^-$ is given by

$$\psi\left(x,y,z=0^-;t\right)=\psi_p\left(x,y;z=0^-\right)e^{j\omega_0 t}. \tag{3.4-1}$$

The quantity $\psi_p\left(x,y;z=0^-\right)$ is referred to as the *complex amplitude* in optics. $\psi_p\left(x,y;z=0^-\right)$ is a complex function containing the amplitude and phase information and is called a *phasor* in electrical engineering. Let us definite the *amplitude transmittance function*, also simply called *transparency function*, of the aperture as the ratio of the transmitted complex field $\psi_p\left(x,y;z=0^+\right)$ to the incident complex field $\psi_p\left(x,y;z=0^-\right)$ in the $z=0$ plane,

$$t\left(x,y\right)=\frac{\psi_p\left(x,y;z=0^+\right)}{\psi_p\left(x,y;z=0^-\right)}. \tag{3.4-2a}$$

Then

$$\psi_p\left(x,y;z=0^+\right)=\psi_p\left(x,y;z=0^-\right)t\left(x,y\right)=\psi_{p0}\left(x,y\right). \tag{3.4-2b}$$

This complex amplitude is the initial condition or the boundary condition of the situation.

For a given incident field and the transparency function, we seek to find the diffracted field $\psi(x,y,z;t)$.

To find the field distribution at z away from the aperture, we model the solution in the form of

$$\psi\left(x,y,z,t\right)=\psi_p\left(x,y;z\right)e^{j\omega_0 t}, \tag{3.4-3}$$

where $\psi_p\left(x,y;z\right)$ is the unknown complex field to be found with the given initial condition $\psi_{p0}\left(x,y\right)$. Substituting $\psi(x,y,z,t)=\psi_p\left(x,y;z\right)e^{j\omega_0 t}$ into the 3-D scalar wave in Eq. (3.3-1), we obtain

$$\frac{\partial^2\psi_p}{\partial x^2}e^{j\omega_0 t}+\frac{\partial^2\psi_p}{\partial y^2}e^{j\omega_0 t}+\frac{\partial^2\psi_p}{\partial z^2}e^{j\omega_0 t}=\frac{1}{v^2}\left(j\omega_0\right)^2\psi_p e^{j\omega_0 t},$$

which becomes the *Helmholtz equation* for $\psi_p(x,y;z)$,

$$\frac{\partial^2 \psi_p}{\partial x^2} + \frac{\partial^2 \psi_p}{\partial y^2} + \frac{\partial^2 \psi_p}{\partial z^2} + k_0^2 \psi_p = 0. \tag{3.4-4}$$

Note that the Helmholtz equation contains no time dependence. We have accomplished dimensionality reduction in the process.

We now use the Fourier transform technique to find the solution to Eq. (3.4-4). By taking the 2-D Fourier transform of Eq. (3.4-4) and using the transform pair $\dfrac{\partial^n f(x,y)}{\partial x^n} \Leftrightarrow (-jk_x)^n F(k_x,k_y)$ from Table 2.2, we obtain

$$(-jk_x)^2 \Psi_p(k_x,k_y;z) + (-jk_y)^2 \Psi_p(k_x,k_y;z) + \frac{d^2 \Psi_p(k_x,k_y;z)}{dz^2} + k_0^2 \Psi_p(k_x,k_y;z) = 0,$$

where $\mathcal{F}\{\psi_p(x,y;z)\} = \Psi_p(k_x,k_y;z)$. The above equation is simplified to

$$\frac{d^2 \Psi_p}{dz^2} + \left(k_0^2 - k_x^2 - k_y^2\right)\Psi_p = 0, \tag{3.4-5}$$

which is to be solved subject to the initial condition $\mathcal{F}\{\psi_{p0}(x,y)\} = \Psi_p(k_x,k_y;z=0^+) = \Psi_{p0}(k_x,k_y)$. $\psi_{p0}(x,y)$ is given by Eq. (3.4-2b). Note that by using the Fourier transform, we have managed to reduce the second-order partial differential equation of the Helmholtz equation into a second-order ordinary differential equation in $\Psi_p(k_x,k_y;z)$.

Recognizing the differential equation of the form

$$\frac{d^2 y(z)}{dz^2} + \alpha^2 y(z) = 0$$

has the solution given by

$$y(z) = y(0)\exp[-j\alpha z],$$

we can write the solution to Eq. (3.4-5) as

$$\Psi_p(k_x,k_y;z) = \Psi_{p0}(k_x,k_y)e^{-jk_0\sqrt{\left(1-k_x^2/k_0^2-k_y^2/k_0^2\right)}\,z} \tag{3.4-6}$$

for waves propagating along the z-direction. Note that conceptually this equation is identical to Eq. (2.2-13a) when we describe linear time-invariant systems in terms of the transfer function $H(\omega)$ in Chapter 2. Hence, we can define some kind of transfer function for diffraction and the system involved is being a linear and space-invariant system. We, therefore, define the *spatial transfer function of propagation* $\mathcal{H}(k_x,k_y;z)$ as

$$\mathcal{H}\left(k_x,k_y;z\right)=\Psi_p\left(k_x,k_y;z\right)/\Psi_{p0}\left(k_x,k_y\right)=e^{-jk_0\sqrt{\left(1-k_x^2/k_0^2-k_y^2/k_0^2\right)}\,z}. \qquad (3.4\text{-}7)$$

This transfer function relates the input spectrum $\Psi_{p0}\left(k_x,k_y\right)$ at $z=0^+$ and output spectrum $\Psi_p\left(k_x,k_y;z\right)$ at z. To find the complex field $\psi_p\left(x,y;z\right)$, we simply take the inverse Fourier transform of Eq. (3.4-7) to give

$$\psi_p\left(x,y;z\right)=\mathcal{F}^{-1}\left\{\Psi_p\left(k_x,k_y;z\right)\right\}=\mathcal{F}^{-1}\left\{\Psi_{p0}\left(k_x,k_y\right)\mathcal{H}\left(k_x,k_y;z\right)\right\}$$

or

$$=\mathcal{F}^{-1}\left\{\Psi_{p0}\left(k_x,k_y\right)e^{-jk_0\sqrt{\left(1-k_x^2/k_0^2-k_y^2/k_0^2\right)}\,z}\right\}$$

$$=\frac{1}{4\pi^2}\iint_{-\infty}^{\infty}\Psi_{p0}\left(k_x,k_y\right)e^{-jk_0\sqrt{\left(1-k_x^2/k_0^2-k_y^2/k_0^2\right)}\,z}\,e^{-jk_xx-jk_yy}\,dk_x\,dk_y. \qquad (3.4\text{-}8)$$

This is a very important result as it is the foundation to *Fresnel diffraction* and *Fraunhofer diffraction*, to be discussed next. Let us now bring out the physical interpretation of Eq. (3.4-8).

For a given field distribution over the $z=0^+$ plane, that is, $\psi_p\left(x,y;z=0^+\right)=\psi_{p0}\left(x,y\right)$, we can find the field distribution across a plane parallel to the x–y plane but at a distance z from it by calculating Eq. (3.4-8). The term $\Psi_{p0}\left(k_x,k_y\right)$ in Eq. (3.4-8) is a Fourier transform of $\psi_{p0}\left(x,y\right)$, and according to Eq. (2.3-1b), we have

$$\psi_{p0}\left(x,y\right)=\mathcal{F}^{-1}\left\{\Psi_{p0}\left(k_x,k_y\right)\right\}=\frac{1}{4\pi^2}\iint_{-\infty}^{\infty}\Psi_{p0}\left(k_x,k_y\right)e^{-jk_xx-jk_yy}\,dk_x\,dk_y.$$

The physical meaning of the above integral is that we first recognize that a unit-amplitude plane wave propagating with propagation vector $\boldsymbol{k}_0$ is given by $e^{j\left(\omega_0t-k_{0x}x-k_{0y}y-k_{0z}z\right)}$ [see Eq. (3.3-9)]. The complex field of the plane wave, according to Eq. (3.4-3), is

$$e^{-jk_{0x}x-jk_{0y}y-jk_{0z}z}.$$

The field distribution or the plane-wave component at $z=0$ is then

$$e^{-jk_{0x}x-jk_{0y}y}.$$

By comparing this equation with the one given by $\psi_{p0}\left(x,y\right)$ above and recognizing that the spatial radian frequency variables k_x and k_y of the field distribution $\psi_{p0}\left(x,y\right)$ can be regarded as k_{0x} and k_{0y} of the plane wave, respectively. Therefore, $\Psi_{p0}\left(k_x,k_y\right)e^{-jk_xx-jk_yy}$ is the plane wave with amplitude $\Psi_{p0}\left(k_x,k_y\right)$ and by summing over various directions of k_x and k_y, we have the field distribution $\psi_{p0}\left(x,y\right)$ at $z=0^+$. For this reason, $\Psi_{p0}\left(k_x,k_y\right)$ is also called the *angular plane wave spectrum* of the field distribution $\psi_{p0}\left(x,y\right)$.

To find the field distribution a distance of z away, we simply let the various plane waves to propagate over a distance z, which means acquiring a phase shift of $e^{-jk_z z}$ or $e^{-jk_{0z} z}$ by noting that the variable k_z is regarded as k_{0z} of the plane wave:

$$\Psi_{p0}\left(k_x, k_y\right) e^{-jk_x x - jk_y y} e^{-jk_{0z} z}. \tag{3.4-9}$$

Now by summing over various directions of k_x and k_y, we have

$$\psi_p\left(x, y; z\right) = \frac{1}{4\pi^2} \iint\limits_{-\infty}^{\infty} \Psi_{p0}\left(k_x, k_y\right) \ e^{-jk_x x - jk_y y - jk_z z} dk_x dk_y$$

$$= \mathcal{F}^{-1}\left\{\Psi_{p0}\left(k_x, k_y\right) e^{-jk_{0z} z}\right\}.$$

Recall that $k_0 = \sqrt{k_{0x}^2 + k_{0y}^2 + k_{0z}^2}$ and hence $k_z = k_{0z} = \pm k_0 \sqrt{1 - k_x^2 / k_0^2 - k_y^2 / k_0^2}$. We keep the + sign in the above relation to represent waves traveling in the positive z-direction. Hence, we have

$$\psi_p\left(x, y; z\right) = \mathcal{F}^{-1}\left\{\Psi_{p0}\left(k_x, k_y\right) \exp\left(-jk_{0z} z\right)\right\}$$

$$= \mathcal{F}^{-1}\left\{\Psi_{p0}\left(k_x, k_y\right) e^{-jk_0 \sqrt{\left(1 - k_x^2 / k_0^2 - k_y^2 / k_0^2\right)} \, z}\right\},$$

and we immediately recover Eq. (3.4-8) and provide physical meaning to the equation. Note that the angular plane wave spectrum of the field distribution $\psi_p\left(x, y; z\right)$ is given by

$$\Psi_{p0}\left(k_x, k_y\right) e^{-jk_{0z} z}$$

or

$$\Psi_{p0}\left(k_x, k_y\right) e^{-jk_0 \sqrt{\left(1 - k_x^2 / k_0^2 - k_y^2 / k_0^2\right)} \, z},$$

which is for the propagation of plane waves, that is, we must have $1 - k_x^2 / k_0^2 - k_y^2 / k_0^2 \geq 0$ or $k_0^2 \geq k_x^2 + k_y^2$.

The range of super high spatial frequencies k_x, k_y covers the *evanescent waves* when $1 - k_x^2 / k_0^2 - k_y^2 / k_0^2 < 0$ or $k_x^2 + k_y^2 \geq k_0^2$. It follows that $k_{0z} = \pm k_0 \sqrt{1 - k_x^2 / k_0^2 - k_y^2 / k_0^2} = \pm jk_0 \sqrt{k_x^2 / k_0^2 + k_y^2 / k_0^2 - 1}$. Physically, we simply choose the negative sign so that the wave amplitude decays with distance from the source rather than growing exponentially. Hence,

$$\Psi_{p0}\left(k_x, k_y\right) e^{-jk_{0z} z} = \Psi_{p0}\left(k_x, k_y\right) e^{-j\left(-jk_0 \sqrt{k_x^2 / k_0^2 + k_y^2 / k_0^2 - 1}\right) z}$$

$$= \Psi_{p0}\left(k_x, k_y\right) e^{-k_0 z \sqrt{k_x^2 / k_0^2 + k_y^2 / k_0^2 - 1}}.$$

This represents a field propagating on the x y plane with an exponentially decreasing amplitude in the z-direction. In other words, the feature of the evanescent waves is absent or not observable at a distance $z \gg \lambda_0$ due to the damping effect in the z-direction. The ability to observe or measure evanescent waves is one of the important problems in optics for super-resolution imaging. We want to point out that the use of Eq. (3.4-8) is commonly referred to as *the angular spectrum method (ASM)* to calculate numerically the diffraction of a wave field though a distance z using *the fast Fourier transform (FFT)*.

Since $\mathcal{H}(k_x, k_y; z)$ is the spatial transfer function of propagation, its inverse transform $G(x, y; z)$ is called the *spatial impulse response of propagation* in the linear, homogeneous, and isotropic medium:

$$G(x, y; z) = \mathcal{F}^{-1}\left\{\mathcal{H}(k_x, k_y; z)\right\}$$

$$= \frac{1}{4\pi^2} \iint\limits_{-\infty}^{\infty} e^{-jk_0\sqrt{\left(1 - k_x^2/k_0^2 - k_y^2/k_0^2\right)}\,z}\; e^{-jk_x x - jk_y y}\, dk_x dk_y. \tag{3.4-10a}$$

The solution to the above equation is formidable and its solution has been provided in the literature [Stark (1982)]:

$$G(x, y; z) = \frac{jk_0 e^{-jk_0\sqrt{x^2 + y^2 + z^2}}}{2\pi\sqrt{x^2 + y^2 + z^2}}\; \frac{z}{\sqrt{x^2 + y^2 + z^2}}\left(1 + \frac{1}{jk_0\sqrt{x^2 + y^2 + z^2}}\right). \tag{3.4-10b}$$

While Eq. (3.4-8) allows us to find the diffracted field through the Fourier transform, we can also write the field distribution at z in the spatial domain as

$$\psi_p(x, y; z) = \psi_{p0}(x, y) * G(x, y; z). \tag{3.4-11}$$

Once we have $\psi_p(x, y; z)$, the final solution is then given by Eq. (3.4-3), and for physical real quantity, we simple take the real part of the complex function, that is, $\mathrm{Re}\left[\psi_p(x, y; z)e^{j\omega_0 t}\right]$.

3.4.1   Huygens' Principle

The convolution integral involving $G(x, y; z)$ in Eq. (3.4-11) is very complicated and, often, we make some realistic physical assumptions to simplify the functional form of $G(x, y; z)$. With the assumption that we observe the field distribution many wavelengths away from the diffracting aperture, that is, $z \gg \lambda_0 = 2\pi / k_0$, the term

$$1 + \frac{1}{jk_0\sqrt{x^2 + y^2 + z^2}} = 1 + \frac{1}{j\frac{2\pi}{\lambda_0}z\sqrt{1 + \frac{x^2 + y^2}{z^2}}} \approx 1.$$

Now, for small angles (*paraxial approximation*), that is, $z \gg \sqrt{x^2 + y^2}$, we have. $\sqrt{x^2 + y^2 + z^2} = z\sqrt{1 + (x^2 + y^2)/z^2} \approx z$. Physically, it means we only consider small excursions around the axis of propagation along z. The factor $z / \sqrt{x^2 + y^2 + z^2}$ in $G(x, y; z)$ can then be replaced by 1. The spatial impulse response of propagation now becomes

$$G(x, y; z) = \frac{jk_0 e^{-jk_0 \sqrt{x^2 + y^2 + z^2}}}{2\pi\sqrt{x^2 + y^2 + z^2}}. \tag{3.4-12}$$

Let us state the spherical wave given in Eq. (3.3-13) and re-write it terms of its complex amplitude as follows:

$$\psi(R; t) = \frac{1}{R} e^{j(\omega_0 t - k_0 R)} = \psi_p(R) e^{j\omega_0 t}.$$

We then notice that the complex amplitude of a spherical wave is

$$\psi_p(R) = \frac{1}{R} e^{-jk_0 R}. \tag{3.4-13}$$

Since $\sqrt{x^2 + y^2 + z^2} = R$ [see Figure 3.3-5], we observe that $G(x, y; z)$ in Eq. (3.4-12) is basically a spherical wave with amplitude of $\frac{jk_0}{2\pi}$ upon comparing with Eq. (3.4-13). We now write out the convolution integral in Eq. (3.4-11) by using Eq. (3.4-12), and we have the well-known *Huygens–Fresnel diffraction integral*:

$$\psi_p(x, y; z) = \psi_{p0}(x, y) * G(x, y; z) = \psi_{p0}(x, y) * \frac{jk_0 e^{-jk_0 R}}{2\pi R}$$

$$= \frac{jk_0}{2\pi} \iint_{-\infty}^{\infty} \psi_{p0}(x', y') \frac{e^{-jk_0 R'}}{R'} dx' dy', \tag{3.4-14}$$

where $R' = \sqrt{(x - x')^2 + (y - y')^2 + z^2}$. The Huygens–Fresnel diffraction integral is basically a mathematical statement of the *Huygens' principle*, which states that every unobstructed point of a wave front from the diffracting aperture serves as a source of spherical waves. The complex amplitude at any point beyond is the superposition of all these spherical waves. The situation is depicted in Figure 3.4-2. As we want to find the complex field at point P on the observation x–y plane that is at a distance z from the source (aperture) $x' - y'$ plane, we first consider a differential area $dx' dy'$ on the source plane. The field at point P due to spherical waves emanating from the differential area is

$$\frac{jk_0}{2\pi} \psi_{p0}(x', y') \frac{e^{-jk_0 R'}}{R'} dx' dy',$$

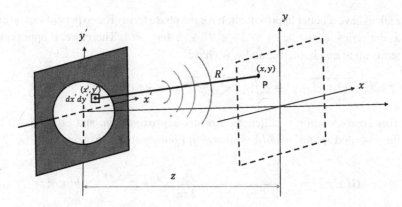

Figure 3.4-2 Diffraction geometry illustrating the source and observation planes

where the term $\dfrac{jk_0}{2\pi}\psi_{p0}(x',y')$ is the amplitude of the spherical wave, which is a weighting factor due to the aperture and light illumination on the aperture as $\psi_{p0}(x,y)=\psi_p(x,y;z=0^-)t(x,y)$ [see Eq. (3.4-2)]. To find the total field at point P, we simply sum over all the differential areas in the aperture to obtain

$$\iint\limits_{-\infty}^{\infty}\frac{jk_0}{2\pi}\psi_{p0}(x',y')\frac{e^{-jk_0R'}}{R'}dx'dy',$$

which is Eq. (3.4-14).

3.4.2 Fresnel Diffraction and Fraunhofer Diffraction

Fresnel Diffraction

The Huygens' principle, derived from Eq. (3.4-14) along with Eq. (3.4-12), is fairly complicated for practical applications. However, we can further reduce $G(x,y;z)$ to obtain a more simple and usable expression. Using the paraxial approximation, that is, $\sqrt{x^2+y^2+z^2}=z\sqrt{1+(x^2+y^2)/z^2}\approx z$, we can replace the denominator factor by z to get

$$G(x,y;z)=\frac{jk_0e^{-jk_0\sqrt{x^2+y^2+z^2}}}{2\pi\sqrt{x^2+y^2+z^2}}\approx\frac{jk_0e^{-jk_0\sqrt{x^2+y^2+z^2}}}{2\pi z}. \qquad (3.4\text{-}15)$$

However, the phase term is more sensitive to the approximation. The reason is that the phase term $k_0\sqrt{x^2+y^2+z^2}$ is multiplied by a very large wavenumber $k_0=2\pi/\lambda_0$. For a typical value of $\lambda_0=0.6\times10^{-6}$m in the visible region of the light spectrum, $k_0\approx10^7$rad/m. Phase changes due to wave propagation of as little as several wavelengths change the value of the phase significantly. For example, for $z=5\lambda_0$, $k_0\sqrt{x^2+y^2+z^2}\approx k_0z=10^7\times5\times0.6\times10^{-6}=30$rad. In light of this, we seek to have a better approximation for the phase term. The exponent can be expanded as the series $\sqrt{1+\varepsilon} \approx 1+\varepsilon/2-\varepsilon^2/8+\ldots$ for $\varepsilon \ll 1$. Therefore, if approximated by terms up to quadratic in x and y, we have

$$\sqrt{x^2+y^2+z^2} = z\sqrt{1+\left(x^2+y^2\right)/z^2} \approx z+\left(x^2+y^2\right)/2z.$$

This approximation is called the *Fresnel approximation*, and Eq. (3.4-15) becomes the so-called *spatial impulse response in Fourier optics*, $h(x,y;z)$, as

$$G(x,y;z) = \frac{jk_0 e^{-jk_0\sqrt{x^2+y^2+z^2}}}{2\pi z} \approx \frac{jk_0}{2\pi z}e^{-jk_0 z}e^{\frac{-jk_0}{2z}\left(x^2+y^2\right)} = h(x,y;z). \qquad (3.4\text{-}16)$$

Using $h(x,y;z)$, Eq. (3.4-11) becomes

$$\psi_p\left(x,y;z\right) = \psi_{p0}\left(x,y\right) * G\left(x,y;z\right) \approx \psi_{p0}\left(x,y\right) * h\left(x,y;z\right)$$

$$= e^{-jk_0 z}\frac{jk_0}{2\pi z}\iint_{-\infty}^{\infty}\psi_{p0}\left(x',y'\right)e^{\frac{-jk_0}{2z}\left[(x-x')^2+(y-y')^2\right]}dx'dy'. \qquad (3.4\text{-}17)$$

Equation (3.4-17) is called the *Fresnel diffraction formula* and describes *Fresnel diffraction*, often known as *near field diffraction*, of a beam during propagation and having the initial complex profile $\psi_{p0}\left(x,y\right)$. By taking the Fourier transform of the impulse response $h(x,y;z)$, we have the *spatial frequency transfer function in Fourier optics* $H\left(k_x,k_y;z\right)$:

$$H\left(k_x,k_y;z\right) = \mathcal{F}\left\{h\left(x,y;z\right)\right\}$$

$$= e^{-jk_0 z}e^{\frac{jz}{2k_0}\left(k_x^2+k_y^2\right)}. \qquad (3.4\text{-}18)$$

To perform the calculation of the Fresnel diffraction formula in the frequency domain, we take the Fourier transform of Eq. (3.4-17):

$$\mathcal{F}\left\{\psi_p\left(x,y;z\right)\right\} = \mathcal{F}\left\{\psi_{p0}\left(x,y\right)\right\} \times \mathcal{F}\left\{h\left(x,y;z\right)\right\},$$

where we have used the convolution theorem in Eq. (2.3-27). Expressing the above using their transforms, we have

$$\Psi_p\left(k_x,k_y;z\right) = \Psi_{p0}\left(k_x,k_y\right)H\left(k_x,k_y;z\right)$$

and finally in the spatial domain, we have

$$\psi_p\left(x,y;z\right) = \mathcal{F}^{-1}\left\{\Psi_p\left(k_x,k_y;z\right)\right\} = \mathcal{F}^{-1}\left\{\Psi_{p0}\left(k_x,k_y\right)H\left(k_x,k_y;z\right)\right\}, \qquad (3.4\text{-}19)$$

which is the Fresnel diffraction formulas written in terms of Fourier transform.

Alternatively, the calculation of the diffracted field can be performed using a single Fourier transform. We re-write Eq. (3.4-17) by expanding the quadratic terms in the exponent inside the integral and grouping all the primed variables within the integral:

$$\psi_p(x,y;z) = e^{-jk_0z}\frac{jk_0}{2\pi z}\int\int_{-\infty}^{\infty}\psi_{p0}(x',y')\,e^{\frac{-jk_0}{2z}\left[(x^2+y^2)+(x'^2+y'^2)-2xx'-2yy'\right]}dx'dy'$$

$$= e^{-jk_0z}\frac{jk_0}{2\pi z}e^{\frac{-jk_0}{2z}(x^2+y^2)}\int\int_{-\infty}^{\infty}\psi_{p0}(x',y')\,e^{\frac{-jk_0}{2z}(x'^2+y'^2)}e^{\frac{jk_0}{z}(xx'+yy')}dx'dy'. \quad (3.4\text{-}20)$$

By comparing this equation with the definition of the Fourier transform from Eq. (2.3-1a) re-stated below for convenience:

$$\mathcal{F}\{f(x,y)\} = F(k_x,k_y) = \int\int_{-\infty}^{\infty}f(x,y)\,e^{jk_xx+jk_yy}dxdy,$$

we can write the Fresnel diffraction formulas using a single Fourier transform operation as follows:

$$\psi_p(x,y;z) = e^{-jk_0z}\frac{jk_0}{2\pi z}e^{\frac{-jk_0}{2z}(x^2+y^2)}\mathcal{F}\left\{\psi_{p0}(x,y)e^{\frac{-jk_0}{2z}(x^2+y^2)}\right\}\Bigg|_{k_x=\frac{k_0x}{z},\,k_y=\frac{k_0y}{z}}. \quad (3.4\text{-}21)$$

Fraunhofer Diffraction

From Eq. (3.4-20) to Eq. (3.4-21), we see that Fresnel diffraction can be obtained from a Fourier transform of the product of the source distribution $\psi_{p0}(x,y)$ and a quadratic phase factor $e^{\frac{-jk_0}{2z}(x^2+y^2)}$. If we now impose the so-called *Fraunhofer approximation*

$$\frac{k_0}{2}(x'^2+y'^2)\Big|_{max} = z_R \ll z, \quad (3.4\text{-}22)$$

the quadratic phase factor within the integral of Eq. (3.4-20) is approximated unity over the entire aperture, giving Eq. (3.4-21) to become

$$\psi_p(x,y;z) = e^{-jk_0z}\frac{jk_0}{2\pi z}e^{\frac{-jk_0}{2z}(x^2+y^2)}\mathcal{F}\left\{\psi_{p0}(x,y)\right\}\Bigg|_{k_x=\frac{k_0x}{z},\,k_y=\frac{k_0y}{z}}. \quad (3.4\text{-}23)$$

When this happens, we are in the region of *Fraunhofer diffraction* or *far field diffraction*. z_R is called the *Rayleigh length* and if we express it in terms of the wavelength λ_0, we have

$$z_R = \frac{\pi}{\lambda_0}(x'^2+y'^2)\Big|_{max}.$$

The term $\pi(x'^2+y'^2)\big|_{max}$ defines the maximum area of the source and if this area divided by the wavelength is much less than the observation distance z, we observe diffracted field under Fraunhofer diffraction. Equation (3.4-23) is the *Fraunhofer diffraction formula.*

Equation (3.4-23) says that we can compute optically the spectrum of $\psi_{p0}(x,y)$, that is, $\mathcal{F}\{\psi_{p0}(x,y)\}$, by simply observing the diffracted field much greater than the Rayleigh distance. That is good news for *optical image processing* as we could perform image processing optically. The notion of optical image processing comes from our previous experiences with signal processing discussed in Chapter 2. From Eq. (2.2-13b), we have

$$y(t) = \mathcal{F}^{-1}\{Y(\omega)\} = \mathcal{F}^{-1}\{X(\omega)H(\omega)\}.$$

We need to compute the input spectrum of $x(t)$, that is, $X(\omega)$, so that the spectrum is "processed" by the transfer function, $H(\omega)$, of the system to obtain the processed output $y(t)$. In order to explore such way of processing the input, we need the spectrum of the input as a start. From Eq. (3.4-23), we do have the spectrum of the input $\psi_{p0}(x,y)$. However, the spectrum is corrupted by the quadratic phase term, $e^{\frac{-jk_0}{2z}(x^2+y^2)}$, and that introduces errors into the spectrum. We need, somehow, to eliminate the quadratic phase errors. In the next section, we introduce a lens in an optical system. As it turns out, an ideal lens carries a phase transformation property that will allow us to achieve an exact Fourier transformation in the process.

Example: Fraunhofer Diffraction of a Square Aperture

The complex amplitude of a square aperture illuminated by a plane wave of unity amplitude is represented by $\psi_{p0}(x,y) = \text{rect}\left(\dfrac{x}{l_x}, \dfrac{y}{l_x}\right)$ at the exit of the aperture, where $l_x \times l_x$ is the area of the opening. According to the *Fraunhofer diffraction formula*, that is, Eq. (3.4-23), the complex field at a distance of z away from the aperture is given by

$$\psi_p(x,y;z) = e^{-jk_0 z}\frac{jk_0}{2\pi z}e^{\frac{-jk_0}{2z}(x^2+y^2)}\mathcal{F}\{\psi_{p0}(x,y)\}\Big|_{k_x=\frac{k_0 x}{z},\, k_y=\frac{k_0 y}{z}}$$

$$= e^{-jk_0 z}\frac{jk_0}{2\pi z}e^{\frac{-jk_0}{2z}(x^2+y^2)}\mathcal{F}\left\{\text{rect}\left(\frac{x}{l_x},\frac{y}{l_x}\right)\right\}\Big|_{k_x=\frac{k_0 x}{z},\, k_y=\frac{k_0 y}{z}}.$$

Using the result of Eq. (2.3-6) or through the use of Table 2.1, the above equation becomes

$$\psi_p(x,y;z) = e^{-jk_0 z}\frac{jk_0}{2\pi z}e^{\frac{-jk_0}{2z}(x^2+y^2)}l_x l_x\, \text{sinc}\left(\frac{k_x l_x}{2\pi},\frac{k_y l_x}{2\pi}\right)\Big|_{k_x=\frac{k_0 x}{z},\, k_y=\frac{k_0 y}{z}}$$

$$= e^{-jk_0 z}\frac{jk_0}{2\pi z}e^{\frac{-jk_0}{2z}(x^2+y^2)}l_x l_x\, \text{sinc}\left(\frac{k_0 x}{z}\frac{l_x}{2\pi},\frac{k_0 y}{z}\frac{l_x}{2\pi}\right).$$

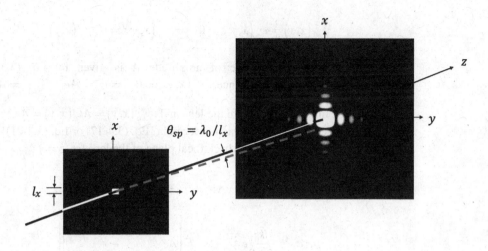

Figure 3.4-3 Fraunhofer diffraction of a square

The intensity pattern, according to Eq. (3.3-22), is

$$I = \psi_p \psi_p^* = |\psi_p|^2 \propto \text{sinc}^2\left(\frac{x\, l_x}{z\, \lambda_0}, \frac{y\, l_x}{z\, \lambda_0}\right).$$

The first zeros of the *sinc* function along the x-direction occur at $x = \pm z\lambda_0\,/\,l_x$, and the *angle of spread* or *spread angle* can be defined as $\theta_{sp} = \lambda_0\,/\,l_x$. The situation is shown in Figure 3.4-3. Note that the spread angle θ_{sp} obtained from diffraction is consistent with that derived from the uncertainty principle discussed in Chapter 1 for a slit aperture [see Section 1.3].

3.4.3 Phase Transforming Property of an Ideal Lens

We have discussed light diffraction by aperture in the previous section. In this section, we will discuss the passage of light through an *ideal lens*. An ideal lens is a pure phase optical element of infinite extent and has no thickness. Hence, the ideal lens only modifies the phase of the incoming wave fronts. The transparency function of the ideal lens is given by

$$t_f(x,y) = e^{j\frac{k_0}{2f}(x^2+y^2)}, \tag{3.4-24}$$

where f is the focal length of the lens. For a converging (or positive) lens, $f > 0$ and $f < 0$ is for a diverging (or negative) lens.

For a plane wave of amplitude A incident on the lens, let us find the complex field distribution in the back focal plane of the lens. We assume that the lens is located on the plane $z = 0$. The situation is shown in Figure 3.4-4. According to Eq. (3.4-2b), the field distribution after the lens is

$$\psi_{p0}(x,y) = \psi_p(x,y;z=0^+) = \psi_p(x,y;z=0^-)t_f(x,y).$$

Since the incident plane wave of amplitude A is given by Eq. (3.3-8) as $\psi(z;t) = Ae^{j(\omega_0 t - k_0 z)} = \psi_p e^{j\omega_0 t}$. Hence, $\psi_p(x,y;z=0^-) = \psi_p = Ae^{-jk_0 z}|_{z=0^-} = A$ is the amplitude just immediately before the lens and $\psi_{p0}(x,y) = At_f(x,y) = Ae^{j\frac{k_0}{2f}(x^2+y^2)}$. We can now use the Fresnel diffraction formula [Eq. (3.4-17) or Eq. (3.4-21)] to calculate the field distribution in the back focal plane of the lens for $z = f$:

$$\psi_p(x,y;z=f) = \psi_{p0}(x,y)*h(x,y;z=f) = Ae^{j\frac{k_0}{2f}(x^2+y^2)}*h(x,y;z=f)$$

$$= e^{-jk_0 f}\frac{jk_0}{2\pi f}e^{\frac{-jk_0}{2f}(x^2+y^2)}\mathcal{F}\left\{Ae^{j\frac{k_0}{2f}(x^2+y^2)}e^{\frac{-jk_0}{2f}(x^2+y^2)}\right\}\Bigg|_{k_x=\frac{k_0 x}{f},\,k_y=\frac{k_0 y}{f}}$$

$$= e^{-jk_0 f}\frac{jk_0}{2\pi f}e^{\frac{-jk_0}{2f}(x^2+y^2)}\mathcal{F}\{A\}\Bigg|_{k_x=\frac{k_0 x}{f},\,k_y=\frac{k_0 y}{f}}. \qquad (3.4\text{-}25)$$

We observe that the lens phase function cancels out exactly the quadratic phase function associated with Fresnel diffraction and the Fourier transform of A, using Table 2.1, is $4\pi^2 A\delta(k_x,k_y)$. Therefore, Eq. (3.4-25) finally becomes

$$\psi_p(x,y;z=f) = e^{-jk_0 f}\frac{jk_0}{2\pi f}e^{\frac{-jk_0}{2f}(x^2+y^2)}A4\pi^2\delta\left(\frac{k_0 x}{f},\frac{k_0 y}{f}\right) \propto \delta(x,y).$$

The result makes perfect sense as it is consistent with geometrical optics in that all input rays parallel to the optical axis converges behind the lens to a point called the back focal point. What we have just demonstrated is that the functional form of the phase function given by Eq. (3.4-24) is indeed correct for the ideal lens.

From Eq. (3.4-25), we recognize that A is the field distribution just in front of the lens. Therefore, we can generalize Eq. (3.4-25) to

$$\psi_p(x,y;z=f) = e^{-jk_0 f}\frac{jk_0}{2\pi f}e^{\frac{-jk_0}{2f}(x^2+y^2)}\mathcal{F}\{\psi_{fl}(x,y)\}\Bigg|_{k_x=\frac{k_0 x}{f},\,k_y=\frac{k_0 y}{f}}, \qquad (3.4\text{-}26)$$

where $\psi_{fl}(x,y)$ denotes the field distribution just in front of the lens. For example, if the physical situation is changed to include a transparency function $t(x,y)$ just in front of the lens as shown in Figure 3.4-5, the field distribution just in front of the lens becomes $\psi_{fl}(x,y) = At(x,y)$. The transparency function under discussion could represent some kind of aperture in front of the lens or simply as an input image.

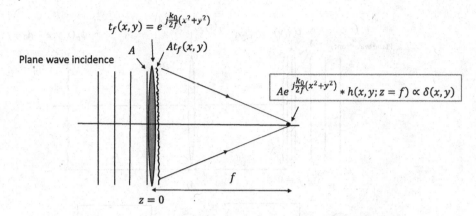

Figure 3.4-4 Plane wave of amplitude A incident on the ideal lens of focal length f

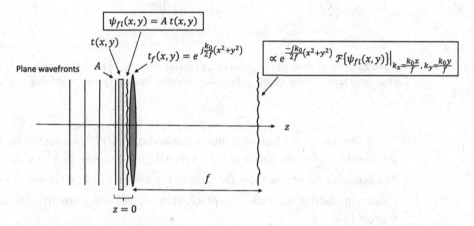

Figure 3.4-5 Field distribution relationship between the field just in front of the lens, that is, $\psi_{fl}(x,y)$, and the field on the focal plane of the lens

Example: Transparency Function as an Input Placed in Front of the Ideal Lens

Let us consider a more general situation where a transparency function $t(x,y)$ is placed a distance d_0 in front of the idea lens. The transparency is illuminated by a plane wave with amplitude A, as shown in Figure 3.4-6a. Therefore, $\psi_p(x,y;z=0^-)=A$ is the complex amplitude just immediately before $t(x,y)$ and $A \times t(x,y)$ is the field distribution after the transparency. In general, if the transparency is illuminated by some general complex field $\psi_p^i(x,y)$, that is, $\psi_p(x,y;z=0^-)=\psi_p^i(x,y)$, the field distribution immediately behind the transparency is then given by $\psi_p^i(x,y) \times t(x,y)$. For the current situation, we simply have $\psi_p^i(x,y)=A$.

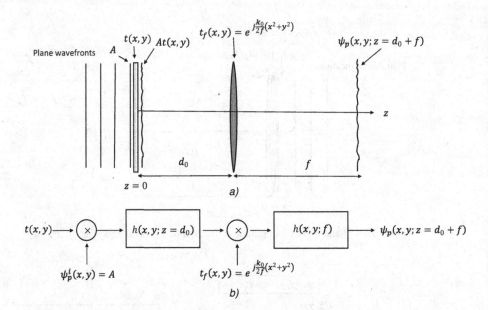

We first use the Fresnel diffraction formula [Eq. (3.4-17)] to calculate the field distribution just before the lens at $z = d_0$, which is given by $[A \times t(x,y)] * h(x,y;z = d_0)$. The field after the lens is, therefore, given by $\{[A \times t(x,y)] * h(x,y;z = d_0)\} \times t_f(x,y)$. Finally, the field at the back focal plane, after Fresnel diffraction of a distance of f, is given by

$$\psi_p(x,y;z = d_0 + f) = \{\{[A \times t(x,y)] * h(x,y;z = d_0)\} \times t_f(x,y)\} * h(x,y;f). \quad (3.4\text{-}27)$$

This equation can be represented by a system block diagram shown in Figure 3.4-6b.

Equation (3.4-27) is a precise way to represent the physical situation in Figure 3.4-6a, but it is complicated to evaluate owing to the repeated convolution operations. Let us employ the result of Eq. (3.4-26) to facilitate the calculation to obtain the field distribution on the focal plane. In order to do it, we must first find the complex field just in front of the lens, which is the first step in Eq. (3.4-27):

$$\psi_{fl}(x,y) = A \times t(x,y) * h(x,y;z = d_0)$$

$$= A\mathcal{F}^{-1}\{T(k_x,k_y)H(k_x,k_y;d_0)\}$$

$$= A\mathcal{F}^{-1}\left\{T(k_x,k_y)e^{-jk_0 d_0}e^{j\frac{d_0}{2k_0}(k_x^2+k_y^2)}\right\},$$

$$(3.4\text{-}28)$$

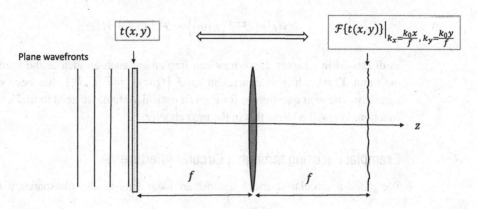

Figure 3.4-7 Ideal lens as an optical Fourier transformer

where we have used the transfer function approach to wave propagation [see Eqs. (3.4-18) and (3.4-19)] and $\mathcal{F}\{t(x,y)\} = T(k_x, k_y)$ is the spectrum of the input. Now, the complex field at the focal plane is given by substituting Eq. (3.4-28) into Eq. (3.4-26), and we have

$$
\psi_p(x,y;z = f + d_0) = e^{-jk_0 f} \frac{jk_0}{2\pi f} e^{\frac{-jk_0}{2f}(x^2+y^2)} \mathcal{F}\{\psi_{fl}(x,y)\}\Big|_{k_x=\frac{k_0 x}{f}, k_y=\frac{k_0 y}{f}}
$$

$$
= e^{-jk_0 f} \frac{jk_0}{2\pi f} A e^{\frac{-jk_0}{2f}(x^2+y^2)} \mathcal{F}\left\{\mathcal{F}^{-1}\left\{T(k_x,k_y) e^{-jk_0 d_0} e^{\frac{jd_0}{2k_0}(k_x^2+k_y^2)}\right\}\right\}\Big|_{k_x=\frac{k_0 x}{f}, k_y=\frac{k_0 y}{f}}
$$

$$
= e^{-jk_0 f} \frac{jk_0}{2\pi f} A e^{\frac{-jk_0}{2f}(x^2+y^2)} T(k_x,k_y) e^{-jk_0 d_0} e^{\frac{jd_0}{2k_0}(k_x^2+k_y^2)}\Big|_{k_x=\frac{k_0 x}{f}, k_y=\frac{k_0 y}{f}}
$$

$$
= e^{-jk_0(f+d_0)} \frac{jk_0}{2\pi f} A e^{\frac{-jk_0}{2f}(x^2+y^2)} T(k_x,k_y)\Big|_{k_x=\frac{k_0 x}{f}, k_y=\frac{k_0 y}{f}} \times e^{i\frac{d_0}{2k_0}\left[\left(\frac{k_0 x}{f}\right)^2+\left(\frac{k_0 y}{f}\right)^2\right]}
$$

or finally,

$$
\psi_p(x,y;z = f + d_0) = e^{-jk_0(f+d_0)} \frac{jk_0}{2\pi f} A e^{-j\frac{k_0}{2f}\left[1-\frac{d_0}{f}\right](x^2+y^2)} \mathcal{F}\{t(x,y)\}\Big|_{k_x=\frac{k_0 x}{f}, k_y=\frac{k_0 y}{f}},
$$

$$
(3.4\text{-}29)
$$

where we have grouped the $(x^2 + y^2)$ terms in the exponents to simplify in order to obtain the above result. Note that, as in Eq. (3.4-23), a phase curvature factor again precedes the Fourier transform but vanished when $d_0 = f$. Thus, when the input transparency is placed in the front focal plane of the lens, aside from some constants we have the exact Fourier transform of the input on the back focal plane. This important result is illustrated in Figure 3.4-7. Fourier processing on an input transparency located on the front focal plane may now be performed on the back focal plane, as will be seen in the next chapter. With reference to signal processing where

$$y(t) = \mathcal{F}^{-1}\{Y(\omega)\} = \mathcal{F}^{-1}\{X(\omega)H(\omega)\}$$

as discussed in Chapter 2, we now can implement such an idea as the spectrum of the input $X(\omega)$, which is equivalent to $\mathcal{F}\{t(x,y)\} = T(k_x, k_y)$, has been obtained optically. The next question is, for a given optical system, we need to find its transfer function. We will address this in the next chapter.

Example: Focusing through a Circular Aperture

We place a circular aperture against an ideal lens of the transparency function given by $t_f(x,y) = e^{j\frac{k_0}{2f}(x^2+y^2)}$. The complex amplitude of a circular aperture illuminated by a plane wave of unity amplitude is then represented by

$$\psi_{p0}(x,y) = \text{circ}\left(\frac{r}{r_0}\right) t_f(x,y) = \text{circ}\left(\frac{r}{r_0}\right) e^{j\frac{k_0}{2f}(x^2+y^2)} \quad \text{at the exit of the lens, where}$$

r_0 is radius of the aperture. According to the Fresnel diffraction formula, that is, Eq. (3.4-17) or Eq. (3.4-21), we write the complex field at a distance of $f+z$ away from the aperture as

$$\psi_p(x,y;f+z)$$
$$= e^{-jk_0(f+z)} \frac{jk_0}{2\pi(f+z)} e^{\frac{-jk_0}{2(f+z)}(x^2+y^2)} \mathcal{F}\left\{\psi_{p0}(x,y) e^{\frac{-jk_0}{2(f+z)}(x^2+y^2)}\right\}\Bigg|_{k_x=\frac{k_0 x}{f+z}, k_y=\frac{k_0 y}{f+z}}$$

$$= e^{-jk_0(f+z)} \frac{jk_0}{2\pi(f+z)} e^{\frac{-jk_0}{2(f+z)}(x^2+y^2)} \mathcal{F}\left\{\text{circ}\left(\frac{r}{r_0}\right) e^{j\frac{k_0}{2f}(x^2+y^2)} e^{\frac{-jk_0}{2(f+z)}(x^2+y^2)}\right\}\Bigg|_{k_x=\frac{k_0 x}{f+z}, k_y=\frac{k_0 y}{f+z}}.$$

Since the problem has circular symmetry, we can re-write the above equation in terms of the Fourier–Bessel transform as

$$\psi_p(r;f+z) = e^{-jk_0(f+z)} \frac{jk_0}{2\pi(f+z)} e^{\frac{-jk_0}{2(f+z)}r^2} \mathcal{B}\left\{\text{circ}\left(\frac{r}{r_0}\right) e^{j\frac{k_0}{2f}r^2} e^{\frac{-jk_0}{2(f+z)}r^2}\right\}\Bigg|_{k_r=\frac{k_0 r}{f+z}}. \quad (3.4\text{-}30)$$

Focused Circular Beam: The Airy Pattern

To find the complex field at the focal plane, we let $z=0$ in the above equation:

$$\psi_p(r;f) = e^{-jk_0 f} \frac{jk_0}{2\pi f} e^{\frac{-jk_0}{2f}r^2} \mathcal{B}\left\{\text{circ}\left(\frac{r}{r_0}\right)\right\}\Bigg|_{k_r=\frac{k_0 r}{f}}.$$

Since the Fourier–Bessel transform of a circular function is given by Eq. (2.3-15) as

$$\mathcal{B}\left\{\text{circ}\left(\frac{r}{r_0}\right)\right\} = 2\pi r_0^2 \,\text{jinc}(r_0 k_r),$$

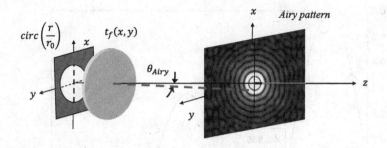

Figure 3.4-8 Focusing through a circular aperture

the complex field in the focal plane is then

$$\psi_p(r;f) = e^{-jk_0 f} \frac{jk_0}{2\pi f} e^{\frac{-jk_0}{2f} r^2} 2\pi r_0^2 \text{jinc}\left(r_0 \frac{k_0 r}{f}\right),$$

and the intensity on the focal plane is

$$I(r,f) = |\psi_p(r;f)|^2 = \left(\frac{k_0 r_0^2}{f}\right)^2 \text{jinc}^2\left(r_0 \frac{k_0 r}{f}\right) = I_0 \text{jinc}^2\left(r_0 \frac{k_0 r}{f}\right), \qquad (3.4\text{-}31)$$

which is known as the *Airy pattern*. The Airy pattern consists of a central disc, known as the *Airy disk*, surrounded by rings, as illustrated in Figure 2.3-5. Figure 3.4-8 shows the physical situation of the current example.

Let us find the radius of the central disc. Since $\text{jinc}(x) = \dfrac{J_1(x)}{x}$, the first zero of J_1, therefore, determines the radius. In Figure 3.4-9, we plot $J_1(x)$ and estimate its first zero, which happens at $x \approx 3.835$. By letting the argument of the jinc function in Eq. (3.4-31) equal to 3.835, we have the radius of the disk $r = r_{\text{Airy}}$:

$$r_0 \frac{k_0 r_{\text{Airy}}}{f} \approx 3.835,$$

or

$$r_{\text{Airy}} \approx 1.22 f \frac{\lambda_0}{2r_0} = f\theta_{\text{Airy}}, \qquad (3.4\text{-}32)$$

where

$$\theta_{\text{Airy}} = 1.22 \frac{\lambda_0}{2r_0}$$

is the spread angle, which is the half angle subtended by the Airy disk. Note that the radius of the central peak r_{Airy} is a measure of lateral resolution. From the uncertainty principle discussed in Chapter 1, the lateral resolution of a circular aperture, according to Eq. (1.2-3), is

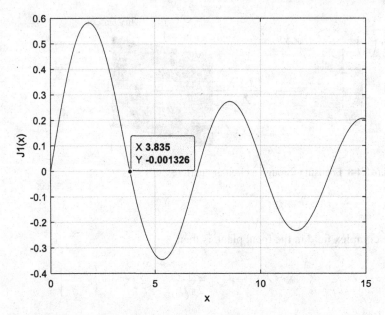

Figure 3.4-9 Plot of the first-order Bessel function of the first kind

$$\Delta r \approx \frac{\lambda_0}{2\sin(\theta_{im}/2)}, \tag{3.4-33}$$

where $\sin(\theta_{im}/2)$ is given by r_0/f for small angles. Δr from Eq. (3.4-33) then becomes

$$\Delta r \approx f\frac{\lambda_0}{2r_0},$$

which is consistent to that given by Eq. (3.4-32).

Figure 3.4-9 is generated using the m-file below.

```
%besselroot.m
x=0:0.001:15;
plot (x, besselj(1,x)), grid on
j1=besselj(1,x);
xlabel ('x')
ylabel('J1(x)')
```

Depth of Focus

Let us find the on-axis intensity. From Eq. (3.4-30), we have

$$\psi_p\left(r;f+z\right)=e^{-jk_0(f+z)}\frac{jk_0}{2\pi\left(f+z\right)}e^{\frac{-jk_0}{2(f+z)}r^2}\mathcal{B}\left\{\text{circ}\left(\frac{r}{r_0}\right)e^{j\frac{k_0}{2f}r^2}e^{\frac{-jk_0}{2(f+z)}r^2}\right\}\Bigg|_{k_r=\frac{k_0 r}{f+z}}$$

$$= e^{-jk_0(f+z)} \frac{jk_0}{2\pi(f+z)} e^{\frac{-jk_0}{2(f+z)}r^2} 2\pi \int_0^{r_0} r' e^{j\frac{k_0}{2f}r'^2} e^{\frac{-jk_0}{2(f+z)}r'^2} J_0(k_r r') dr' \Bigg|_{k_r = \frac{k_0 r}{f+z}}$$

$$= e^{-jk_0(f+z)} \frac{jk_0}{2\pi(f+z)} e^{\frac{-jk_0}{2(f+z)}r^2} 2\pi \int_0^{r_0} r' e^{j\frac{k_0}{2}\epsilon r'^2} J_0\left(\frac{k_0 r}{f+z} r'\right) dr',$$

where $\varepsilon = z/f(f+z)$. Hence, the on-axis or the axial complex field is

$$\psi_p(0; f+z) = e^{-jk_0(f+z)} \frac{jk_0}{(f+z)} \int_0^{r_0} r' e^{j\frac{k_0}{2}\epsilon r'^2} dr'.$$

After integrating, we have

$$\psi_p(0; f+z) = e^{-jk_0(f+z)} \frac{jk_0 r_0^2}{2(f+z)} \frac{\sin\left(k_0 \epsilon r_0^2 / 4\right)}{k_0 \epsilon r_0^2 / 4} e^{jk_0 \epsilon r_0^2 / 4}.$$

The axial intensity is then given by

$$I(0, f+z) = \psi_p(0; f+z)\psi_p^*(0; f+z)$$

$$= I_0 \left[\frac{f}{(f+z)}\right]^2 \text{sinc}^2\left(\frac{\epsilon r_0^2}{2\lambda_0}\right) = I_0 \left[\frac{f}{(f+z)}\right]^2 \text{sinc}^2\left[\frac{z r_0^2}{2\lambda_0 f(f+z)}\right], \quad (3.4\text{-}34)$$

where $I_0 = \left(\frac{\pi r_0^2}{\lambda_0 f}\right)^2$ is the intensity of the geometrical focus at $z = 0$ due to the

circular aperture of radius r_0. Let us find the axial distance $z = \Delta z$ at which the *sinc* function finds its first zero:

$$\frac{\Delta z r_0^2}{2\lambda_0 f(f+\Delta z)} = 1.$$

From this, we solve for

$$\Delta z = \frac{2\lambda_0 f^2}{r_0^2 - 2\lambda_0 f} \approx \frac{2\lambda_0 f^2}{r_0^2} = \frac{2\lambda_0}{NA^2} \quad (3.4\text{-}35)$$

as $r_0 \gg \lambda_0$, $r_0 \sim f$ and $r_0 / f \approx NA$ for small angles. Note that Δz is the depth of focus [see Eq. (1.2-9)], derived in Chapter 1 through the use of the uncertainty principle.

3.5　Gaussian Beam Optics

Optical beams from lasers typically have a Gaussian amplitude distribution. There-fore, the study of the diffraction of a *Gaussian beam* is of great importance. Let us consider a Gaussian beam propagating along the z-direction with planar wave fronts at $z = 0$. Hence,

$$\psi_{p0}(x,y) = e^{-(x^2+y^2)/\omega_0^2},$$ (3.5-1)

where ω_0 is called the *beam waist* of the Gaussian beam as it is at this distance from the z-axis, that the amplitude of the beam falls by a factor $1/e$. We can now find Fresnel diffraction of the beam using the frequency domain equation from Eq. (3.4-19):

$$\psi_p(x,y;z) = \mathcal{F}^{-1}\left\{\Psi_p(k_x,k_y;z)\right\} = \mathcal{F}^{-1}\left\{\Psi_{p0}(k_x,k_y)H(k_x,k_y;z)\right\}.$$ (3.5-2)

The Fourier transform of the Gaussian beam can be found by using item #5 of Table 2.1, and the result is

$$\Psi_{p0}(k_x,k_y) = \mathcal{F}\left\{e^{-(x^2+y^2)/\omega_0^2}\right\} = \pi\omega_0^2 e^{-\omega_0^2(k_x^2+k_y^2)/4}.$$ (3.5-3)

Using the spatial frequency transfer function in Fourier optics from Eq. (3.4-18), Eq. (3.5-2) becomes

$$\psi_p(x,y;z) = \mathcal{F}^{-1}\left\{\pi\omega_0^2 e^{-\omega_0^2(k_x^2+k_y^2)/4} \times e^{-jk_0 z} e^{jz(k_x^2+k_y^2)/2k_0}\right\}$$

$$= e^{-jk_0 z}\mathcal{F}^{-1}\left\{\pi\omega_0^2 e^{j\frac{q}{2k_0}(k_x^2+k_y^2)}\right\},$$ (3.5-4)

where q is the *q-parameter of the Gaussian beam*, given by

$$q(z) = z + jz_{RG}$$ (3.5-5)

with z_{RG} defined as the *Raleigh range of the Gaussian beam*:

$$z_{RG} = \frac{k_0\omega_0^2}{2}.$$ (3.5-6)

The inverse transform of the complex Gaussian in Eq. (3.5-4) can be found by using item #5 of Table 2.1 again to give

$$\psi_p(x,y;z) = e^{-jk_0 z}\frac{jk_0\omega_0^2}{2q} e^{-j\frac{k_0}{2q}(x^2+y^2)}.$$ (3.5-7)

This is an important result for a diffracted Gaussian beam expressed in terms of the q-parameter.

Substituting Eq. (3.5-5) into Eq. (3.5-7), we have, after some manipulations,

$$\psi_p(x,y;z) = \frac{\omega_0}{\omega(z)} e^{\frac{-(x^2+y^2)}{\omega^2(z)}} e^{\frac{-jk_0}{2R(z)}(x^2+y^2)} e^{-j\phi(z)} e^{-jk_0 z},$$ (3.5-8a)

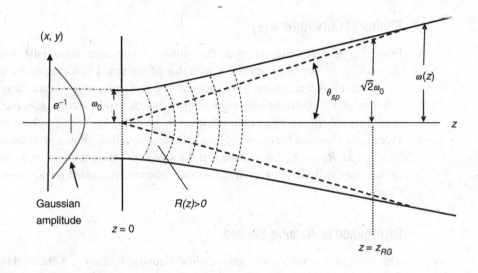

Figure 3.5-1 Spreading of a Gaussian beam upon diffraction

where

$$\omega(z) = \omega_0 \sqrt{\left[1 + \left(\frac{z}{z_{RG}}\right)^2\right]} \qquad (3.5\text{-}8b)$$

is the waist of the Gaussian beam at a distance z, and

$$R(z) = z\left[1 + \left(\frac{z_{RG}}{z}\right)^2\right] \qquad 3.5\text{-}8c$$

is the *radius of curvature of the Gaussian beam* at a distance z, and finally

$$\phi(z) = -\tan^{-1}\left(\frac{z}{z_{RG}}\right). \qquad (3.5\text{-}8d)$$

Figure 3.5-1 illustrates the propagation of a Gaussian beam with initial profile given by Eq. (3.5-1). In what follows, we will discuss some Gaussian beam properties upon diffraction with reference to Figure 3.5-1.

Beam Waist $\omega(z)$

From Eq. (3.5-8b), we observe that the beam waist $\omega(z)$ stays approximately constant when it has been diffracted well within the Raleigh range z_{RG}. At $z = z_{RG}$ the beam waist increases to $\omega_0\sqrt{2}$. In the far field when $z \gg z_{RG}$, the beam waist increases linearly according to $\omega(z) \approx \omega_0 z / z_{RG}$.

Radius of Curvature $R(z)$

From Eq. (3.5-8c), we see that the radius of curvature is initially infinite as $R(z=0)=\infty$, which is consistent with that of the initial Gaussian beam when we assume that it is with planar wave fronts at $z=0$. The wave fronts are curved when $z>0$ as $R(z)>0$. Therefore, we anticipate that the equiphase surfaces (i.e., wavefronts) are spherical with a radius of curvature given by $R(z)$ for $z>0$ as $R(z)>0$. Indeed, the Gaussian beam resembles a spherical wavefront from a point source when $z\gg z_{RG}$ as $R(z\gg z_{RG})\approx z$. Note that the sign convention is the same as the radius of curvature defined for a spherical surface when we discussed refraction matrix in Chapter 1.

Divergence or Angular Spread, θ_{sp}

The *divergence* or the *angular spread* of the Gaussian beam, θ_{sp}, is obtained by $\tan\theta_{sp} = \omega(z)/z$, as it is found from the geometry shown in Figure 3.5-1. In the far field, that is, $z\gg z_{RG}$, the divergence can be determined by

$$\theta_{sp} \approx \frac{\omega(z)}{z} = \frac{\lambda_0}{\pi\omega_0},\qquad(3.5\text{-}9)$$

as $\omega(z)\approx 2z/k_0\omega_0$ and λ_0 is the wavelength in the medium of refractive index n in which the Gaussian beam propagates.

Gouy Phase $\phi(z)$

The term $e^{-j\phi(z)}e^{-jk_0z}$ in Eq. (3.5-8a) represents the phase acquired as the beam propagates along the z-axis, where $\phi(z)=-\tan^{-1}(z/z_{RG})$ is called the *Gouy phase*. This phase gives a small deviation from that of a plane wave, which is simply given by k_0z as we have discussed this in Eq. (3.3-8). The reason is that we have a beam, which consists of superposition of plane waves propagating in different directions. The overall phase shift is the superposition of all these plane waves, giving rise to the Gouy phase.

3.5.1   q-Transformation and Bilinear Transformation

q-Transformation

The q-parameter of the Gaussian beam, introduced by Eq. (3.5-5), is a very useful quality to know for a Gaussian beam as it contains all the important information, namely its curvature $R(z)$ and its waist $\omega(z)$, about the beam. Once we know how the q transforms as the beam going through an optical system, we know how the Gaussian beam behaves. From the definition of q, we write

$$\frac{1}{q(z)} = \frac{1}{z+jz_{RG}} = \frac{z-jz_{RG}}{z^2+z_{RG}^{2}} = \frac{z}{z^2+z_{RG}^{2}} - j\frac{z_{RG}}{z^2+z_{RG}^{2}}$$

$$= \frac{1}{R(z)} - j\frac{2}{k_0\omega^2(z)} , \qquad (3.5\text{-}10)$$

where we have used Eqs. (3.5-8b) and (3.5-8c) to simplify the above results. Clearly, we see that the q-parameter contains the information on $R(z)$ and $\omega(z)$ of the Gaussian beam. For example, the real part of q, that is, $\mathrm{Re}\{1/q\}$, is equal to $1/R$, and for this reason q is often called the *complex curvature* of a Gaussian beam. The imaginary part of q, that is, $\mathrm{Im}\{1/q\}$, is a measure of $1/\omega^2(z)$.

Because of the importance of q, we shall develop how q evolves as a Gaussian beam is propagating along an optical system.

Bilinear Transformation

From Gaussian optics discussed in Chapter 1, we have described the behavior of light rays in optical systems using the formalism of the ray transfer matrix under the paraxial approximation. In this chapter, we have described wave propagation using the formalism of the Fresnel diffraction formula, which is also developed under the paraxial approximation. As it turns out, the two formalisms are related in that the diffraction formula can be written in terms of the elements of the ray transfer matrix, providing a connection between Gaussian optics and diffraction theory. The connection is given by the following integral, often known as *Collins' integral*:

$$\psi_p(x,y;z) = e^{-jk_0z} \frac{jk_0}{2\pi B} \iint\limits_{-\infty}^{\infty} \psi_{p0}(x',y') \; e^{-j\frac{k_0}{2B}\left[A\left(x'^2+y'^2\right)+D\left(x^2+y^2\right)-2\left(xx'+yy'\right)\right]}dx'dy', \quad (3.5\text{-}11)$$

where the input $x' - y'$ plane and the output $x - y$ plane are defined according to Figure 3.4-2 with z being the distance between the two planes, and the parameters A, B, C, and D are the matrix elements of the system matrix discussed in Chapter 1.

Example: Translation Matrix in Air

According to Eq. (1.1-3b), the translation matrix in air for a distance z is given by

$$T_z = \begin{pmatrix} A & B \\ C & D \end{pmatrix} = \begin{pmatrix} 1 & z \\ 0 & 1 \end{pmatrix}.$$

Substituting the *ABCD* parameters into Eq. (3.5-11) accordingly, we have

$$\psi_p(x,y;z) = e^{-jk_0z} \frac{jk_0}{2\pi z} \iint\limits_{-\infty}^{\infty} \psi_{p0}(x',y') \; e^{-j\frac{k_0}{2z}\left[\left(x'^2+y'^2\right)+\left(x^2+y^2\right)-2\left(xx'+yy'\right)\right]}dx'dy'$$

$$= e^{-jk_0z} \frac{jk_0}{2\pi z} \iint\limits_{-\infty}^{\infty} \psi_{p0}(x',y') \; e^{-j\frac{k_0}{2z}\left[\left(x-x'\right)^2+\left(y-y'\right)^2\right]}dx'dy',$$

which is identical to Eq. (3.4-17) for $\psi_{p0}(x,y)$ undergoing Fresnel diffraction for a distance z.

Example: Input Transparency Function in Front of the Ideal Lens

With reference to the physical situation in Figure 3.4-6a, from the input plane, which is a distance d_0 in front of the lens to the output plane on the back focal plane, we have the system matrix as, according to the process given by Eq. (1.1-12),

$$\mathcal{S} = \begin{pmatrix} A & B \\ C & D \end{pmatrix} = \begin{pmatrix} 1 & f \\ 0 & 1 \end{pmatrix}\begin{pmatrix} 1 & 0 \\ -1/f & 1 \end{pmatrix}\begin{pmatrix} 1 & d_0 \\ 0 & 1 \end{pmatrix} = \begin{pmatrix} 0 & f \\ -1/f & 1-d_0/f \end{pmatrix}.$$

Substituting the *ABCD* parameters into Eq. (3.5-11) accordingly with $z = f + d_0$ and $\psi_{p0}(x,y) = t(x,y)$ for unit amplitude plane wave incidence, we have

$$\psi_p(x,y;z=f+d_0) = e^{-jk_0(f+d_0)}\frac{jk_0}{2\pi f}\int\!\!\!\int_{-\infty}^{\infty} t(x',y')\, e^{-j\frac{k_0}{2f}\left[\left(1-\frac{d_0}{f}\right)(x^2+y^2)-2(xx'+yy')\right]}dx'dy'$$

$$= e^{-jk_0(f+d_0)}\frac{jk_0}{2\pi f}e^{-j\frac{k_0}{2f}\left[1-\frac{d_0}{f}\right](x^2+y^2)}\int\!\!\!\int_{-\infty}^{\infty} t(x',y')\, e^{j\frac{k_0}{f}(xx'+yy')}dx'dy'$$

$$= e^{-jk_0(f+d_0)}\frac{jk_0}{2\pi f}e^{-j\frac{k_0}{2f}\left[1-\frac{d_0}{f}\right](x^2+y^2)}\,\mathcal{F}\{t(x,y)\}\Big|_{k_x=\frac{k_0 x}{f},\,k_y=\frac{k_0 y}{f}},$$

which is identical to Eq. (3.4-29). As we can see, using Collins' integral along with the knowledge of the system matrix, we can obtain the complex field along the optical system easily.

Now that we have discussed Collins' integral, we use it to derive the *bilinear transformation* of a Gaussian beam propagating through an optical system characterized by its system matrix. The bilinear transformation is a mathematical mapping of variables and in our case, the variable in question is the q parameter.

We assume we have an initial Gaussian beam $e^{-(x^2+y^2)/\omega_0^2}$ at $z = 0$. From Eq. (3.5-5), $q(z) = z + jz_{RG}$. Therefore, at input plane $z = 0$, we have

$$q(z=0) = q_0 = jz_{RG} = j\frac{k_0\omega_0^2}{2}\,, \quad \text{and we can express } \psi_p(x,y;z=0) = \psi_{p0}(x,y) =$$

$e^{-(x^2+y^2)/\omega_0^2} = e^{-j\frac{k_0}{2q_0}(x^2+y^2)}$ in terms of the q-parameter.

Applying Collins' integral with z being the distance between the input plane and output plane, we write

$$\psi_p(x,y;z) = e^{-jk_0 z}\frac{jk_0}{2\pi B}\int\!\!\!\int_{-\infty}^{\infty} \psi_{p0}(x',y')\, e^{-j\frac{k_0}{2B}\left[A(x'^2+y'^2)+D(x^2+y^2)-2(xx'+yy')\right]}dx'dy',$$

$$= e^{-jk_0 z} \frac{jk_0}{2\pi B} \int\!\!\!\int_{-\infty}^{\infty} e^{-j\frac{k_0}{2q_0}\left(x'^2+y'^2\right)} e^{-j\frac{k_0}{2B}\left[A\left(x'^2+y'^2\right)+D\left(x^2+y^2\right)-2\left(xx'+yy'\right)\right]} dx'dy'$$

$$= e^{-jk_0 z} \frac{jk_0}{2\pi B} e^{-j\frac{k_0}{2B}\left[D\left(x^2+y^2\right)\right]} \int\!\!\!\int_{-\infty}^{\infty} e^{-j\frac{k_0}{2q_0}\left(x'^2+y'^2\right)} e^{-j\frac{k_0}{2B}\left[A\left(x'^2+y'^2\right)\right]} e^{\frac{jk_0}{B}\left[\left(xx'+yy'\right)\right]} dx'dy'$$

$$= e^{-jk_0 z} \frac{jk_0}{2\pi B} e^{-\frac{jk_0}{2B}\left[D\left(x^2+y^2\right)\right]} \int\!\!\!\int_{-\infty}^{\infty} e^{-j\left(\frac{k_0}{2q_0}+\frac{k_0 A}{2B}\right)\left(x'^2+y'^2\right)} e^{\frac{jk_0}{B}\left[\left(xx'+yy'\right)\right]} dx'dy',$$

which can be written in terms of the Fourier transform as

$$\psi_p\left(x,y;z\right) = e^{-jk_0 z} \frac{jk_0}{2\pi B} e^{-j\frac{k_0}{2B}\left[D\left(x^2+y^2\right)\right]} \mathcal{F}\left\{ e^{-jk_0\left(\frac{1}{2q_0}+\frac{A}{2B}\right)\left(x^2+y^2\right)} \right\} \Bigg|_{k_x=\frac{k_0 x}{B},\, k_y=\frac{k_0 y}{B}}.$$

The complex Gaussian function can be Fourier transformed, according to Table 2.1, to give

$$\psi_p\left(x,y;z\right) = e^{-jk_0 z} \frac{q_0}{Aq_0+B} e^{-jk_0 \frac{Cq_0+D}{2\left(Aq_0+B\right)}\left(x^2+y^2\right)} = e^{-jk_0 z} \frac{q_0}{Aq_0+B} e^{-j\frac{k_0}{2q_1}\left(x^2+y^2\right)} \tag{3.5-12a}$$

where we recognize that the term $e^{-j\frac{k_0}{2q_1}\left(x^2+y^2\right)}$ is a Gaussian beam with

$$q_1 = \frac{Aq_0+B}{Cq_0+D}. \tag{3.5-12b}$$

This is called the *bilinear transformation* for the q-parameter as it relates q_0 to q_1 for a given system matrix *ABCD*.

3.5.2 Examples on the Use of the Bilinear Transformation

Propagation of a Gaussian Beam with a Distance z in Air

The translation matrix in air for a distance z is given by

$$\mathcal{T}_z = \begin{pmatrix} A & B \\ C & D \end{pmatrix} = \begin{pmatrix} 1 & z \\ 0 & 1 \end{pmatrix}.$$

Assume $\psi_p\left(x,y;z=0\right) = e^{-j\frac{k_0}{2q_0}\left(x^2+y^2\right)}$. According to Eq. (3.5-12b) along with $\mathcal{T}_z$,

$$q_1\left(z\right) = z + q_0, \tag{3.5-13}$$

which is consistent with Eq. (3.5-5). Equation (3.5-13) is the *translation law for a Gaussian beam* propagating in air. Now, the complex field is given by Eq. (3.5-12a) (for $A=1$, $B=z$):

$$\psi_p(x,y;z) = e^{-jk_0 z} \frac{q_0}{q_0 + z} e^{-j\frac{k_0}{2q_1}(x^2+y^2)}$$

$$= e^{-jk_0 z} \frac{q_0}{q_1} e^{-j\frac{k_0}{2q_1}(x^2+y^2)} = e^{-jk_0 z} \frac{jk_0 \omega_0^2}{2q} e^{-j\frac{k_0}{2q}(x^2+y^2)},$$

which is the same result as in Eq. (3.5-7) as q_1 is the output q.

Lens Law for a Gaussian Beam

To analyze the effects of an ideal lens of focal length f on a Gaussian beam, we recall the thin-lens matrix, $\mathcal{L}_f$ [see Eq. (1.1-13)]:

$$\mathcal{L}_f = \begin{pmatrix} 1 & 0 \\ -\dfrac{1}{f} & 1 \end{pmatrix}.$$

Using the bilinear transformation from Eq. (3.5-12b) along with $\mathcal{L}_f$, we obtain

$$q_1 = \frac{q_0}{-\dfrac{1}{f} q_0 + 1} = \frac{f q_0}{f - q_0}$$

or

$$\frac{1}{q_1} = \frac{1}{q_0} - \frac{1}{f}. \tag{3.5-14}$$

This is the *lens law for a Gaussian beam*, which relates the qs before and after the lens.

Gaussian Beam Focusing

We consider the propagation of a Gaussian beam through an optical system that consists of two transfer matrices, such as the focusing of a Gaussian beam through a lens. The general situation involving two matrices is illustrated in Figure 3.5-2. Let us relate the input beam parameter, q_1, to the output beam parameter, q_3.

Using Eq. (3.5-12b), we can write the relations between the various qs as follows:

$$q_2 = \frac{A_1 q_1 + B_1}{C_1 q_1 + D_1} \text{ and } q_3 = \frac{A_2 q_2 + B_2}{C_2 q_2 + D_2}.$$

Combining the two equations, we have

$$q_3 = \frac{A_2 \dfrac{A_1 q_1 + B_1}{C_1 q_1 + D_1} + B_2}{C_2 \dfrac{A_1 q_1 + B_1}{C_1 q_1 + D_1} + D_2} = \frac{A_2 (A_1 q_1 + B_1) + B_2 (C_1 q_1 + D_1)}{C_2 (A_1 q_1 + B_1) + D_2 (C_1 q_1 + D_1)}$$

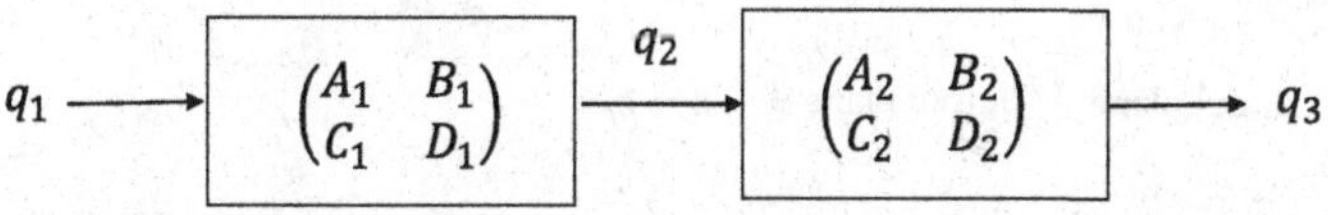

Figure 3.5-2 q-transformation involving more than one transfer matrix

$$= \frac{(A_2A_1 + B_2C_1)q_1 + A_2B_1 + B_2D_1}{(C_2A_1 + D_2C_1)q_1 + C_2B_1 + D_2D_1} = \frac{A_Tq_1 + B_T}{C_Tq_1 + D_T}, \qquad (3.5\text{-}15)$$

where (A_T, B_T, C_T, D_T) are the elements of the system matrix relating q_3 on the output plane to q_1 on the input plane, which is given conveniently by

$$\begin{pmatrix} A_T & B_T \\ C_T & D_T \end{pmatrix} = \begin{pmatrix} A_2 & B_2 \\ C_2 & D_2 \end{pmatrix}\begin{pmatrix} A_1 & B_1 \\ C_1 & D_1 \end{pmatrix}.$$

Let us now consider a Gaussian beam of planar wave fronts at $z = 0$, that is incident at its waist ω_0 on an ideal lens of focal length f, as shown in Figure 3.5-3. We will find the location of the waist of the output beam after the lens. The input plane is the immediate left of the lens and the output plane is at a distance z away from the lens. With reference to Eq. (3.5-10), we have

$$\frac{1}{q(z)} = \frac{1}{R(z)} - j\frac{2}{k_0\omega^2(z)}.$$

Note that the minimum waist of the input beam, ω_0, is when $R(z = 0) = \infty$. Therefore, the input q is $q_0 = \left(-j\frac{2}{k_0\omega^2(z=0)}\right)^{-1} = \frac{jk_0\omega_0^2}{2} = jz_{\text{RG}}$ and it is purely imaginary.

The output q, according to Eq. (3.5-15), is

$$q_1 = \frac{A_Tq_0 + B_T}{C_Tq_0 + D_T},$$

where

$$\begin{pmatrix} A_T & B_T \\ C_T & D_T \end{pmatrix} = T_z\mathcal{L}_f = \begin{pmatrix} 1 & z \\ 0 & 1 \end{pmatrix}\begin{pmatrix} 1 & 0 \\ -\dfrac{1}{f} & 1 \end{pmatrix} = \begin{pmatrix} 1 - \dfrac{z}{f} & z \\ -\dfrac{1}{f} & 1 \end{pmatrix}.$$

Therefore,

$$q_1(z) = \frac{(1 - \dfrac{z}{f})q_0 + z}{-\dfrac{1}{f}q_0 + 1} = \frac{fq_0}{f - q_0} + z.$$

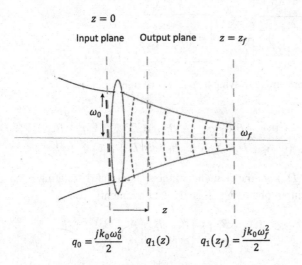

Figure 3.5-3 Focusing of a Gaussian beam

The Gaussian beam is defined to be at focus (minimum waist) at $z = z_f$ when its q becomes purely imaginary, that is, when the beam having planar wave fronts. Let ω_f be the waist at focus; we, therefore, can write

$$q_1\left(z_f\right) = \frac{f q_0}{f - q_0} + z_f = \frac{j k_0 \omega_f^2}{2},$$

Substituting $q_0 = j z_{RG}$ into the above equation, and after some manipulations, we obtain

$$\left(z_f - \frac{f z_{RG}^2}{f^2 + z_{RG}^2}\right) + j \frac{f^2 z_{RG}}{f^2 + z_{RG}^2} = \frac{j k_0 \omega_f^2}{2}. \tag{3.5-16}$$

Equating the real part of Eq. (3.5-16), we have

$$z_f - \frac{f z_{RG}^2}{f^2 + z_{RG}^2} = 0,$$

or equivalently

$$z_f = \frac{f z_{RG}^2}{f^2 + z_{RG}^2} = \frac{f}{1 + \left(\dfrac{f}{z_{RG}}\right)^2} \leq f. \tag{3.5-17}$$

Since $z_{RG} = \dfrac{k_0 \omega_0^2}{2}$ is always a positive number, $z_f \leq f$. In other words, the Gaussian beam does not focus exactly at the geometrical back focus of the lens. Instead, the focus is shifted close to the lens. This phenomenon is called the *focal shift* [Poon (1988)] and it is crucial for high laser power applications in which precise focus is of paramount importance. Note that for a plane wave incidence, that is, $\omega_0 \to \infty$, $z_{RG} \gg f$, and $z_f = f$.

Now, equating the imaginary part of Eq. (3.5-16), we have

$$\frac{f^2 z_{RG}}{f^2 + z_{RG}^2} = \frac{k_0 \omega_f^2}{2}.$$

With $z_{RG} = \dfrac{k_0 \omega_0^2}{2}$ in the nominator, we solve for ω_f:

$$\omega_f = \frac{f \omega_0}{\sqrt{f^2 + z_{RG}^2}} \approx \frac{\lambda_0 f}{\pi \omega_0} \tag{3.5-18}$$

for a large ω_0 as $z_{RG} \gg f$. For $\omega_0 = 10$ mm, $f = 10$ cm, and $\lambda_0 = 0.633\ \mu\text{m}$ (red light), the focused spot size $\omega_f \approx 2\ \mu\text{m}$.

Problems

3.1 Show that the wave equation for H in a linear, homogeneous, and isotopic medium characterized by ε and μ is given by

$$\nabla^2 H - \mu \varepsilon \frac{\partial^2 H}{\partial t^2} = -\nabla \times J_c.$$

3.2 Show that the 3-D scalar wave equation

$$\nabla^2 \psi = \mu \varepsilon \frac{\partial^2 \psi}{\partial t^2} = \frac{1}{v^2} \frac{\partial^2 \psi}{\partial t^2}$$

is a linear differential equation. If a differential equation is linear, the principle of superposition holds, which provides the basis for interference and diffraction.

3.3 Determine which of the following functions describe travelling waves

$$\psi(z,t) = \sin\big[(3z - 4t)(3z + 4z)\big]$$

$$\psi(z,t) = \cos^3\left[\left(\frac{z}{3} + \frac{t}{b}\right)^2\right]$$

$$\psi(z,t) = \tan(3t - 4z).$$

Also, find the phase velocity and direction ($+z$ or $-z$ direction) for those travelling waves.

3.4 Employ the Fresnel diffraction formula,
(a) find the complex field at a distance z away from a point source [Hint: you can model a point source by $\delta(x, y)$], and
(b) find the complex field at a distance z away from a plane wave. [Hint: You might want to take advantage of the Fourier transform to find your final answer.]

3.5 Using the result of problem 3.4a, show that the transparency function of the ideal lens of focal length f is given by $t_f(x,y) = e^{j\frac{k_0}{2f}(x^2+y^2)}$ when a point object is imaged to give an image point, as shown in Figure P.3.5.

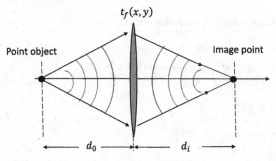

Figure P.3.5

3.6 For a plane wave of unit amplitude incident on a circular aperture $t(x,y) = \mathrm{circ}\left(\dfrac{r}{r_0}\right)$, show that the spread angle is given by

$$\theta_{sp} = 1.22\frac{\lambda_0}{2r_0}.$$

3.7 With reference to Figure P.3.7, employ Collins' integral and show that the complex field on the focal plane is given by Eq. (3.4-29), that is,

$$\psi_p(x,y;z=f+d_0) = e^{-jk_0(f+d_0)}\frac{jk_0}{2\pi f}Ae^{-j\frac{k_0}{2f}\left[1-\frac{d_0}{f}\right](x^2+y^2)}\mathcal{F}\{t(x,y)\}\Big|_{k_x=\frac{k_0 x}{f},\,k_y=\frac{k_0 y}{f}}.$$

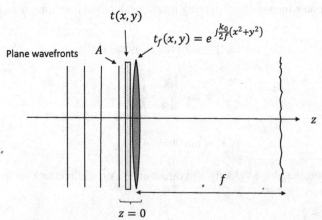

Figure P.3.7

3.8 Figure P.3.8 shows a 4-f imaging system. The input, $t(x,y)$, is illuminated by a plane wave of amplitude A.

    (a) Give an expression involving convolution and multiplications similar to that in Eq. (3.4-27) for the complex field on the image plane.

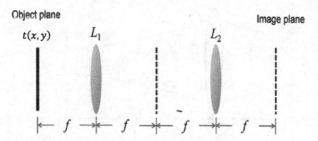

Figure P.3.8

    (b) From the result in part a, draw a block diagram of the optical system.

    (c) Show that on the image plane, the complex field distribution is proportional to $t(-x,-y)$ (Figure P.3.8).

3.9 For a Gaussian beam with initial profile given by

$$\psi_p(x,y;z=0)=\psi_{p0}(x,y)=e^{-j\frac{k_0}{2q_0}(x^2+y^2)} \text{ with } q_0=jz_{RG}=j\frac{k_0\omega_0^2}{2},$$

we have shown that its diffracted complex field at a distance z is

$$\psi_p(x,y;z)=e^{-jk_0z}\frac{jk_0\omega_0^2}{2q}e^{-j\frac{k_0}{2q}(x^2+y^2)}.$$

By substituting $q(z)=z+jz_{RG}$ into the above diffracted field, show that $\psi_p(x,y;z)$ is given by Eq. (3.5-8), which is given below:

$$\psi_p(x,y;z)=\frac{\omega_0}{\omega(z)}e^{\frac{-(x^2+y^2)}{\omega^2(z)}}e^{-j\frac{k_0}{2R(z)}(x^2+y^2)}e^{-j\phi(z)}e^{-jk_0z}.$$

3.10 With reference to Figure P.3.10, we assume that a Gaussian beam with a waist ω_0,

$$\psi_{p0}(x,y)=e^{-(x^2+y^2)/\omega_0^2},$$

travels at a distance d_0 to an ideal lens of focal length f. At a distance d_f, the Gaussian beam is transformed to a beam of planar wave fronts with waist ω_f.

    (a) Find ω_f and d_f.

    (b) When $\omega_0 \to \infty$, that is, the Gaussian has become a plane wave, find an equation relating d_0, d_f, and f from the result you have obtained in part (a).

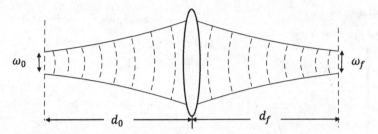

Figure P.3.10

3.11   A Gaussian beam of q_1 enters an optical system specified by $\begin{pmatrix} A & B \\ C & D \end{pmatrix}$. At the exit of the optical system, the Gaussian beam has q_2. Show that the ratio of the two waists is

$$\frac{\omega_2}{\omega_1} = \left| A + \frac{B}{q_1} \right|,$$

where ω_1 and ω_2 are the waists of the input and output beams, respectively.

Bibliography

Banerjee, P. P. and T.-C. Poon (1991). *Principles of Applied Optics*. Irwin, Illinois.

Gerard, A. and J. M. Burch (1975). *Introduction to Matrix Methods in Optics*. Wiley, New York.

Li, J. and Y. Wu (2016). *Diffraction Calculation and Digital Holography I*. Science Press, Beijing, China.

Poon, T.-C. (1988). "Focal shift in focused annular beams," *Optics Communications* 6, pp. 401–406.

Poon, T.-C. (2007). *Optical Scanning Holography with MATLAB®*. Springer, New York.

Poon, T.-C. and T. Kim (2018). *Engineering Optics with MATLAB®*, 2nd ed., World Scientific, New Jersey.

Stark, H., ed. (1982). *Application of Optical Transforms*. Academic Press, Florida.

4 Spatial Coherent and Incoherent Optical Systems

4.1 Temporal Coherence and Spatial Coherence

The formal treatment of coherence of light is a rather complicated topic. In this section, we shall describe optical coherence qualitatively to have some basic understanding. In *temporal coherence*, we are concerned with the ability of a light field to interfere with a time delayed version of itself. Temporal coherence tells us how monochromatic a source is, and it is a measure of the correlation of light wave's phase at different points along the direction of propagation. Therefore, it is often referred to as *longitudinal coherence*. In spatial coherence, the ability of a light field to interfere with a spatially shifted version of itself is considered. *Spatial coherence* is a measure of the correlation of a light wave's phase at different points transverse to the direction of propagation. It is often referred to as *transverse coherence*. Spatial coherence tells us how uniform the phase of a wave front is. It is never possible to separate completely the two effects, but it is important to point out their significance.

Temporal Coherence

A discussion of wave optics from the last chapter is incomplete without considering the conditions that light waves exist for *interference*. Interference is a phenomenon as a result of summing two light waves. Light sources can be categorized into laser sources and thermal sources. A good example of a thermal source is a gas-discharge lamp such as a light bulb. In such a light source, light is emitted by excited atoms. According to modern physics, electrons in atoms and molecules can make transitions in energy levels by emitting or absorbing a photon, whose energy must be exactly equal to the energy difference between the two energy levels. The atoms of a light source do not continuously send out waves but rather they send out "wave trains" of a particular frequency. The frequency ω_0 is related to the energy difference between the atomic levels involved in the transition according to $\hbar\omega_0 = E_2 - E_1$, where E_2 and E_1 are energy levels and $\hbar = h/2\pi$ with h being Plank's constant.

We can measure the temporal coherence of the light wave by splitting the light wave into two paths of different distances and then recombine them to form an interference fringe pattern. Figure 4.1 shows the *Michelson interferometer* for the measure of temporal coherence.

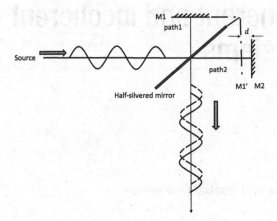

Figure 4.1-1 (a) Michelson interferometer: The path difference $2d$ is less than the length of the wave train emitted by the source. Interference occurs

At the half-silvered mirror, the incoming wave train from the source splits into two waves of equal amplitude, one travelling along path 1 to be reflected by mirror M1. The other wave taking path 2 travels a slightly larger distance and reflected by mirror M2. The two reflected wave trains are shown emerging from the interferometer with some displacement. The displacement is equal to the difference in path length of $2d$, where d is defined in the figure. If $2d$ is much smaller than the length of the wave trains as shown in Figure 4.1-1a, the two wave trains are almost superposed and are able to interfere. The created interference pattern will be sharp. We have temporal coherence. As we increase the path difference $2d$ by translating M2 away from M1′, the two wave trains emerging from the interferometer will overlap less and created interference pattern will be less sharp. When $2d$ is larger than the length of the wave trains, which come from the same initial wave train from the source, the two wave trains can no longer overlap and no interference can occur. The situation is shown in Figure 4.1-1b. For this simple case, we can define the *coherence time* as the duration of the wave train, τ_c. The *coherence length* l_c, therefore, equals $c\tau_c$, where c is the speed of light in vacuum. However, it is still possible to have wave trains overlap as they emerge from the interferometer. The reason is that there might be another atom emitting a wave train of random length and at random times with slightly different frequencies or different wavelengths (e.g., due to the thermal motion of the atom). In practice, there are enormous number of wave trains emitted from different atoms within the source.

Let us assume the source emits waves with wavelengths ranging from λ to $\lambda + \Delta\lambda$, and $\Delta\lambda$ is the *spectral width* of the source. Remember that the phase of a propagating wave is $\varphi = \omega_0 t - k_0 z$. We monitor the wave pattern in space at some time $t = t_0$. At some distance l, the phase difference between two waves with wavenumbers k_{01} and k_{02} which are in phase at some $z = z_0$ becomes out of phase by $\Delta\varphi = (k_{01} - k_{02})l$. As a rule of thumb, when $\Delta\varphi \approx 1$ rad, we say that the light is no longer considered coherent,

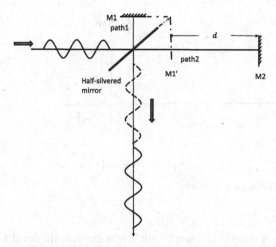

Figure 4.1-1 (b) Michelson interferometer: The path difference $2d$ is greater than the length of the wave train emitted by the source. Interference cannot occur

which means interference loose its fringe contrast completely. For $\Delta\varphi = 1\,\mathrm{rad}$, l becomes l_c. We, therefore, have

$$\Delta\varphi = 1 = \left(k_{01} - k_{02}\right)l_c$$

$$= \left(\frac{2\pi}{\lambda} - \frac{2\pi}{\lambda + \Delta\lambda}\right)l_c,$$

or

$$l_c = \frac{\lambda(\lambda + \Delta\lambda)}{2\pi\Delta\lambda} \approx \frac{\lambda^2}{2\pi\Delta\lambda}. \tag{4.1-1}$$

Since $\lambda f = \lambda\left(\dfrac{\omega}{2\pi}\right) = c$, we can establish that

$$\left|\frac{\Delta\omega}{\omega}\right| = \left|\frac{\Delta\lambda}{\lambda}\right|.$$

Using this result, Eq. (4.1-1) becomes

$$l_c \approx \frac{\lambda^2}{2\pi\Delta\lambda} = \frac{c}{\Delta\omega} = \frac{c}{2\pi\Delta f},$$

or the *coherence time* of a quasi-monochromatic field of bandwidth Δf is

$$\tau_c = \frac{l_c}{c} = \frac{1}{2\pi\Delta f}. \tag{4.1-2}$$

White light has a spectral width (often called *line width*) $\Delta\lambda \approx 300$ nm, ranging about from 400 to 700 nm. If we take the average wavelength at $\lambda = 550$ nm, Eq. (4.1-1)

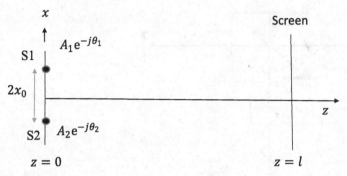

Figure 4.1-2 Young's interference experiment

gives $\ell_c \sim 0.16\,\mu m$, a very short coherence length. Light-emitting diodes (LEDs) have a spectral width $\Delta\lambda$ of about 50 nm and have a coherence length of about 1 μm for red color with wavelength of 632 nm. As for the green line of mercury at $\lambda = 546$ nm having a line width $\Delta\lambda$ about 0.025 nm, its coherence length is about 0.2 cm. Lasers typically have a long coherence length. Helium–neon lasers can produce light with coherence lengths greater than 5 m but 20 cm is typical. Some industrial CO_2 lasers of line width $\Delta\lambda$ of around 10^{-5} nm producing emission at the infrared wavelength of $\lambda = 10.6\,\mu m$ would give a coherence length of around a couple of kilometers. When we discuss holography in subsequent chapters, the coherence length places an upper limit on the allowable path difference between the object and reference wave for interference. If interference fringes are to be observed, the path separation must be less than the coherence length of the source used.

Spatial Coherence

We first consider the well-known Young's interference experiment shown in Figure 4.1-2. Two point sources S1 and S2 separated by a distance $2x_0$ generate spherical waves that are interfered on a screen at a distance l.

According to Fresnel diffraction [see Eq. (3.4-17)], the complex field of point source S1 of amplitude A_1 with phase θ_1 on the screen is given by

$$\psi_{S1} = A_1 e^{-j\theta_1}\delta(x-x_0,y)*h(x,y;z=l) = A_1 e^{-j\theta_1}h(x-x_0,y;z=l)$$

$$= A_1 e^{-j\theta_1}\frac{jk_0}{2\pi l}e^{-jk_0 l}e^{-jk_0\left[(x-x_0)^2+y^2\right]/2l}.$$

Similarly, for point source S2 of amplitude A_2 with phase θ_2, the complex field on the screen is given by

$$\psi_{S2} = A_2 e^{-j\theta_2}\delta(x+x_0,y)*h(x,y;z=l) = A_2 e^{-j\theta_2}\frac{jk_0}{2\pi l}e^{-jk_0 l}e^{-jk_0\left[(x+x_0)^2+y^2\right]/2l}.$$

The total field ψ_t on the screen, according to the principle of superposition, is

$$\psi_t = \psi_{S1} + \psi_{S2},$$

and the corresponding intensity $I(x,y)$ is the square magnitude of the total field [see Eq. (3.3-22)]:

$$I(x,y) = |\psi_{S1} + \psi_{S2}|^2 = |\psi_{S1}|^2 + |\psi_{S2}|^2 + \psi_{S1}\psi_{S2}^* + \psi_{S1}^*\psi_{S2} = \left(A_1\frac{k_0}{2\pi l}\right)^2 + \left(A_2\frac{k_0}{2\pi l}\right)^2$$

$$+ A_1 e^{-j\theta_1}\frac{jk_0}{2\pi l}e^{-jk_0 l}e^{-jk_0\left[(x-x_0)^2+y^2\right]/2l} A_2 e^{+j\theta_2}\frac{-jk_0}{2\pi l}e^{+jk_0 l}e^{-jk_0\left[(x+x_0)^2+y^2\right]/2l} + \text{c.c.}$$

$$= \left(A_1\frac{k_0}{2\pi l}\right)^2 + \left(A_2\frac{k_0}{2\pi l}\right)^2 + A_1 A_2 e^{-j(\theta_1-\theta_2)}\left(\frac{k_0}{2\pi l}\right)^2 e^{-jk_0\left[(x-x_0)^2+y^2\right]/2l}e^{-jk_0\left[(x+x_0)^2+y^2\right]/2l} + \text{c.c.}$$

$$= |\psi_{S1}|^2 + |\psi_{S2}|^2 + A_1 A_2 e^{-j(\theta_1-\theta_2)}\left(\frac{k_0}{2\pi l}\right)^2 e^{-jk_0(-4xx_0)/2l} + \text{c.c.}$$

where c.c. denotes the complex conjugate. We can express the above equation in terms of a real function:

$$I(x,y) = |\psi_{S1}|^2 + |\psi_{S2}|^2 + 2A_1 A_2\left(\frac{k_0}{2\pi l}\right)^2 \cos\left[\frac{k_0}{l}2x_0 x + (\theta_2 - \theta_1)\right]$$

$$= |\psi_{S1}|^2 + |\psi_{S2}|^2 + 2A_1 A_2\left(\frac{k_0}{2\pi l}\right)^2 \cos\left[\frac{2\pi}{\Lambda}x + (\theta_2 - \theta_1)\right]. \qquad (4.1\text{-}3)$$

There is a fringe pattern along the x-direction on the screen with the fringe period $\Lambda = \lambda_0 l / 2x_0$. The first peak x_p near the optical axis (z) is shifted according to the phase difference $\theta_1 - \theta_2$ between the two point sources:

$$x_p = \frac{\Lambda}{2\pi}(\theta_1 - \theta_2). \qquad (4.1\text{-}4)$$

The last term of Eq. (4.1-3) is the interference term. If the fields are *spatially coherent*, there is a fixed phase relationship between θ_1 and θ_2, and we have the fringe pattern on the screen given by Eq. (4.1-3) as the intensity pattern is calculated according to

$$I(x,y)\big|_{\text{coh}} = |\psi_{S1} + \psi_{S2}|^2, \qquad (4.1\text{-}5)$$

that is, the intensity is computed as the modulus-squared of the sum of the two complex fields. In the spatially incoherent case, $\theta_1 - \theta_2$ is randomly fluctuating and, hence, x_p, or the fringes, moves randomly. The $\cos(\cdot)$ fluctuations average out to zero. Hence, Eq. (4.1-3) becomes

$$I(x,y)\big|_{\text{incoh}} = |\psi_{S1}|^2 + |\psi_{S2}|^2. \qquad (4.1\text{-}6)$$

Therefore, in the perfectly incoherent case, the intensity is computed as the sum of the moduli-squared of the complex fields of the interfering fields. Equations (4.1-5) and (4.1-6) are the bases when we consider coherent imaging and incoherent imaging in subsequent sections. In coherent imaging, we sum up all the complex fields and then find the overall intensity, whereas in incoherent imaging, we manipulate the intensities directly.

4.2   Spatial Coherent Image Processing

With reference to our earlier discussion on signal filtering in Chapter 1, from Eq. (2.2-13b), we have

$$y(t) = \mathcal{F}^{-1}\{Y(\omega)\} = \mathcal{F}^{-1}\{X(\omega)H(\omega)\}.$$

The input spectrum $X(\omega)$ is modified or filtered by the system's transfer function $H(\omega)$ as the input signal, $x(t)$, goes through the system to give an output $y(t)$, which is the filtered version of the input. To effectuate "image" filtering similar to that in the above equation, we analyze a standard two-lens coherent image processing shown in Figure 4.2-1.

4.2.1   Pupil Function, Coherent Point Spread Function, and Coherent Transfer Function

The two-lens system shown in Figure 4.2-1 is known as the $4f$-system as Lens 1 and Lens 2 both have the same focal length, f. An input of the form of a transparency, $t(x,y)$, is illuminated by a plane wave (which is a coherent wave) of unit amplitude. A transparency function, $p(x,y)$, is placed on the confocal plane of the optical system. The confocal plane is often called the *Fourier plane* of the processing system. The reason is that the back focal plane of Lens 1 is the plane where the spectrum or the

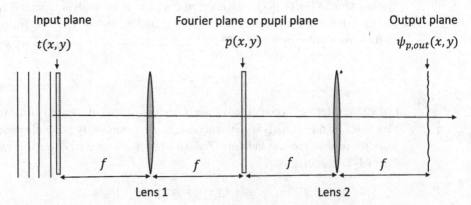

Figure 4.2-1 Standard 4-f optical image processing system with plane-wave illumination

Fourier transform of $t(x,y)$ appears [see Figure 3.4-7], which is $T(k_0 x/f, k_0 y/f)$ with T being the Fourier transform of $t(x,y)$. The Fourier plane is also called the *pupil plane* of the processing system. For this reason, $p(x,y)$ is called the *pupil function*. We can see that the spectrum of the input is now modified by $p(x,y)$ and the field immediately after the pupil function is $T\left(\dfrac{k_0 x}{f}, \dfrac{k_0 y}{f}\right) p(x,y)$. This field again will be Fourier transformed by lens 2 to give the complex field on the output plane as

$$\psi_{p,\text{out}}(x,y) = \mathcal{F}\left\{ T\left(\frac{k_0 x}{f}, \frac{k_0 y}{f}\right) p(x,y) \right\}_{k_x = \frac{k_0 x}{f}, \, k_y = \frac{k_0 y}{f}} \tag{4.2-1}$$

Since

$$T\left(\frac{k_0 x}{f}, \frac{k_0 y}{f}\right) = \mathcal{F}\{t(x,y)\}_{k_x = \frac{k_0 x}{f}, \, k_y = \frac{k_0 y}{f}} = \iint t(x',y') e^{\frac{jk_0}{f}(xx'+yy')} dx' dy', \tag{4.2-2}$$

we rewrite Eq. (4.2-1) as

$$\psi_{p,\text{out}}(x,y) = \mathcal{F}\left\{ \iint t(x',y') e^{\frac{jk_0}{f}(xx'+yy')} dx' dy' \, p(x,y) \right\}_{k_x = \frac{k_0 x}{f}, \, k_y = \frac{k_0 y}{f}}$$

$$= \iint \left[\iint t(x',y') e^{\frac{jk_0}{f}(x''x'+y''y')} dx' dy' \, p(x'',y'') \right] e^{\frac{jk_0}{f}(xx''+yy'')} dx'' dy''. \tag{4.2-3}$$

Let us group the double-primed coordinates and integrate over these coordinates first, and we have

$$\iint p(x'', y'') e^{\frac{jk_0}{f}(x''x'+y''y')} e^{\frac{jk_0}{f}(xx''+yy'')} dx'' dy'' = \iint p(x'', y'') e^{\frac{jk_0}{f}[x''(x'+x)+y''(y'+y)]} dx'' dy''$$

$$= \mathcal{F}\{p(x, y)\}_{k_x = \frac{k_0}{f}(x'+x), \, k_y = \frac{k_0}{f}(y'+y)}$$

$$= P\left[\frac{k_0}{f}(x'+x), \frac{k_0}{f}(y'+y) \right].$$

Substituting this result back into Eq. (4.2-3), we have

$$\psi_{p,\text{out}}(x,y) = \iint t(x',y') P\left[\frac{k_0}{f}(x'+x), \frac{k_0}{f}(y'+y) \right] dx' dy',$$

which can be expressed in terms of convolution as

$$\psi_{p,\text{out}}(x,y) = t(-x,-y) * P\left(\frac{k_0 x}{f}, \frac{k_0 y}{f} \right) = t(-x,-y) * h_c(x,y), \tag{4.2-4}$$

where

$$h_c(x,y) = P\left(\frac{k_0 x}{f}, \frac{k_0 y}{f}\right) = \mathcal{F}\{p(x, y)\}_{k_x = \frac{k_0 x}{f}, k_y = \frac{k_0 y}{f}}$$

is the impulse response of the system and defined as the *coherent point spread function (CPSF)* in optics as the *geometrical image* $t(-x,-y)$ of the input $t(x,y)$ and the output $\psi_{p,\text{out}}(x,y)$ are related through the convolution integral. Note that the geometrical image in general is given by $t(x/M, y/M)$, where M is the lateral magnification of the optical system. In the optical system presently under consideration [see Figure 4.2-1], we have $M = -1$. Therefore, the optical system shown in Figure 4.2-1 is a *linear and space-invariant coherent system*. Image processing capabilities can be varied by carefully designing the pupil function, $p(x,y)$, and the design of pupils to accommodate certain intended applications such as long depth of field for 3D imaging or enhanced resolution for precision measurement, etc., is called *pupil engineering*. Since the pupil function is on the Fourier plane of the input, it actually modifies the spectrum of the input. Hence, the pupil function is also called a *spatial filter* – physical filter capable of modifying the spatial frequencies of the input. Such a process is often called *spatial filtering* in the literature.

By definition, the Fourier transform of the impulse response is the transfer function. The transform of the *CPSF* is called the *coherent transfer function (CTF)*:

$$H_c(k_x, k_y) = \mathcal{F}\{h_c(x,y)\} = \mathcal{F}\left\{P\left(\frac{k_0 x}{f}, \frac{k_0 y}{f}\right)\right\} = p(x, y)\Big|_{x = \frac{-fk_x}{k_0}, y = \frac{-fk_y}{k_0}}$$

$$= p\left(\frac{-fk_x}{k_0}, \frac{-fk_y}{k_0}\right). \tag{4.2-5}$$

If we take the Fourier transform of Eq. (4.2-4), we have

$$\mathcal{F}\{\psi_{p,\text{out}}(x,y)\} = \mathcal{F}\left\{t(-x,-y) * P\left(\frac{k_0 x}{f}, \frac{k_0 y}{f}\right)\right\}$$

$$= \mathcal{F}\{t(-x,-y)\}\mathcal{F}\left\{P\left(\frac{k_0 x}{f}, \frac{k_0 y}{f}\right)\right\}$$

$$= \mathcal{F}\{t(-x,-y)\}H_c(k_x, k_y) = T(-k_x, -k_y)H_c(k_x, k_y),$$

or

$$\psi_{p,\text{out}}(x,y) = \mathcal{F}^{-1}\{T(-k_x, -k_y)H_c(k_x, k_y)\}. \tag{4.2-6}$$

This equation has the same connotation in terms of filtering as we have discussed earlier in Chapter 2 when comparing to the equation below [see Eq. (2.2-13b)]:

$$y(t) = \mathcal{F}^{-1}\{Y(\omega)\} = \mathcal{F}^{-1}\{X(\omega)H(\omega)\}.$$

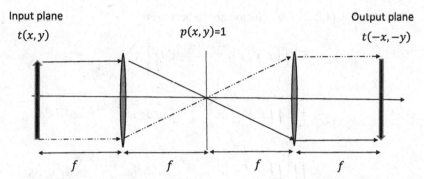

Input plane
$t(x, y)$

$p(x, y) = 1$

Output plane
$t(-x, -y)$

$f \qquad f \qquad f \qquad f$

Figure 4.2-2 Two-lens imaging system

However, in coherent image processing, we process the flipped and inverted spectrum of $t(x, y)$, that is, $T(-k_x, -k_y)$, owing to the double Fourier transformation naturally occurring in the two-lens system as the lenses only perform forward Fourier transform. Finally, once the output complex field is found, the corresponding image intensity is

$$I(x, y) = \psi_{p,\text{out}}(x, y)\psi_{p,\text{out}}^{*}(x, y) = \left|t(-x, -y) * h_c(x, y)\right|^{2}, \qquad (4.2\text{-}7)$$

which is the basis for *coherent image processing*.

4.2.2 Coherent Image Processing Examples

Example: All-Pass Filtering, $p(x, y) = 1$

From Eq. (4.2-4), the CPSF is then given by

$$h_c(x, y) = \mathcal{F}\{1\}_{k_x = \frac{k_0 x}{f}, \, k_y = \frac{k_0 y}{f}} = 4\pi^2 \delta\left(\frac{k_0 x}{f}, \frac{k_0 y}{f}\right) = 4\pi^2 \left(\frac{f}{k_0}\right)^2 \delta(x, y),$$

where we have used the scaling property of the delta function to obtain the last step of the equation [see Eq. (2.1-6)]. The output, according to Eq. (4.2-4), is

$$\psi_{p,\text{out}}(x, y) = t(-x, -y) * h_c(x, y) = 4\pi^2 \left(\frac{f}{k_0}\right)^2 t(-x, -y) * \delta(x, y) \propto t(-x, -y). \quad (4.2\text{-}8)$$

This represents an imaging system with the output being flipped and inverted on the output plane, which is consistent with the ray optics diagram shown in Figure 4.2-2.

According to Eq. (4.2-5), the coherent transfer function is

$$H_c(k_x, k_y) = p(x, y)\Big|_{x = \frac{-fk_x}{k_0}, \, y = \frac{-fk_y}{k_0}} = 1,$$

representing all-pass filtering.

An alternative way to analyze the problem is through Eq. (4.2-1), where the input has gone through double Fourier transforms; we, therefore, have

$$\psi_{p,\text{out}} = \mathcal{F}\left\{T\left(\frac{k_0 x}{f}, \frac{k_0 y}{f}\right)p(x, y)\right\}_{k_x = \frac{k_0 x}{f}, \, k_y = \frac{k_0 y}{f}} = \mathcal{F}\left\{T\left(\frac{k_0 x}{f}, \frac{k_0 y}{f}\right)\right\}_{k_x = \frac{k_0 x}{f}, \, k_y = \frac{k_0 y}{f}}.$$

Using Eq. (4.2-2), the equation above becomes

$$\psi_{p,\text{out}} = \mathcal{F}\left\{ \iint t(x',y')e^{\frac{jk_0}{f}(xx'+yy')}dx'dy' \right\}_{k_x=\frac{k_0 x}{f},\, k_y=\frac{k_0 y}{f}}$$

$$= \iint \left[\iint t(x',y')e^{\frac{jk_0}{f}(x''x'+y''y')}dx'dy' \right]e^{j(k_x x''+k_y y'')}dx''dy''$$

$$= \iint \left[\iint t(x',y')e^{\frac{jk_0}{f}(x''x'+y''y')}dx'dy' \right]e^{j\left(\frac{k_0 x}{f}x''+\frac{k_0 y}{f}y''\right)}dx''dy'' , \tag{4.2-9}$$

where we have used the definition of Fourier transform [see Eq. (2.3-1a)] to set up the above integral. Let us now perform the integration involving the double primed coordinates first:

$$\iint e^{\frac{jk_0}{f}(x''x'+y''y')}e^{j\left(\frac{k_0 x}{f}x''+\frac{k_0 y}{f}y''\right)}dx''dy'' = \iint e^{\frac{jk_0}{f}[x''(x'+x)+y''(y'+y)]}dx''dy''$$

$$= 4\pi^2\delta\left(\frac{k_0}{f}(x'+x),\frac{k_0}{f}(y'+y)\right) = 4\pi^2\left(\frac{f}{k_0}\right)^2\delta(x'+x,y'+y), \tag{4.2-10}$$

where we have recalled the integral definition of the delta function as [see Eq (2.3-4)]

$$\delta(x,y) = \frac{1}{4\pi^2}\int_{-\infty}^{\infty}\!\!\int e^{\pm jxx' \pm jyy'}dx'dy'.$$

Putting Eq. (4.2-10) into Eq. (4.2-9), we have

$$\psi_{p,\text{out}} = 4\pi^2\left(\frac{f}{k_0}\right)^2 \iint t(x',y')\delta(x'+x,y'+y)dx'dy'$$

$$= 4\pi^2\left(\frac{f}{k_0}\right)^2 t(-x,-y) \propto t(-x,-y),$$

which is the same result in Eq. (4.2-8).

Example: Low-Pass Filtering, $p(x,y) = \text{circ}\left(\frac{r}{r_0}\right)$

According to Eq. (4.2-5), the coherent transfer function is

$$H_c(k_x,k_y) = p(x,y)\Big|_{x=-\frac{fk_x}{k_0},\, y=-\frac{fk_y}{k_0}}$$

$$= \text{circ}\left(\frac{r}{r_0}\right)\Big|_{x=-\frac{fk_x}{k_0},\, y=-\frac{fk_y}{k_0}} = \text{circ}\left(\frac{\sqrt{x^2+y^2}}{r_0}\right)\Big|_{x=-\frac{fk_x}{k_0},\, y=-\frac{fk_y}{k_0}} = \text{circ}\left(\frac{k_r}{r_0 k_0 / f}\right),$$

where $k_r = \sqrt{k_x^2 + k_y^2}$. This translates to *low-pass filtering* as the opening of the circle on the pupil plane only allows spatial frequencies up to $r_0 k_0 / f$ to transmit. Figure 4.2-3 shows an example of low-pass filtering. In Figure 4.2-3a and b, we show

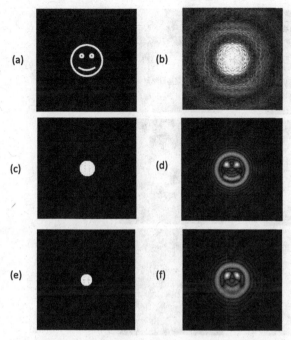

Figure 4.2-3 Low-pass filtering example: (a) original image, (b) spectrum of (a), (c) low-pass filtered spectrum, (d) low-pass filtered image of (a) for the low-pass filtered spectrum in (c), (e) low-pass filtered spectrum with $r_0 k_0 / f$ designed to be smaller than that in (c), (f) low-pass filtered image of (a) for the low-pass filtered spectrum in (e)

the original image and its spectrum, respectively. In Figure 4.2-3c and e, we show the low-pass filtered spectra for different values of r_0, and their corresponding filtered images are shown in Figure 4.2-3d and f, respectively. The filtered spectra are obtained by multiplying the original spectrum by $\mathrm{circ}\left(\dfrac{k_r}{r_0 k_0 / f}\right)$ [see Eq. (4.2-6)] as the radius of the "white" circle is $r_0 k_0 / f$. Note that the radius in Figure 4.2-3c is larger than that in Figure 4.2-3e, meaning we let more spatial frequencies pass through the optical system. In general, the low-pass filtered images are blurred or have been smooth as compared to the original image. As it turns out, a low-pass filter, also called a "blurring" or "smoothing" filter, averages out any rapid changes in the original image.

Example: High-Pass Filtering, $p(x,y)=1-\mathrm{circ}\left(\dfrac{r}{r_0}\right)$

The coherent transfer function is

$$H_c\left(k_x,k_y\right) = p(x,\ y)\Big|_{x=\frac{-fk_x}{k_0},y=\frac{-fk_y}{k_0}}$$

$$= 1-\mathrm{circ}\left(\frac{r}{r_0}\right)\Bigg|_{x=\frac{-fk_x}{k_0},y=\frac{-fk_y}{k_0}}$$

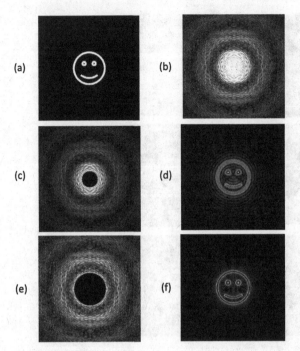

Figure 4.2-4 High-pass filtering example: (a) original image, (b) spectrum of (a), (c) high-pass filtered spectrum, (d) high-pass filtered image of (a) for high-pass filtered spectrum in (c), (e) high-pass filtered spectrum with $r_0 k_0 / f$ designed to be larger than that in (c), (f) high-pass filtered image of (a) for high-pass filtered spectrum in (e)

$$= 1 - \mathrm{circ}\left(\frac{\sqrt{x^2 + y^2}}{r_0}\right)\Bigg|_{x = \frac{-f k_x}{k_0}, y = \frac{-f k_y}{k_0}}$$

$$= 1 - \mathrm{circ}\left(\frac{k_r}{r_0 k_0 / f}\right) = \overline{\mathrm{circ}}\left(\frac{k_r}{r_0 k_0 / f}\right).$$

While $\mathrm{circ}\left(\dfrac{k_r}{r_0 k_0 / f}\right)$ is a low-pass filter, $\overline{\mathrm{circ}}\left(\dfrac{k_r}{r_0 k_0 / f}\right)$ is a high-pass filter as the blocking of the circle on the pupil plane forbids spatial frequencies up to $r_0 k_0 / f$ to transmit. In Figure 4.2-4a and b, we show the original image and its spectrum, respectively. In Figure 4.2-4c and e, we show the high-pass filtered spectra, and their corresponding filtered images are shown in Figure 4.2-4d and f, respectively. The filtered spectra are obtained by multiplying the original spectrum by $\overline{\mathrm{circ}}\left(\dfrac{k_r}{r_0 k_0 / f}\right)$. Note that for sufficient blocking of low spatial frequencies such as the case shown in Figure 4.2-4e, where the "black" circle of radius $r_0 k_0 / f$ is larger than that in Figure 4.2-4c, the high-pass filtered image is edge-extracted. Indeed, *high-pass filtering* is useful for edge detection, that is, locating boundaries in images. *Edge detection* plays a crucial role in *pre-processing* for pattern recognition.

Figures 4.2-3 and 4.2-4 are generated using the m-file shown below.

```
% lowpass_highpass.m, Low-pass and higpass filtering of an
  image
% Adapted from "Introduction to Modern Digital Holography"
% by T.-C. Poon & J.-P. Liu
% Cambridge University Press(2014), Tables 1.5 and 1.6.
clear all;close all;
A=imread('front.jpg');          % read 512 X 512 8-bit image
A=double(A);
A=A/255;
FA=fftshift(fft2(fftshift(A)));
D=abs(FA);
D=D(129:384,129:384);
figure;imshow(A);
% title('Original image')
figure;imshow(30.*mat2gray(D));%magnitude spectrum
% title('Original spectrum')

c=1:512;
r=1:512;
[C, R]=meshgrid(c, r);
CI=((R-257).^2+(C-257).^2);
filter=zeros(512,512);
% produce a circular lowpass filter
for a=1:512;
    for b=1:512;
        if CI(a,b)>=20^2;
%Used "20" and "15" to simulate different r0's for results in
    text
            filter(a,b)=0;
        else
            filter(a,b)=1;
        end
    end
end
G=abs(filter.*FA);
G=G(129:384,129:384);
figure;imshow(30.*mat2gray(G));
% title('Magnitude spectrum of lowpass filtered image')
FLP=FA.*filter;
E=abs(fftshift(ifft2(fftshift(FLP))));
figure;imshow(mat2gray(E));
% title('Lowpass filtered image')

c=1:512;
r=1:512;
[C, R]=meshgrid(c, r);
CI=((R-257).^2+(C-257).^2);
filter=zeros(512,512);
% produce a circular high-pass filter
for a=1:512;
    for b=1:512;
```

```
        if CI(a,b)<=20^2;
%Used "20" and "40" to simulate different r0's for
  results in text
            filter(a,b)=0;
        else
            filter(a,b)=1;
        end
    end
end
G=abs(filter.*FA);
G=G(129:384,129:384);
figure;imshow(2.*mat2gray(G));
% title('Magnitude spectrum of highpass filtered image')
FHP=FA.*filter;
E=abs(fftshift(ifft2(fftshift(FHP))));
figure;imshow(mat2gray(E));
% title('Higpass filtered image')
```

4.3   Spatial Incoherent Image Processing

In the last section, we have discussed spatially coherent illumination of an object transparency, such as the use of a laser producing plane waves for illumination. However, light from extended sources, such as fluorescent tube lamps, is spatially incoherent. The system shown in Figure 4.2-1 becomes an incoherent optical system upon illumination from an incoherent source. A coherent system is linear with respect to the complex amplitudes and hence Eqs. (4.2-4) and (4.2-7) hold for coherent optical systems as we manipulate complex amplitudes [see Eq (4.1-5)]. On the other hand, an incoherent optical system is linear with respect to the intensities. In other words, we manipulate intensities, which is not totally surprising as we have seen from our earlier discussion on spatial coherence in Section 4.1 that we add intensities when dealing with spatial incoherent light [see Eq. (4.1-6)]. To find the image intensity, we perform convolution with the given intensity quantities as follows:

$$I(x,y) = |t(-x,-y)|^2 * |h_c(x,y)|^2.$$ (4.3-1)

Equation (4.3-1) is the basis for *incoherent image processing* in a linear and space-invariant incoherent system.

4.3.1   Intensity Point Spread Function and Optical Transfer Function

By inspecting Eq. (4.3-1), we see that the input and output intensities are related by an impulse response $|h_c(x,y)|^2$, which is known as the *intensity point spread function (IPSF)* in incoherent systems:

$$IPSF(x,y) = |h_c(x,y)|^2.$$ (4.3-2)

The Fourier transform of the *IPSF* gives a transfer function known as the *optical transfer function (OTF)* of the incoherent system:

$$OTF(k_x, k_y) = \mathcal{F}\left\{\left|h_c(x, y)\right|^2\right\}. \tag{4.3-3}$$

According to Eq. (4.2-5), we can write $h_c(x, y) = \mathcal{F}^{-1}\left\{H_c(k_x, k_y)\right\}$. Upon substitution of this expression of $h_c(x, y)$ into Eq. (4.3-3), it can be shown that the *OTF* and the coherent transfer function are related as follows:

$$OTF(k_x, k_y) = H_c(k_x, k_y) \otimes H_c(k_x, k_y)$$

$$= \iint_{-\infty}^{\infty} H_c^*(k_x', k_y') H_c(k_x' + k_x, k_y' + k_y) dk_x' dk_y', \tag{4.3-4}$$

where $\otimes$ denotes correlation [see Eq. (2.3-30)]. The *OTF* is the autocorrelation of H_c and in general it is complex. The modulus of the *OTF* is called the *modulation transfer function (MTF)*.

Some important properties of the *OTF* are as follows:

$$OTF(-k_x, -k_y) = OTF^*(k_x, k_y) \tag{4.3-5a}$$

$$\left|OTF(k_x, k_y)\right| \leq \left|OTF(0,0)\right| \text{ or } MTF(k_x, k_y) \leq MTF(0,0). \tag{4.3-5b}$$

Equation (4.3-5b) is particularly important. The property states that the *MTF* always has a central maximum, which always signifies low-pass filtering regardless of the pupil function used in an incoherent system. This gives rise to a question on how to implement high-pass filtering in incoherent systems, for example. We will address this important question in the last section of this chapter.

By taking the Fourier transform of Eq. (4.3-1), we have

$$\mathcal{F}\left\{I(x, y)\right\} = \mathcal{F}\left\{\left|t(-x, -y)\right|^2\right\}\mathcal{F}\left\{\left|h_c(x, y)\right|^2\right\} = \mathcal{F}\left\{\left|t(-x, -y)\right|^2\right\}OTF(k_x, k_y),$$

or

$$I(x, y) = \mathcal{F}^{-1}\left\{\mathcal{F}\left\{\left|t(-x, -y)\right|^2\right\}OTF(k_x, k_y)\right\}. \tag{4.3-6}$$

Again, this equation has the same spirit in terms of filtering as we have discussed earlier in Chapter 2 when comparing to the equation below [see Eq. (2.2-13b)]:

$$y(t) = \mathcal{F}^{-1}\left\{Y(\omega)\right\} = \mathcal{F}^{-1}\left\{X(\omega)H(\omega)\right\}.$$

Example: Show that $\left|OTF(k_x, k_y)\right| \leq \left|OTF(0,0)\right|$

We use *Schwarz's Inequality*. Given two complex functions $P(k_x, k_y)$ and $Q(k_x, k_y)$, Schwarz's Inequality states that

$$\left|\iint_{-\infty}^{\infty} P(k_x', k_y') Q(k_x', k_y') dk_x' dk_y'\right| \leq \left[\iint_{-\infty}^{\infty}\left|P(k_x', k_y')\right|^2 dk_x' dk_y'\right]^{\frac{1}{2}} \times \left[\iint_{-\infty}^{\infty}\left|Q(k_x', k_y')\right|^2 dk_x' dk_y'\right]^{\frac{1}{2}}.$$

By letting $P(k'_x,k'_y) = H^*_c(k'_x,k'_y)$ and $Q(k'_x,k'_y) = H_c(k'_x + k_x, k'_y + k_y)$, we have

$$\left| \iint\limits_{-\infty}^{\infty} H^*_c(k'_x,k'_y) H_c(k'_x + k_x, k'_y + k_y) dk'_x dk'_y \right|$$

$$\leq \left[\iint\limits_{-\infty}^{\infty} \left| H^*_c(k'_x,k'_y) \right|^2 dk'_x dk'_y \right]^{\frac{1}{2}} \times \left[\iint\limits_{-\infty}^{\infty} \left| H_c(k'_x + k_x, k'_y + k_y) \right|^2 dk'_x dk'_y \right]^{\frac{1}{2}}. \qquad (4.3\text{-}7)$$

However, the first term of the right side of the inequality is

$$\iint\limits_{-\infty}^{\infty} \left| H^*_c(k'_x,k'_y) \right|^2 dk'_x dk'_y = \iint\limits_{-\infty}^{\infty} \left| H_c(k'_x,k'_y) \right|^2 dk'_x dk'_y, \qquad (4.3\text{-}8)$$

and in the second term we can express

$$\iint\limits_{-\infty}^{\infty} \left| H_c(k'_x + k_x, k'_y + k_y) \right|^2 dk'_x dk'_y = \iint\limits_{-\infty}^{\infty} \left| H_c(k'_x,k'_y) \right|^2 dk'_x dk'_y \qquad (4.3\text{-}9)$$

for the reason that a shift of the function, that is, a shift by k_x and k_y cannot change the value of the integral. By substituting the results of Eqs. (4.3-8) and (4.3-9) into Eq. (4.3-7), we obtain

$$\left| \iint\limits_{-\infty}^{\infty} H^*_c(k'_x,k'_y) H_c(k'_x + k_x, k'_y + k_y) dk'_x dk'_y \right|$$

$$\leq \left[\iint\limits_{-\infty}^{\infty} \left| H_c(k'_x,k'_y) \right|^2 dk'_x dk'_y \right]^{\frac{1}{2}} \times \left[\iint\limits_{-\infty}^{\infty} \left| H_c(k'_x,k'_y) \right|^2 dk'_x dk'_y \right]^{\frac{1}{2}} = \iint\limits_{-\infty}^{\infty} \left| H_c(k'_x,k'_y) \right|^2 dk'_x dk'_y.$$

Expressing the above result in terms of the *OTF*, we have

$$\left| OTF(k_x,k_y) \right| \leq \left| OTF(0,0) \right|,$$

which is the important property of the *OTF* in Eq. (4.3-5b).

4.3.2   Incoherent Image Processing Examples

Example: Single-Slit Pupil, $p(x,y) = \text{rect}\left(\dfrac{x}{x_0}\right)$

In this example, we use a slit of width x_0 along the y-direction as the pupil in the standard 4-f optical image processing system shown in Figure 4.2-1. According to Eq. (4.2-5), the coherent transfer function is

$$H_c(k_x,k_y) = p(x,y)\Big|_{x=-\frac{fk_x}{k_0}, y=-\frac{fk_y}{k_0}} = \text{rect}\left(\frac{fk_x}{x_0 k_0}\right) = \text{rect}\left(\frac{k_x}{x_0 k_0 / f}\right).$$

The *OTF* is

$$OTF(k_x,k_y) = H_c(k_x,k_y) \otimes H_c(k_x,k_y) = \text{rect}\left(\frac{k_x}{x_0 k_0 / f}\right) \otimes \text{rect}\left(\frac{k_x}{x_0 k_0 / f}\right)$$

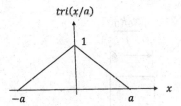

Figure 4.3-1 Triangle function

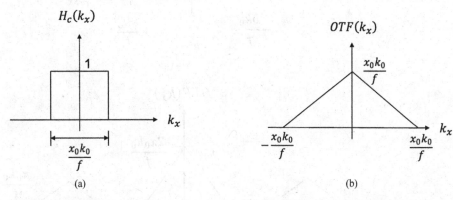

Figure 4.3-2 (a) The coherent transfer function and (b) the *OTF* of the two-lens system with

$$p(x) = \text{rect}\left(\frac{x}{x_0}\right)$$

$$= \frac{x_0 k_0}{f} \text{ tri}\left(\frac{k_x}{x_0 k_0 / f}\right), \tag{4.3-10}$$

where we define a *triangle function* tri(x) as

$$\text{tri}\left(\frac{x}{a}\right) = \begin{cases} 0, & |x/a| \geq 1 \\ 1 - |x/a|, & |x/a| < 1, \end{cases} \tag{4.3-11}$$

with $2a$ being the width of the function. The triangle function is shown in Figure 4.3-1.

We plot $H_c(k_x, k_y)$ along k_x with k_y constant in Figure 4.3-2a along with the *OTF* in Figure 4.3-2b. We observe that both situation perform low-pass filtering along the x-direction on an input image. However, under incoherent illumination, it is possible to transmit twice the range of spatial frequencies of the input image as compared to the use of coherent illumination.

Example: Double-Slit Pupil, $p(x, y) = \text{rect}\left(\dfrac{x-b}{x_0}\right) + \text{rect}\left(\dfrac{x+b}{x_0}\right)$, $b > x_0 / 2$

We use a double-slit pupil of width x_0 along the y-direction in the standard 4-f optical image processing system shown in Figure 4.2-1. The coherent transfer function along the x-direction is

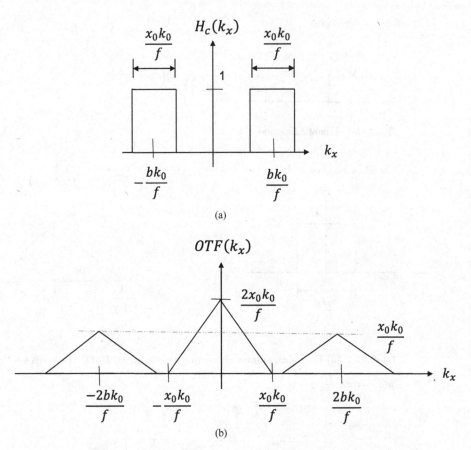

Figure 4.3-3 Two-lens system with double-slit pupil: (a) coherent transfer function illustrating bandpass filtering characteristics and (b) *OTF* illustrating low-pass filtering characteristics

$$H_c\left(k_x,k_y\right)=p(x,y)\Big|_{x=-\frac{fk_x}{k_0},\,y=-\frac{fk_y}{k_0}}=\mathrm{rect}\left(\frac{\frac{-fk_x}{k_0}-b}{x_0}\right)+\mathrm{rect}\left(\frac{\frac{-fk_x}{k_0}+b}{x_0}\right)$$

$$=\mathrm{rect}\left(\frac{-k_x-bk_0\,/\,f}{x_0k_0\,/\,f}\right)+\mathrm{rect}\left(\frac{-k_x+bk_0\,/\,f}{x_0k_0\,/\,f}\right),$$

and its plot is shown in Figure 4.3-3a. The *OTF* is the autocorrelation of $H_c\left(k_x,k_y\right)$ given above and its plot is shown in Figure 4.3-3b. Note that while the coherent transfer function shows bandpass filtering characteristics as the transfer function passes spatial frequencies within a certain range and rejects frequencies outside that range, the *OTF* still exhibits low-pass filtering characteristics with *OTF*(0) being maximum.

Example: MATLAB Examples of Incoherent Image Processing
While we have illustrated in the last two examples that incoherent spatial filtering always leads to low-pass filtering [see Figures 4.3-2b and 4.3-3b]. In this

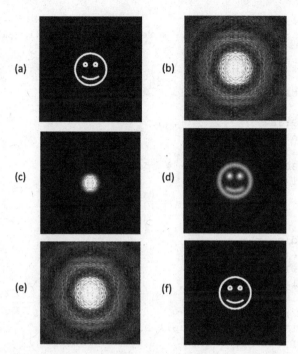

Figure 4.3-4 Incoherent spatial filtering: (a) original image, (b) spectrum of (a), (c) filtered spectrum with $\mathrm{circ}(r/r_0)$ as the pupil, (d) filtered image of (a) with $\mathrm{circ}(r/r_0)$ as the pupil, (e) filtered spectrum with $\overline{\mathrm{circ}}(r/r_0)$ as the pupil, (f) filtered image of (a) with $\overline{\mathrm{circ}}(r/r_0)$ as the pupil

example, we perform some simulations for two types of pupils: $\mathrm{circ}(r/r_0)$ and $\overline{\mathrm{circ}}(r/r_0) = 1 - \mathrm{circ}(r/r_0)$.

In Figure 4.3-4a and b, we show the original image and its spectrum, respectively. In Figure 4.3-4c and e, we show the filtered spectra for pupils $\mathrm{circ}(r/r_0)$ and $\overline{\mathrm{circ}}(r/r_0)$, respectively, and their corresponding filtered images are shown in Figure 4.3-4d and f, respectively. The filtered spectra in Figure 4.3-4c and e are obtained by multiplying the original spectrum by $OTF = \mathrm{circ}\left(\dfrac{k_r}{r_0 k_0 / f}\right) \otimes \mathrm{circ}\left(\dfrac{k_r}{r_0 k_0 / f}\right)$ and $OTF = \overline{\mathrm{circ}}\left(\dfrac{k_r}{r_0 k_0 / f}\right) \otimes \overline{\mathrm{circ}}\left(\dfrac{k_r}{r_0 k_0 / f}\right)$, respectively. Note that even with the pupil of the form of $\overline{\mathrm{circ}}(r/r_0)$, edge detection is not possible with incoherent illumination in the standard two-lens image processing system, as the filtered image is simply a low-pass version of the input image.

```
% Incoherent spatial filtering
% circ(r/r0) and  1-circ(r/r0) as pupil functions
% Adapted from "Introduction to Modern Digital Holography"
% by T.-C. Poon & J.-P. Liu
% Cambridge University Press (2014), Table 1.7 and Table 1.8.
```

```
clear all;close all;
A=imread('front.jpg');              % read image file 512x512
8-bit
A=double(A);
A=A/255;
SP=fftshift(fft2(fftshift(A)));
D=abs(SP);
D=D(129:384,129:384);
figure;imshow(A);
%title('Original image')
figure;imshow(30.*mat2gray(D)); % spectrum
title('Original spectrum')

c=1:512;
r=1:512;
[C, R]=meshgrid(c, r);
CI=((R-257).^2+(C-257).^2);
pup=zeros(512,512);
% produce pupil circ(r/r0)
for a=1:512;
    for b=1:512;
        if CI(a,b)>=15^2;    %pupil size
            pup(a,b)=0;
        else
            pup(a,b)=1;
        end
    end
end
h=ifft2(fftshift(pup));

OTF=fftshift(fft2(h.*conj(h)));
OTF=OTF/max(max(abs(OTF)));
G=abs(OTF.*SP);
G=G(129:384,129:384);
figure;imshow(30.*mat2gray(G));
%title('Spectrum filtered by circ(r/r0)')

I=abs(fftshift(ifft2(fftshift(OTF.*SP))));
figure;imshow(mat2gray(I));
%title('Image filterd by circ(r/r0)')

c=1:512;
r=1:512;
[C, R]=meshgrid(c, r);
CI=((R-257).^2+(C-257).^2);
pup=zeros(512,512);
% produce pupil 1-circ(r/r0)
for a=1:512;
    for b=1:512;
        if CI(a,b)>=15^2;   %pupil size
            pup(a,b)=1;
```

```
        else
            pup(a,b)=0;
        end
    end
end
h=ifft2(fftshift(pup));

OTF=fftshift(fft2(h.*conj(h)));
OTF=OTF/max(max(abs(OTF)));
G=abs(OTF.*SP);
G=G(129:384,129:384);
figure;imshow(30.*mat2gray(G));
%title('Spectrum filtered by 1-circ(r/r0)')

I=abs(fftshift(ifft2(fftshift(OTF.*SP))));
figure;imshow(mat2gray(I));
%title('Image filtered by 1-circ(r/r0)')
```

4.4 Scanning Image Processing

Coherent optical image processing systems use a frequency plane (pupil plane) architecture and processes images in parallel. The technique is conceptually simple and for this reason has received much attention. However, parallel optical processors often lack accuracy and flexibility. The lack of accuracy is owing to the fact that coherent systems are sensitive to phase (as we are dealing with the manipulation of complex amplitudes) and they are extremely susceptible to noise corruption owing to interference. The lack of flexibility comes from the fact that it is difficulty to extend the range of applications beyond the usual Fourier and linear invariant processing. To overcome the lack of accuracy, we can employ incoherent image processing but the main drawback is that the intensity point spread function is real and nonnegative [see Eq. (4.3-2)] and the achievable *OTF*, which is the autocorrelation of the pupil function [see Eq. (4.3-4)] and thus always has a central maximum, is strictly constrained. This means that it is not possible to implement even the simplest but important filtering such as high-pass filtering. In this section, we discuss *optical scanning processing* which offers some advantage over the corresponding parallel system in the performance of certain processing tasks.

In Figure 4.4-1, we show a typical *scanning optical processor*. Briefly, the optical processor scans out an input $t(x, y)$ in a raster fashion with an optical beam by moving the beam through the xy-scanner. A photodetector accepts all the light and gives an electrical signal $i(t)$ as an output. When this electrical signal, which contains the processed information of the scanned input, is digitally stored (i.e., in a computer) in synchronization with 2-D scan signals of the xy-scanner, we have a 2-D digital record, $i(x, y)$, as a processed image. Here, the processed image is built sequentially to give $i(x, y)$ to be stored on a computer or real-time displayed on a monitor.

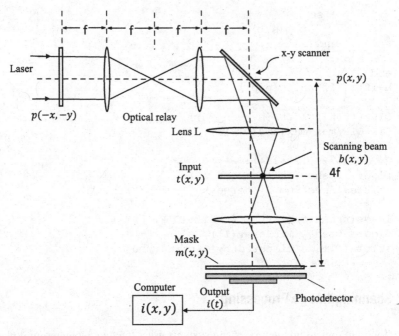

Figure 4.4-1 Optical scanning processing

The scanning position is described by the following relationships:

$$x = x(t) = V_x t, \quad y = y(t) = V_y t, \tag{4.4-1}$$

where $x(t)$ and $y(t)$ are a function of the scan with a uniform scan at speeds V_x and V_y, respectively. Hence, by optical scanning, we translate spatial frequencies, $f_x = k_x / 2\pi \left[\text{mm}^{-1}\right]$, of an image into temporal frequencies, f_t [1/s], according to

$$f_t = V_x f_x,$$

where $V_x = l / T$ with l being the linear range of the scan length and T the scan time over the scan length along x. For example, if $f_x = 2$ mm^{-1} with $l = 20$ mm and $T = 1$ s, $f_t = 40$ Hz.

Let us now further discuss the scanning system in detail. A pupil function $p(x, y)$ is first optically relayed to the surface of the xy-scanner, which is in the front focal plane of lens L. We assume the pupil on the xy-scanner is given by $p(x, y)$ and this can be done by simply putting $p(-x, -y)$ in front of the optical relay system shown in Figure 4.4-1. We, therefore, have a scanning beam with amplitude $b(x, y)$ on the input plane, and $b(x, y) = \mathcal{F}\left\{p(x, y)\right\}_{k_x = \frac{k_0 x}{f}, k_y = \frac{k_0 y}{f}}$.

For brevity, let us now perform analysis in one dimension. The complex field after the input $t(x)$ is now given by $b(x' - x)t(x')$ as the term $b(x' - x)$ simply means that the optical beam is translating (or scanning) according to $x = x(t)$. Now, this field is Fourier transform onto the mask $m(x)$, and the complex field exiting from the mask is

$$\psi(x, x_m) \propto \left[\int b(x' - x)t(x')e^{\frac{jk_0}{f}x_m x'} dx'\right] m(x_m), \tag{4.4-2}$$

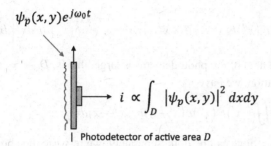

Figure 4.4-2 Optical photodetection

where x_m is the coordinate on the plane of the mask. Immediately behind the mask is a photodetector that converts optical signals into electrical signals, in the form of either current or voltage.

All photodetectors are *square-law detectors* that respond to the intensity, rather than the complex field, of an optical signal. Assume that the complex field just in front of the photodetector's surface is $\psi_p(x,y)e^{j\omega_0 t}$, where ω_0 is the frequency of the light. Since the photodetector could only respond to intensity, that is, $\left|\psi_p(x,y)e^{j\omega_0 t}\right|^2$, it gives out a current i as an output by spatially integrating over the active area, D, of the detector:

$$i(x,y) \propto \int_D \left|\psi_p(x,y)e^{j\omega_0 t}\right|^2 dxdy = \int_D \left|\psi_p(x,y)\right|^2 dxdy. \quad (4.4\text{-}3)$$

The situation of *photodetection* is illustrated in Figure 4.4-2.

We can now use Eq. (4.4-2) to find the current from the photodetector in Figure 4.4-1 by integrating over the mask's coordinate:

$$i(x) = \int_D \left|\psi(x,x_m)\right|^2 dx_m = \int_D \psi(x,x_m)\psi^*(x,x_m)dx_m$$

$$= \int_D \left[\int b(x'-x)t(x')e^{\frac{jk_0}{f}x_m x'}dx'\right]m(x_m)$$

$$\times \left[\int b^*(x''-x)t^*(x'')e^{\frac{-jk_0}{f}x_m x''}dx''\right]m^*(x_m)dx_m$$

$$= \int_D \left[\iint t(x')t^*(x'')b(x'-x)b^*(x''-x)e^{\frac{jk_0}{f}x_m(x'-x'')}dx'dx''\left|m(x_m)\right|^2\right]dx_m. \quad (4.4\text{-}4)$$

In Eq. (4.4-4), $i(x)$ represents a processed image on a computer. The scanning processor can perform all the usual coherent and incoherent image processing operations achievable by optical means under a parallel architecture.

4.4.1 Coherent Imaging

Let us consider a *point detector*. We take $m(x) = \delta(x)$. Physically, this is accomplished by putting a pinhole in front of the detector. Hence, the detector is termed as a point detector. Eq. (4.4-4) becomes

$$i(x) = \int_D \left| \iint t(x')t^*(x'')b(x'-x)b^*(x''-x)e^{\frac{jk_0}{f}x_m(x'-x'')}\delta(x_m)dx'dx'' \right| dx_m.$$

Assuming that the active area of the photodetector is large, that is, $D \to \infty$, and after the delta function integration, we have

$$i(x) = \iint t(x')t^*(x'')b(x'-x)b^*(x''-x)dx'dx''.$$

Since the x' and x'' integrations can be done separately, we re-write the above equation to become

$$i(x) = \int t(x')b(x'-x)dx' \times \int t^*(x'')b^*(x''-x)dx'' = \left| \int t(x')b(x'-x)dx' \right|^2$$

$$= \left| \int t(x')b[-(x-x')]dx' \right|^2 = |t(x)*b(-x)|^2. \tag{4.4-5}$$

Comparing to Eq. (4.2-7), we can see that corresponds to coherent image processing with a coherent point spread function in the scanning system given by $b(-x)$ as we manipulate (or process) complex amplitudes first and then compute the intensity.

4.4.2   Incoherent Imaging

Let us consider an *integrating detector*. We take $m(x) = 1$. Equation (4.4-4) becomes

$$i(x) = \int_D \left| \iint t(x')t^*(x'')b(x'-x)b^*(x''-x)e^{\frac{jk_0}{f}x_m(x'-x'')}dx'dx'' \right| dx_m. \tag{4.4-6}$$

Along with $m(x) = 1$ and assuming the active area of the photodetector is large, we accept all the light incident on the detector (hence, the detector is termed as an *integrating detector*). The integral in Eq. (4.4-6) involving x_m can be evaluated first to get [see Eqs. (2.3-4) and (2.1-6)]

$$\int_{-\infty}^{\infty} e^{\frac{jk_0}{f}x_m(x'-x'')}dx_m = 2\pi\delta\left(\frac{k_0}{f}(x'-x'')\right) \propto \delta(x'-x'').$$

With this result, Eq. (4.4-6) becomes

$$i(x) \propto \iint t(x')t^*(x'')b(x'-x)b^*(x''-x)\delta(x'-x'')dx'dx''.$$

Using the sampling property of a delta function [see Eq. (2.1-5)], we perform the integration over x' to obtain

$$i(x) \propto \int t(x'')t^*(x'')b(x''-x)b^*(x''-x)dx'' = \int |t(x'')|^2 |b(x''-x)|^2 dx''$$

$$= |t(x)|^2 * |b(-x)|^2. \tag{4.4-7}$$

Comparing to Eq. (4.3-1), we can see that this is incoherent image processing with the intensity point spread function in the scanning system given by $|b(-x)|^2$.

We have seen that the scanning system can perform coherent or incoherent process-ing by simply manipulating the mask in front of the photodetector. Using $m(x) = \delta(x)$, we have coherent image processing. For $m(x) = 1$, we have incoherent image process-ing. The use of a delta function and a uniform function is two extreme cases on the size of the mask. It is not hard to envision *partial coherent image processing* by using a mask of some finite size but the topic is beyond the scope of the present book. To end this subsection, we want to point out some important potential applications with the scanning approach to imaging.

In the scanning approach, it is the coherent optical beam $b(x, y)$ that is generated and scanned over the input. The size of the mask in front of the photodetector deter-mines the coherency of the scanning system. Some of the advantages of the scanning technique over the parallel counterpart are, therefore, a better utilization of the light energy and the fact that the output signal is an electronic form, which is ready to be processed or transmitted serially. In many instances, the input already exists in a scanned format (e.g., from a scanner) and in such case it seems natural to process the signal in that form rather than put it on film or use an expensive spatial light modulator for parallel processing (we will discuss spatial light modulators in Chapter 7). Another advantage of optical scanning as compared to non-scanning optical techniques is the capability of efficiently performing image processing of large-scale objects such as in remote sensing applications.

The scanning system is much more flexible than the parallel counterpart. The function of the xy-scanner represented by Eq. (4.4-1) defines the spatial to temporal conversion. For $x = x(t) = V_x t$, we have a linear mapping between the spatial and temporal frequencies. The mapping can, in principle, be nonlinear if $x(t)$ and t are related nonlinearly, thus representing some unusual manipulating operation through *nonlinear scanning*. Also, since image processing is performed via scanning, we can envision that the scanning beam can be modified continuously as it scans over the input simply by updating the pupil function $p(x, y)$, giving the possibility of *space-variant filtering* – a task that is difficult with the parallel counterpart.

4.5   Two-Pupil Synthesis of Optical Transfer Functions

As we have seen from Eqs. (4.3-2) and (4.3-3), the incoherent point spread func-tion (*IPSF*) of incoherent systems is a real and nonnegative function which imposes restrictions on both the magnitude and the phase of the *OTF*. Consequently, it limits the range of processing as we have discussed before. Many important processing oper-ations, such as high-pass filtering, edge extraction, etc., require point spread functions that are bipolar (having positive as well as negative values). In order to overcome the limitation of a positive and real *IPSF*, the so-called *two-pupil synthesis* of the optical transfer function has been introduced [Lohmann and Rhodes (1978)]. Considerable work has been done in past decades on two-pupil synthesis owing to its practical importance. We shall limit ourselves to discuss one approach called the *non-pupil interaction synthesis*. Let us discuss this concept in the frequency domain first.

In incoherent systems, the *OTF* is given by Eq. (4.3-3) or equivalently by Eq. (4.3-4):

$$OTF(k_x,k_y) = \mathcal{F}\left\{\left|h_c(x,y)\right|^2\right\} = H_c(k_x,k_y) \otimes H_c(k_x,k_y)$$

$$= \iint\limits_{-\infty}^{\infty} H_c^*(k_x',k_y') H_c(k_x'+k_x,k_y'+k_y) dk_x' dk_y'. \qquad (4.5\text{-}1)$$

Now, according to Eq. (4.2-5), the coherent transfer function is expressed in terms of the pupil $p(x,y)$ as

$$H_c(k_x,k_y) = p\left(\frac{-fk_x}{k_0},\frac{-fk_y}{k_0}\right).$$

Therefore, we can re-write the *OTF* in terms of the pupil function as

$$OTF(k_x,k_y) = p\left(\frac{-fk_x}{k_0},\frac{-fk_y}{k_0}\right) \otimes p\left(\frac{-fk_x}{k_0},\frac{-fk_y}{k_0}\right). \qquad (4.5\text{-}2)$$

We can see that the autocorrelation of the pupil function is the *OTF* of the optical system. This is called *one-pupil synthesis* of the optical transfer functions. As any autocorrelation function always has a central maximum [see Eq. (4.3-5b)], we could not obtain, for example, a bandpass filter function. In order to circumvent the limitations of one-pupil syntheses, *two-pupil synthesis* is introduced.

According to Eq. (4.5-1), we first define the normalized *OTF* by

$$\overline{OTF}(k_x,k_y) = \frac{H_c(k_x,k_y) \otimes H_c(k_x,k_y)}{OTF(0,0)}$$

$$= \frac{\displaystyle\iint\limits_{-\infty}^{\infty} H_c^*(k_x',k_y') H_c(k_x'+k_x,k_y'+k_y) dk_x' dk_y'}{\displaystyle\iint\limits_{-\infty}^{\infty} \left|H_c(k_x',k_y')\right|^2 dk_x' dk_y'}. \qquad (4.5\text{-}3)$$

Therefore, $\overline{OTF}(0,0) = 1$. By the same token, in terms of the pupil function, we have

$$\overline{OTF}(k_x,k_y) = \frac{p\left(\dfrac{-fk_x}{k_0},\dfrac{-fk_y}{k_0}\right) \otimes p\left(\dfrac{-fk_x}{k_0},\dfrac{-fk_y}{k_0}\right)}{OTF(0,0)}. \qquad (4.5\text{-}4)$$

In the non-pupil interaction synthesis, we can obtain a bipolar *IPSF* by subtracting the *IPSF*s of two different $\overline{OTF}$ s, $\overline{OTF}_1$ and $\overline{OTF}_2$:

$$OTF_{\text{eff}} = \overline{OTF}_1 - \overline{OTF}_2$$

$$= \frac{p_1\left(\dfrac{-fk_x}{k_0},\dfrac{-fk_y}{k_0}\right) \otimes p_1\left(\dfrac{-fk_x}{k_0},\dfrac{-fk_y}{k_0}\right)}{OTF_1(0,0)} - \frac{p_2\left(\dfrac{-fk_x}{k_0},\dfrac{-fk_y}{k_0}\right) \otimes p_2\left(\dfrac{-fk_x}{k_0},\dfrac{-fk_y}{k_0}\right)}{OTF_2(0,0)}, \qquad (4.5\text{-}5)$$

(a)   (b)

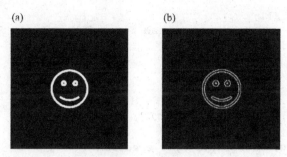

Figure 4.5-1 (a) Original input image and (b) bipolar filtered output image

where p_1 and p_2 are the different pupil functions. Note that OTF_{eff} is no longer maximum at (0,0) and in fact, $OTF_{eff}(0,0) = 0$, giving the optical system of non-low-pass filtering characteristics. The processed intensity, according to Eq. (4.3-6), is

$$I(x,y) = \mathcal{F}^{-1}\left\{\mathcal{F}\left\{|t(-x,-y)|^2\right\}OTF_{eff}(k_x,k_y)\right\}. \tag{4.5-6}$$

To implement $OTF_{eff}(k_x,k_y)$, we need to first capture the processed intensity using p_1, store it, and then capture another processed intensity using p_2. Finally we perform subtraction of the two acquired stored intensities to obtain *bipolar incoherent image processing*. The term "bipolar" is used here to represent that the achievable point spread function of the system is bipolar. Because we cannot subtract intensities directly, this method requires special techniques such as, for instance, the use of computer storage. In *pupil interaction synthesis* of the *OTF*s, interference of light from p_1 and p_2 is involved. These methods are relatively complicated because of the interferometric nature of their operation. They are nevertheless attractive because of the greater flexibility they allow in the synthesis of relatively general bipolar point spread functions. For an in-depth review of the research area, the interested reader is referred to the cited references at the end of this chapter for detailed description of the method [Indebetouw and Poon (1992)].

Example: MATLAB Example of Bipolar Incoherent Image Processing

We take two different circular functions as an example to perform bipolar incoherent image processing: $p_1(x,y) = \text{circ}(r/r_1)$ and $p_2(x,y) = \text{circ}(r/r_2)$. The simulation is performed according to Eq. (4.5-6). Figure 4.5-1a shows the original image, and Figure 4.5-1b shows the filtered image and it clearly illustrates non-low-pass filtering, and indeed edge extraction has been performed with bipolar incoherent filtering.

The images produced in Figure 4.5-1 are generated using the m-file shown below.

```
% Bipolar_incoherent_image_processing
% circ(r/r1) and  circ(r/r2) as pupil functions
% Adapted from "Introduction to Modern Digital Holography"
% by T.-C. Poon & J.-P. Liu
% Cambridge University Press (2014), Table 1.7 .
```

```
clear all;close all;
A=imread('front.jpg');            % read image file 512x512 8-bit
A=double(A);
A=A/255;
SP=fftshift(fft2(fftshift(A)));
D=abs(SP);
D=D(129:384,129:384);
figure;imshow(A);
%title('Original image')
figure;imshow(30.*mat2gray(D)); % spectrum
title('Original spectrum')

c=1:512;
r=1:512;
[C, R]=meshgrid(c, r);
CI=((R-257).^2+(C-257).^2);
pup=zeros(512,512);
% produce pupil circ(r/r1)
for a=1:512;
    for b=1:512;
        if CI(a,b)>=100^2;  %pupil size
            pup(a,b)=0;
        else
            pup(a,b)=1;
        end
    end
end
h=ifft2(fftshift(pup));

OTF1=fftshift(fft2(h.*conj(h)));
OTF1=OTF1/max(max(abs(OTF1)));
G=abs(OTF1.*SP);
G=G(129:384,129:384);
figure;imshow(30.*mat2gray(G));
%title('Spectrum filtered by circ(r/r1)')

I=abs(fftshift(ifft2(fftshift(OTF1.*SP))));
figure;imshow(mat2gray(I));
%title('Image filterd by circ(r/r1)')

c=1:512;
r=1:512;
[C, R]=meshgrid(c, r);
CI=((R-257).^2+(C-257).^2);
pup=zeros(512,512);
% produce pupil circ(r/r2)
for a=1:512;
    for b=1:512;
        if CI(a,b)>=105^2;  %pupil size
            pup(a,b)=0;
        else
            pup(a,b)=1;
```

```
        end
     end
  end
  h=ifft2(fftshift(pup));

  OTF2=fftshift(fft2(h.*conj(h)));
  OTF2=OTF2/max(max(abs(OTF2)));
  G=abs(OTF2.*SP);
  G=G(129:384,129:384);
  figure;imshow(30.*mat2gray(G));
  %title('Spectrum filtered by circ(r/r2)')

  I=abs(fftshift(ifft2(fftshift(OTF2.*SP))));
  figure;imshow(mat2gray(I));
  %title('Image filterd by circ(r/r2)')
  %-------------------
  TOTF=OTF1-OTF2;
  TOTF=TOTF/max(max(abs(TOTF)));
  G=abs(TOTF.*SP);
  G=G(129:384,129:384);
  figure;imshow(10.*mat2gray(G));
  title('Spectrum filtered by TOTF')

  I=abs(fftshift(ifft2(fftshift(TOTF.*SP))));
  figure;imshow(mat2gray(I));
```

Problems

4.1 In the coherent image processing system shown in Figure 4.2-1, the coherent point spread function is $h_c(x,y) = \mathcal{F}\{p(x, y)\}_{k_x = \frac{k_0 x}{f}, \, k_y = \frac{k_0 y}{f}}$. Show that the coherent transfer function

is given by $H_c(k_x, k_y) = p\left(\dfrac{-fk_x}{k_0}, \dfrac{-fk_y}{k_0}\right)$.

4.2 With reference to the standard 4-f optical image processing system shown in Figure 4.2-1, design $p(x,y)$ to implement the following mathematical operations:

(a) $\dfrac{\partial t(x,y)}{\partial x}, \, \dfrac{\partial t(x,y)}{\partial y}$

(b) $\dfrac{\partial t(x,y)}{\partial x} + \dfrac{\partial t(x,y)}{\partial y}$

(c) $\dfrac{\partial^2 t(x,y)}{\partial x^2} + \dfrac{\partial^2 t(x,y)}{\partial y^2}$

(d) $\dfrac{\partial^2 t(x,y)}{\partial x \partial y}$.

4.3 For $t(x,y) = \text{rect}\left(\dfrac{x}{a}, \dfrac{y}{a}\right)$ in a 4-f coherent optical image processing system, write MATLAB codes to implement mathematical operations in P.4.2. Display the intensity of the output.

4.4 The two-dimensional *Hilbert transform* of $t(x,y)$ is defined as follows:

$$\mathcal{H}\{t(x,y)\} = \frac{1}{\pi^2} \iint\limits_{-\infty}^{\infty} \frac{t(x',y')}{(x-x')(y-y')} \, dx' dy'.$$

(a) For a standard coherent image processing system shown in Figure 4.2-1, find the pupil function that is able to implement the 2-D Hilbert transform.

Hint: The 1-D Fourier transform of $\dfrac{1}{\pi x}$ is $\mathcal{F}\left\{\dfrac{1}{\pi x}\right\} = j\,\text{sgn}(k_x)$, where $\text{sgn}(x)$ is the *signum function* defined by $\text{sgn}(x) = \begin{cases} 1, & x > 0 \\ -1, & x < 0. \end{cases}$

(b) Write a MATLAB code to implement the Hilbert transform of $t(x,y) = \text{rect}\left(\dfrac{x}{a}, \dfrac{y}{a}\right)$. Display the intensity of the output.

4.5 Suppose we have a pure phase object $t(x,y) = e^{j\phi(x,y)}$, where $\phi(x,y) \ll 1$, such as biological cells, in the standard 4-f optical image processing system shown in Figure 4.2-1.

(a) For $p(x,y) = 1$, find the intensity distribution $\left|\psi_{p,\text{out}}(x,y)\right|^2$ on the output plane.

(b) For $p(x, y) = p\left(r = \sqrt{x^2 + y^2}\right) = \begin{cases} ja, & r = \varepsilon \approx 0 \\ 1, & \text{otherwise}, \end{cases}$

where a is a real and positive constant, show that the intensity distribution on the output plane is given by

$$\left|\psi_{p,\text{out}}(x,y)\right|^2 \approx a^2 + 2\,\phi(-x,-y).$$

This is the basis of the *phase contrast microscope*, invented by Frits Zernike in 1935. To implement the phase shift of $\pi/2$, we can put a glass plate of a correct thickness at the origin of the pupil plane, and in this case, $a = 1$.

4.6 With reference to the 4-f optical image processing system shown in Figure 4.2-1, find and plot the coherent transfer function. Also find the corresponding coherent point spread function for the pupil given below:

(a) $p(x, y) = \text{rect}\left(\dfrac{x - x_c}{x_0}\right) + \text{rect}\left(\dfrac{x + x_c}{x_0}\right)$, a double slit.

(b) $p(x, y) = \text{rect}\left(\dfrac{x - x_c}{x_0}\right)$, a single slit offset from the optical axis.

4.7 In Figure P.4.7, we have an input $t(x,y)$ illuminated by a plane wave of unit amplitude.

(a) Show that the complex amplitude on the output plane is given by

$$\psi_{p,\text{out}}(x,y)=t\left(\frac{x}{M},\frac{y}{M}\right)*h_c(x,y),$$

where the coherent point spread function $h_c(x,y)=\mathcal{F}\{p(x,\,y)\}_{k_x=\frac{k_0 x}{f_2},\,k_y=\frac{k_0 y}{f_2}}$, and $M=-f_2/f_1$.

(b) Show that the coherent transfer function is given by $H_c(k_x,k_y)=p(x,\,y)\big|_{x=\frac{-f_2 k_x}{k_0},\,y=\frac{-f_2 k_y}{k_0}}$.

Input plane
$t(x,y)$

$p(x,y)$

Output plane

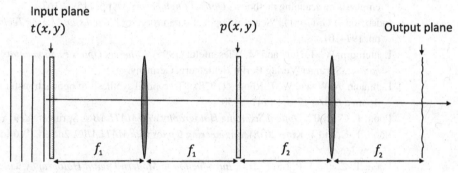

Figure P.4.7

4.8 Referring to the 4-f optical image processing system shown in Figure 4.2-1, find the complex field on the output plane for

(a) $p(x,\,y)=T_2\left(\dfrac{k_0 x}{f},\dfrac{k_0 y}{f}\right),$

(b) $p(x,\,y)=T_2^*\left(-\dfrac{k_0 x}{f},-\dfrac{k_0 y}{f}\right),$

where $\mathcal{F}\{t_2(x,y)\}=T_2(k_x,k_y)$. What type of mathematical operations does this system represent?

4.9 The coherent transfer function is given by

$$H_c(k_x,k_y)=\text{rect}\left(\frac{k_x}{K}\right),$$

show that the OTF is

$$OTF(k_x,k_y)=K\,\text{tri}\left(\frac{k_x}{K}\right).$$

4.10　The optical transfer function of incoherent optical systems is given by

$$OTF\left(k_x,k_y\right) = \mathcal{F}\left\{\left|h_c\left(x,y\right)\right|^2\right\},$$

where $h_c\left(x,y\right) = \mathcal{F}^{-1}\left\{H_c\left(k_x,k_y\right)\right\}$ with $H_c\left(k_x,k_y\right)$ being the coherent transfer function. Show that

$$OTF\left(k_x,k_y\right) = \iint\limits_{-\infty}^{\infty} H_c^*\left(k_x',k_y'\right) H_c\left(k_x'+k_x,k_y'+k_y\right) dk_x' dk_y'.$$

Bibliography

Francon, M. (1974). *Holography*. Academic Press, New York and London.

Indebetouw, G. and T.-C. Poon (1992). "Novel approaches of incoherent image processing with emphasis on scanning methods," *Optical Engineering* 31, pp. 2159–2167.

Indebetouw, G. (1981). "Scanning optical data processor," *Optics and Laser Technology* 13, pp. 197–201.

Lauterborn, W., T. Hurz and M. Wiesenfeldt (1995). *Coherent Optics Fundamentals and Applications*. Springer-Verlag Berlin Heidelberg, Germany.

Lohmann, A. W. and W. T. Rhodes (1978). "Two-pupil synthesis of optical transfer functions," *Applied Optics* 17, pp. 1141–1150.

Poon, T.-C. (2007) *Optical Scanning Holography with MATLAB®*. Springer, New York.

Poon, T.-C. and T. Kim (2018). *Engineering Optics with MATLAB®*, 2nd ed., World Scientific, New Jersey.

Poon, T.-C. and J.-P. Liu (2014). *Introduction to Modern Digital Holography with MATLAB*. Cambridge University Press, Cambridge, United Kingdom.

Zhang, Y., T.-C. Poon, P. W. M. Tsang, R. Wang, and L. Wang (2019). "Review on feature extraction for 3-D incoherent image processing using optical scanning holography," *IEEE Transactions on Industrial Informatics* 15 (11), pp. 6146–6154.

5 Principles of Coherent Holography

In photography, the intensity of a 3-D object is imaged and recorded in a 2-D recording medium such as a photographic film or a charge-coupled device (CCD) camera, which responds only to light intensity. Since there is no interference during recording, the phase information of the wave field is not preserved. The loss of the phase information of the light field from the object destroys the 3-D characteristics of the recorded scene, and therefore *parallax* and depth information of the 3-D object cannot be observed by viewing a photograph. Parallax is a difference in the apparent position of an object viewed from different angles. *Holography*, invented by Dennis Gabor in 1948, is a technique in which the amplitude and phase information of the light field of the object are recorded through interference. The phase is coded in the interference pattern. The recorded interference pattern is a *hologram*. It is reminiscent of Young's interference experiment in which the position of the interference fringes depends on the phase difference between the two sources [see Eqs. (4.1-3) and (4.1-4), where the first peak of the interference pattern x_p is a function of the phase difference of the two point sources, $\theta_1 - \theta_2$]. Once the hologram of a 3-D object has been recorded, we can reconstruct the 3-D image of the object by simply illuminating the hologram (if the interference pattern is recorded on a film) or through digital reconstruction (if the interference pattern is captured by a CCD). We record the complex amplitude of the 3-D object in *coherent holography*, whereas in *incoherent holography*, we record the intensity distribution of the 3-D object. In this chapter, we discuss the principles of coherent holography. Incoherent holography will be discussed in Chapter 6.

5.1 Fresnel Zone Plate as a Point-Source Hologram

In this section, we discuss the holographic recording of a point source. Once we understand how a point source behaves in an optical system, we will get much insight into how to deal with a 3-D object as any object is considered as a collection of many point sources.

5.1.1 On-axis Recording

Holography is a two-step process: recording and reconstruction. We will discuss each process separately.

Recording

Figure 5.1-1a illustrates holographic recording of a point source. A collimated laser is split into two plane waves and then recombined by using two mirrors (M) and two beam splitters (BS). One of the plane waves is used to illuminate a pinhole, which acts as our point source object. The point source gives a diverging spherical wave toward the recording medium. This spherical wave is known as the *object wave* in holography. The plane wave that directly illuminates the recording medium is known as the *reference wave*. Figure 5.1-1b shows the spherical object wave fronts and the plane reference wave fronts toward the recording medium. Since both of the wave fronts propagate along the same direction, that is, along the z-direction, this recording condition is called *on-axis recording*, giving what is known as *on-axis holography*.

Let ψ_0 and ψ_r represent the complex fields of the object wave and the reference wave on the plane of the recording medium, respectively. The interference of the two waves on the recording plane is given by $\psi_0 + \psi_r$. Since the recording medium only records intensity, what is recorded is given by the intensity $I = |\psi_0 + \psi_r|^2$, provided the reference wave and the object wave are mutually coherent over the recording medium. The coherence of the light waves is guaranteed by the use of a laser with the difference between the two light paths being less than the coherence length of the laser. This kind of recording is known as *holographic recording*. In photographic recording, the reference wave does not exist and hence only the object wave is recorded.

If the intensity is recorded by a photographic film, what is developed is a hologram given by a transparency function

$$t(x,y) \propto I = |\psi_0 + \psi_r|^2. \tag{5.1-1}$$

However, if a CCD is used for recording, the recorded digital file is $t(x,y)$. Let us model the pinhole as a delta function point source and assume that the point source is located at the origin ($z = 0$) with a distance of z_0 away from the recording medium. According to Fresnel diffraction [see Eq. (3.4-17)], on the recording medium the object wave is

$$\psi_0(x,y;z_0) = \delta(x,y) * h(x,y;z_0) = \delta(x,y) * e^{-jk_0 z_0} \frac{jk_0}{2\pi z_0} e^{-j\frac{k_0}{2z_0}(x^2+y^2)}$$

$$= e^{-jk_0 z_0} \frac{jk_0}{2\pi z_0} e^{-j\frac{k_0}{2z_0}(x^2+y^2)}. \tag{5.1-2}$$

For the reference plane wave, since it comes from the same laser source and has the same optical path length to the recording medium, it has the same initial phase as the point object at a distance of z_0 away from the recording medium. Therefore, its complex field on the recording medium is

$$\psi_r = ae^{-jk_0 z_0}, \tag{5.1-3}$$

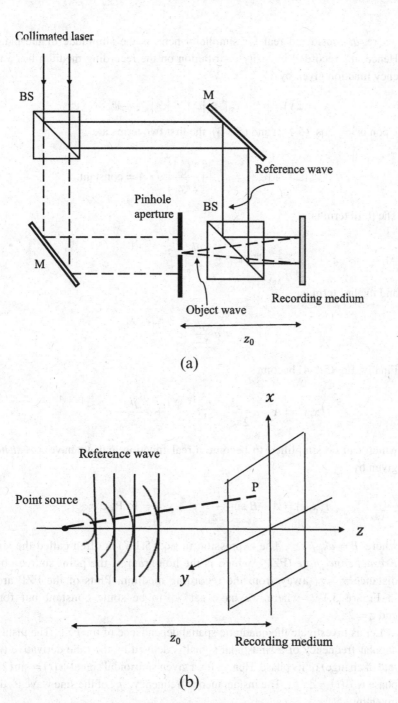

Figure 5.1-1 Holographic recording of a point source object: (a) experimental setup and (b) illustration of wave fronts of plane reference wave and spherical object wave

where a, considered real for simplicity here, is the amplitude of the plane wave. Hence, the recorded intensity distribution on the recording medium has a transparency function given by

$$t(x,y) = |\psi_r + \psi_0|^2 = |\psi_r|^2 + |\psi_0|^2 + \psi_r\psi_0^* + \psi_r^*\psi_0. \qquad (5.1\text{-}4)$$

Upon using Eqs. (5.1-2) and (5.1-3), the first two terms are

$$|\psi_r|^2 + |\psi_0|^2 = a^2 + \left(\frac{k_0}{2\pi z_0}\right)^2 = A = \text{constant}. \qquad (5.1\text{-}5a)$$

The third term is

$$\psi_r\psi_0^* = a\frac{-jk_0}{2\pi z_0}e^{j\frac{k_0}{2z_0}(x^2+y^2)}, \qquad (5.1\text{-}5b)$$

and the last term is

$$\psi_r^*\psi_0 = a\frac{jk_0}{2\pi z_0}e^{-j\frac{k_0}{2z_0}(x^2+y^2)}. \qquad (5.1\text{-}5c)$$

Finally, Eq. (5.1-4) becomes

$$t(x,y) = A + a\frac{-jk_0}{2\pi z_0}e^{j\frac{k_0}{2z_0}(x^2+y^2)} + a\frac{jk_0}{2\pi z_0}e^{-j\frac{k_0}{2z_0}(x^2+y^2)}, \qquad (5.1\text{-}6)$$

which can be simplified to become a real function and we have a *real hologram* given by

$$t(x,y) = A + B\sin\left[\frac{k_0}{2z_0}(x^2+y^2)\right] = \text{FZP}(x,y;z_0), \qquad (5.1\text{-}7)$$

where $B = ak_0/\pi z_0$. The expression in Eq. (5.1-7) is often called the sinusoidal *Fresnel zone plate* (FZP), which is the hologram of the point source object at a distance $z = z_0$ away from the recording medium. Plots of the FZP are shown in Figure 5.1-2, where we have set k_0 to be some constant but for $z = z_0$ and $z = 2z_0$.

Let us investigate the quadratic spatial dependence of the FZP. The instantaneous angular frequency of a sinusoidal signal is defined by the time derivative (i.e., time rate of change) of its phase. Hence, for a given sinusoidal signal $x(t) = \sin(2\pi f_0 t)$, its phase is $\theta(t) = 2\pi f_0 t$. The instantaneous frequency, f_i, of the sine wave is, therefore, given by

$$f_i = \frac{1}{2\pi}\frac{d\theta(t)}{dt} = \frac{1}{2\pi}\frac{d(2\pi f_0 t)}{dt} = f_0. \qquad (5.1\text{-}8)$$

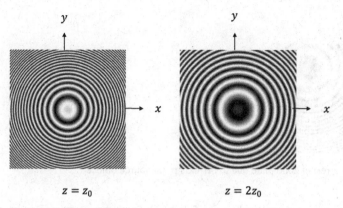

$z = z_0$ $z = 2z_0$

Figure 5.1-2 Plots of Fresnel zone plates

For the FZP, the spatial rate of change of the phase, say, along the x-direction is called the *local frequency* and is given by

$$f_{\text{local}} = \frac{1}{2\pi}\frac{d}{dx}\left(\frac{k_0}{2z_0}x^2\right) = \frac{k_0 x}{2\pi z_0}. \tag{5.1-9}$$

Note that the local fringe frequency increases linearly with the spatial coordinate, x. Note that, for $x < 0$, we take the absolute value of the quantity. In other words, the further we are away from the zone's center, the higher the local spatial frequency, which is obvious from Figure 5.1-2. Note also from the figure, when we double the z value, say from $z = z_0$ to $z = 2z_0$, the local frequency has become less as evident from Eq. (5.1-9) as well. Hence, the local frequency carries the information on depth, that is, z, and from the local frequency, therefore, we can deduce how far the object point source is away from the recording medium. In other words, the depth information is coded by the "fringe density" in the FZP.

For a point source $\delta(x, y)$ which is z_0 away from the recording medium, we have a hologram given by $\mathrm{FZP}(x, y; z_0) = A + B\sin\left[\frac{k_0}{2z_0}(x^2 + y^2)\right]$. It is not hard to see that for an off-axis point source $\delta(x - x_0, y - y_0)$ at a distance of z_0 from the recording medium, its hologram is given by the *off-axis FZP*:

$$\mathrm{FZP}(x - x_0, y - y_0; z_0) = A + B\sin\left[\frac{k_0}{2z_0}\left[(x - x_0)^2 + (y - y_0)^2\right]\right]. \tag{5.1-10}$$

Figure 5.1-3 shows the off-axis FZP. Note that the center of the zone specifies the location, x_0 and y_0, of the point source object, and the depth information, that is, z_0 is coded by the "fringe density." Hence, we see that the off-axis FZP completely contains the complete information of the 3-D coordinates of the delta function, that is, (x_0, y_0, z_0).

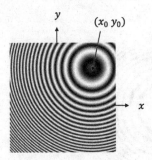

Figure 5.1-3 Fresnel zone plate due to off-axis point source object

The images produced in Figures 5.1-2 and 5.1-3 are generated using the m-file shown below.

```
%Fresnel_zone_plate
%Adapted from "Optical Scanning Holography with MATLAB®,"
%by T.-C. Poon, Springer 2007, p.55.
%display  function  is  1+sin(sigma*((x-x0)^2+(y-y0)^2)).  All
scales are arbitrary.
%sigma=pi/(wavelength*z)
%
clear;

z0=4 %z0 proportional to the distance from the point object to
recording medium
x0=7 % x0=y0, center of the FZP

ROWS=256;
COLS=256;
colormap(gray(256))
sigma=1/z0;
y0=-x0;
y=-12.8;
for r=1:COLS,
 x=-12.8;
    for c=1:ROWS,      %compute Fresnel zone plate
        fFZP(r,c)=exp(j*sigma*(x-x0)*(x-x0)+j*sigma*
        (y-y0)*(y-y0));
        x=x+.1;
        end
  y=y+.1;
end
%normalization
max1=max(fFZP);
max2=max(max1);
scale=1.0/max2;
```

```
fFZP=fFZP.*scale;
image(127*(1+imag(fFZP)));
axis square on
axis off
```

Reconstruction

To reconstruct the original point source object from the hologram $t(x,y)$, we need a *reconstruction* or *decoding* process. This is done simply by illuminating the hologram with the so-called *reconstruction wave*, ψ_{rec}, in holography. Therefore, the complex field immediately after the hologram is $\psi_{rec}t(x,y)$. To find the field after the hologram at some distance z, we just perform Fresnel diffraction as follows [see Eq. (3.4-17)]:

$$\psi_{rec}t(x,y)*h(x,y;z). \tag{5.1-11}$$

Let us take our hologram to be an on-axis FZP as an example with the reconstruction wave being a plane wave normally incident on it. Therefore, we take $\psi_{rec} = 1$ for simplicity. Instead of explicitly calculating Eq. (5.1-11) [which can be done, of course], let us take another approach to find what happens when the hologram is illuminated by a plane wave. The on-axis point-source hologram is

$$t(x,y) = \text{FZP}(x,y;z_0) = |\psi_r + \psi_0|^2 = |\psi_r|^2 + |\psi_0|^2 + \psi_r\psi_0^* + \psi_r^*\psi_0$$

$$= A + B\sin\left[\frac{k_0}{2z_0}(x^2+y^2)\right],$$

where we have calculated that [see Eq. (5.1-5)]

$$|\psi_r|^2 + |\psi_0|^2 = a^2 + \left(\frac{k_0}{2\pi z_0}\right)^2 = A = \text{constant},$$

$$\psi_r\psi_0^* = a\frac{-jk_0}{2\pi z_0}e^{j\frac{k_0}{2z_0}(x^2+y^2)}, \text{ and } \psi_r^*\psi_0 = a\frac{jk_0}{2\pi z_0}e^{-j\frac{k_0}{2z_0}(x^2+y^2)}.$$

These three terms correspond to the effect of a flat glass of negligible thickness, a positive lens and a negative lens, respectively. The first constant term simply means that if a plane wave is incident on it, the plane wave travels straight without any deviation and that gives the so-called *zeroth-order beam*.

Recall that the transparency function of an idea lens of focal length f is $e^{j\frac{k_0}{2f}(x^2+y^2)}$ [see Eq. (3.4-24)]. Hence, the second term $\psi_r\psi_0^*$ is proportional to a transparency function of an idea lens of focal length $f = z_0$. Similarly, the third term represents a negative lens of focal length $f = -z_0$. As a whole, we can consider the FZP as a combination of these three pieces of optical elements as illustrated in Figure 5.1-4. If a plane wave is illuminated onto the FZP, we will have three beams exiting from the hologram. The zeroth-order beam just goes straight without deviation. The focused beam at $z = z_0$ generates a real image, and the virtual image is observed z_0 behind

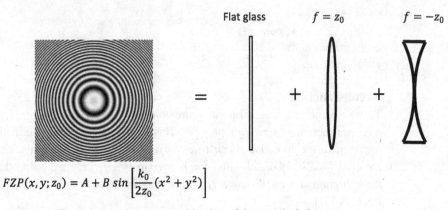

$$FZP(x, y; z_0) = A + B \sin\left[\frac{k_0}{2z_0}(x^2 + y^2)\right]$$

Figure 5.1-4 Fresnel zone plate as a combination of three optical elements

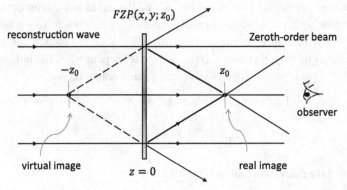

Figure 5.1-5 Ray diagram of reconstruction of a FZP upon illumination by a plane reconstruction wave

the hologram if an observer looks into the hologram as we reconstruct the original point source object at the exact location z_0. The situation is shown in Figure 5.1-5. The reconstructed real image is known as the *twin image*. Besides the zeroth-order beam, the twin image will also create unwanted light affecting the viewing of the virtual image. In the next subsection, we discuss a way to avoid the zeroth-order beam and the twin image disturbances.

Once we understand how a point object is holographically recorded and reconstructed, it is easy to deal with a 3-D object. Figure 5.1-6 shows the holographic recording and reconstruction of a three-point object. The virtual image appears at the correct 3-D location as the original image, and the observer will see a reconstructed image with the same perspective as the original object and the image is called the *orthoscopic image*. The real image (the twin image) is the mirror image of the original object, with the axis of reflection on the plane of the hologram. Such an image is called the *pseudoscopic image*. Since the pseudoscopic image cannot provide the correct parallax to the observer, it is not suitable for 3-D display. It is worthwhile to note that the term

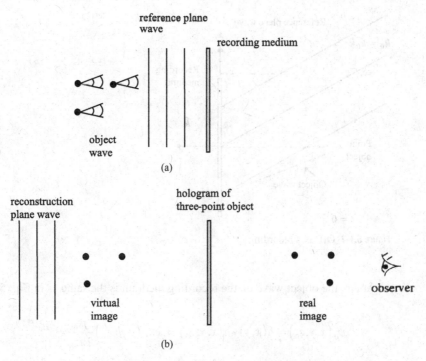

Figure 5.1-6 Three-point object: (a) recording and (b) reconstruction

ψ_0 in the hologram [see Eq. (5.1-4)] gives rise to a virtual image reconstruction, while its complex conjugate ψ_0^* in the hologram corresponds to a real image reconstruction.

5.1.2   Off-axis Recording

We have seen from the last section that the twin image and the zeroth-order beam create an annoying effect when we view the reconstructed virtual image along the viewing direction. In holography, this is infamously known as the *twin image problem*. Much research has been done to solve this problem. *Off-axis holography* is an original method devised by Leith and Upatnieks [1964] to separate the twin image and the zeroth-order beam from the virtual image upon reconstruction. The idea of *off-axis holography* is to set up an off-axis recording geometry.

Recording

With reference to Figure 5.1-1, off-axis recording can be done by simply, for example, rotating the beam splitter between the point source object and the recording medium so that the reference plane wave is incident on the recording medium at an angle. The situation, similar to Figure 5.1-1b in that we show the wave fronts of the reference plane wave and the spherical wave, is shown in Figure 5.1-7. Note that the reference wave and the object wave do not propagate in the same direction along z, and hence it is called off-axis holography. The angle θ is called the *recording angle*.

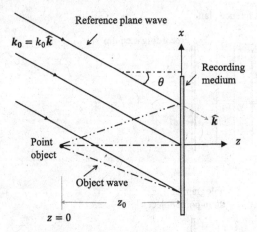

Figure 5.1-7 Off-axis recording

Again, the object wave on the recording medium is the same as in Eq. (5.1-2):

$$\psi_0(x,y;z_0) = \delta(x,y) * h(x,y;z_0) = \exp(-jk_0 z_0)\frac{jk_0}{2\pi z_0}e^{-j\frac{k_0}{2z_0}(x^2+y^2)}$$

$$\propto \frac{jk_0}{2\pi z_0}e^{-j\frac{k_0}{2z_0}(x^2+y^2)}, \tag{5.1-12}$$

where we have neglected the constant phase shift in the last step of the equation for simplicity. As for the reference plane wave, according to Eq. (3.3-9) for a general propagating direction, we have

$$\psi(x,y,z;t) = e^{j(\omega_0 t - k_0 \cdot R)} = e^{j(\omega_0 t - k_{0x}x - k_{0y}y - k_{0z}z)}.$$

Under our current situation shown in Figure 5.1-7, $k_{0x} = k_0\hat{k}\cdot\hat{x} = k_0\cos(\theta+90°) = -\sin\theta$, $k_{0y} = 0$, and $k_{0z} = k_0\hat{k}\cdot\hat{z} = k_0\cos\theta$, where $k_0 = k_0\hat{k}$. The complex amplitude of the reference plane wave with amplitude a on the recording medium at $z = z_0$ is

$$\psi_r = ae^{jk_0\sin\theta x - jk_0\cos\theta z_0} \propto ae^{jk_0\sin\theta x}. \tag{5.1-13}$$

Note that the off-axis reference plane wave has a linear phase shift variation along x on the hologram plane. The recorded intensity distribution on the recording medium gives a transparency function given by

$$t(x,y) = |\psi_r + \psi_0|^2 = |\psi_r|^2 + |\psi_0|^2 + \psi_r\psi_0^* + \psi_r^*\psi_0. \tag{5.1-14}$$

Upon using Eqs. (5.1-12) and (5.1-13), the first two terms are

$$|\psi_r|^2 + |\psi_0|^2 = a^2 + \left(\frac{k_0}{2\pi z_0}\right)^2 = A = \text{constant}. \tag{5.1-15a}$$

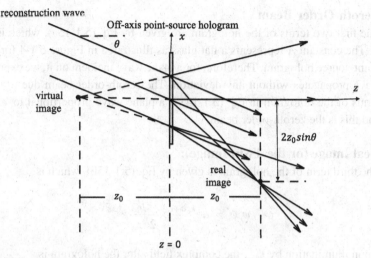

Figure 5.1-8 Holographic reconstruction of off-axis point-source hologram

The third term is

$$\psi_r \psi_0^* = a e^{jk_0 \sin\theta x} \times \frac{-jk_0}{2\pi z_0} e^{j\frac{k_0}{2z_0}(x^2+y^2)}, \tag{5.1-15b}$$

and the last term is

$$\psi_r^* \psi_0 = a e^{-jk_0 \sin\theta x} \times \frac{jk_0}{2\pi z_0} e^{-j\frac{k_0}{2z_0}(x^2+y^2)}. \tag{5.1-15c}$$

Finally, Eq. (5.1-14) becomes

$$t(x,y) = A + B \sin\left[\frac{k_0}{2z_0}(x^2+y^2) + k_0 \sin\theta x\right], \tag{5.1-16}$$

where $B = \frac{ak_0}{\pi z_0}$. $t(x,y)$ given by the above equation is called an off-axis point-source hologram.

Reconstruction

To reconstruct the hologram, we can illuminate the hologram with a reconstruction wave identical to the reference wave. We then have the reconstruction wave $\psi_{rec} = \psi_r = a e^{jk_0 \sin\theta x}$ [see Eq. 5.1-13]. The situation is shown in Figure 5.1-8. The reconstruction, according to Fresnel diffraction, is given by

$$\psi_{rec} t(x,y) * h(x,y;z) = a e^{jk_0 \sin\theta x} t(x,y) * h(x,y;z). \tag{5.1-17}$$

From Eq. (5.1-14), the hologram $t(x,y)$ contains three distinct terms and upon illumination by the reconstruction wave, these three terms create three different waves as follows:

Zeroth-Order Beam

The first two terms of the hologram are given by Eq. (5.1-15a), which is a constant A. The constant A represents a flat glass as illustrated in Figure 5.1-4 for the on-axis point-source hologram. Therefore, for a plane wave incident on it, we expect the plane wave propagates without any deviation. The zeroth-order beam due to the first two terms of the hologram in Eq. (5.1-14) is a plane wave proportional to $e^{jk_0\sin\theta x - jk_0\cos\theta z}$, and this is the zeroth-order beam.

Real Image (or the Twin Image):

The third term of the hologram is given by Eq. (5.1-15b), which is

$$\psi_r \psi_0^* = ae^{jk_0\sin\theta x} \times \frac{-jk_0}{2\pi z_0} e^{j\frac{k_0}{2z_0}(x^2+y^2)}.$$

Upon illumination by ψ_{rec}, the complex field after the hologram is

$$\left[\psi_{rec} \times \psi_r \psi_0^*\right] * h(x,y;z) = \left[\psi_r \times \psi_r \psi_0^*\right] * h(x,y;z) \tag{5.1-18}$$

as ψ_{rec} is identical to $\psi_r = ae^{jk_0\sin\theta x}$ in our situation shown in Figure 5.1-8. Again, instead of explicitly calculating Eq. (5.1-17), we recognize that

$$\psi_r \times \psi_r \psi_0^* \propto e^{jk_0\sin\theta x} \times e^{jk_0\sin\theta x} e^{j\frac{k_0}{2z_0}(x^2+y^2)} = e^{2jk_0\sin\theta x} e^{j\frac{k_0}{2z_0}(x^2+y^2)},$$

which corresponds to a transparency function of an idea lens of focal length z_0, illuminated by complex wave $e^{2jk_0\sin\theta x}$.

Let us recall Figure 3.4-5 and summarize its result in Figure 5.1-9, where $\psi_{fl}(x,y)$ represents the field distribution just in front of the lens. Using this result for $\psi_r \times \psi_r \psi_0^*$, $\psi_{fl} = e^{2jk_0\sin\theta x}$ because $e^{j\frac{k_0}{2z_0}(x^2+y^2)}$ serves as a transparency function of a lens. At a distance z_0 away from the hologram, we have

$$\psi_p(x,y;z=z_0) \propto e^{-j\frac{k_0}{2z_0}(x^2+y^2)} \mathcal{F}\{\psi_{fl}(x,y)\}\big|_{k_x=\frac{k_0 x}{z_0}, k_y=k_0\frac{y}{z_0}}$$

$$= e^{-j\frac{k_0}{2z_0}(x^2+y^2)} \mathcal{F}\{e^{2jk_0\sin\theta x}\}\big|_{k_x=\frac{k_0 x}{z_0}, k_y=\frac{k_0 y}{z_0}}$$

$$= e^{-j\frac{k_0}{2z_0}(x^2+y^2)} 4\pi^2\delta(k_x + 2k_0\sin\theta, k_y)\big|_{k_x=\frac{k_0 x}{z_0}, k_y=\frac{k_0 y}{z_0}}$$

$$\propto e^{-j\frac{k_0}{2f}(x^2+y^2)} \delta\left(\frac{k_0 x}{z_0} + 2k_0\sin\theta, \frac{k_0 y}{z_0}\right) \propto \delta(x + 2z_0\sin\theta, y), \tag{5.1-19}$$

where we have used Table 2.2 and employed the scaling property of the delta function [see Eq. (2.1-6)] to arrive the final result in Eq. (5.1-19). Equation (5.1-19) means that we have a real image formed $2z_0\sin\theta$ away from the z-axis [see Figure 5.1-8]. In summary, the term $\psi_r\psi_0^*$ in the hologram equation contains the

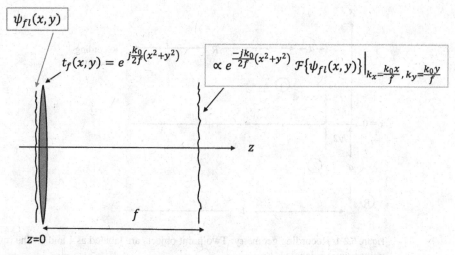

Figure 5.1-9 Fourier transformation of the complex field just in front of an ideal lens

complex conjugate of the original wave front ψ_0, and it gives rise to a real image upon reconstruction.

Virtual Image

The fourth term of the hologram given by Eq. (5.1-15c) is

$$\psi_r^* \psi_0 = ae^{-jk_0\sin\theta x}\, \frac{jk_0}{2\pi z_0}\, e^{-j\frac{k_0}{2z_0}\left(x^2+y^2\right)}.$$

Upon illumination by ψ_{rec}, the complex field after the hologram is

$$\psi_{\text{rec}} \times \psi_r^* \psi_0 * h\left(x,y;z\right) = \psi_r \times \psi_r^* \psi_0 * h\left(x,y;z\right). \qquad (5.1\text{-}20)$$

Again, ψ_{rec} is identical to $\psi_r = ae^{jk_0\sin\theta x}$. As before, instead of explicitly evaluating the above equation, we recognize that

$$\psi_r \times \psi_r^* \psi_0 \propto e^{jk_0\sin\theta x} \times e^{-jk_0\sin\theta x}\, e^{-j\frac{k_0}{2z_0}\left(x^2+y^2\right)} = e^{-j\frac{k_0}{2z_0}\left(x^2+y^2\right)},$$

which corresponds to a transparency function of an idea lens of focal length $-z_0$. So this term in the hologram represents a negative lens, which gives a virtual image on the z-axis (the location of the original point source. See Figure 5.1-8). In summary, the term $\psi_r^* \psi_0$ in the hologram equation contains the original wave front ψ_0, and it gives rise to a virtual image upon reconstruction.

5.2   Three-Dimensional Holographic Imaging

In this section, we study the lateral and longitudinal magnifications in holographic imaging through the recording and reconstruction of two point sources. We consider holographic recording and·reconstruction using spherical waves.

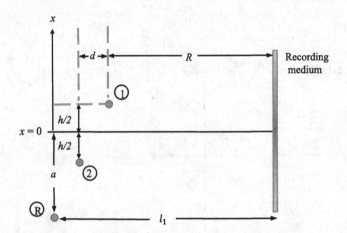

Figure 5.2-1 Recording geometry: Two point objects are labeled as 1 and 2. The reference point source is labeled R

5.2.1 Recording and Reconstruction

Recording

The geometry for recording is shown in Figure 5.2-1. The two point objects, labeled 1 and 2, and the reference wave, labeled R, generate spherical waves that, on the plane of the recording medium, contribute to complex fields, ψ_{p1}, ψ_{p2}, and ψ_{pR}, respectively, given by

$$\psi_{p1}(x,y) = \delta\left(x-\frac{h}{2},y\right) * h(x,y;R) = e^{-jk_0 R}\frac{jk_0}{2\pi R}e^{-j\frac{k_0}{2R}\left[(x-h/2)^2+y^2\right]}$$

$$\propto e^{-j\frac{k_0}{2R}\left[(x-h/2)^2+y^2\right]}, \tag{5.2-1}$$

$$\psi_{p2}(x,y) = \delta\left(x+\frac{h}{2},y\right) * h(x,y;R+d) = e^{-jk_0(R+d)}\frac{jk_0}{2\pi(R+d)}e^{-j\frac{k_0}{2(R+d)}\left[(x+h/2)^2+y^2\right]}$$

$$\propto e^{-j\frac{k_0}{2(R+d)}\left[(x+h/2)^2+y^2\right]} \tag{5.2-2}$$

and

$$\psi_{pR}(x,y) = \delta(x+a,y) * h(x,y;l_1) = e^{-jk_0 l_1}\frac{jk_0}{2\pi l_1}e^{-j\frac{k_0}{2l_1}\left[(x+a)^2+y^2\right]}$$

$$\propto e^{-j\frac{k_0}{2l_1}\left[(x+a)^2+y^2\right]}. \tag{5.2-3}$$

These spherical waves interfere on the recording medium to yield a hologram given by

$$t(x,y) = |\psi_{p1}(x,y)+\psi_{p2}(x,y)+\psi_{pR}(x,y)|^2$$

$$= \left[\psi_{p1}(x,y)+\psi_{p2}(x,y)+\psi_{pR}(x,y)\right]\left[\psi_{p1}^*(x,y)+\psi_{p2}^*(x,y)+\psi_{pR}^*(x,y)\right]. \tag{5.2-4}$$

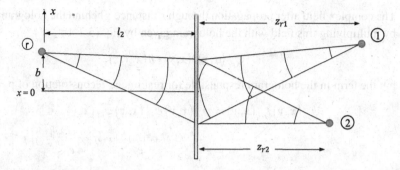

Figure 5.2-2 Reconstruction of two points using a spherical wave

Reconstruction

There are nine terms in the holograms and we shall, on the basis of our previous knowledge regarding point source object recording and reconstruction, pick out some relevant terms responsible for image reconstruction. Again, the terms containing the complex conjugate of the object waves in the hologram, that is, $\psi_{pR}(x,y)\psi_{p1}^{*}(x,y)$ and $\psi_{pR}(x,y)\psi_{p2}^{*}(x,y)$ will give rise to real image reconstructions, and they are

$$t_{rel_1}(x,y) = \psi_{pR}(x,y)\psi_{p1}^{*}(x,y)$$

$$= e^{-j\frac{k_0}{2l_1}\left[(x+a)^2+y^2\right]} \times e^{j\frac{k_0}{2R}\left[(x-h/2)^2+y^2\right]}, \tag{5.2-5}$$

and

$$t_{rel_2}(x,y) = \psi_{pR}(x,y)\psi_{p2}^{*}(x,y)$$

$$= e^{-j\frac{k_0}{2l_1}\left[(x+a)^2+y^2\right]} \times e^{j\frac{k_0}{2(R+d)}\left[(x+h/2)^2+y^2\right]}. \tag{5.2-6}$$

The term containing the original object waves, that is, $\psi_{pR}^{*}(x,y)\psi_{p1}(x,y)$ and $\psi_{pR}^{*}(x,y)\psi_{p2}(x,y)$ will give rise to virtual images, and they are

$$t_{rel_3}(x,y) = \psi_{pR}^{*}(x,y)\psi_{p1}(x,y) = \left[t_{rel_1}(x,y)\right]^{*} \tag{5.2-7}$$

and

$$t_{rel_4}(x,y) = \psi_{pR}^{*}(x,y)\psi_{p2}(x,y) = \left[t_{rel_2}(x,y)\right]^{*}. \tag{5.2-8}$$

On reconstruction, we use a spherical wave from a point source labeled r as our reconstruction wave to illuminate the hologram, shown in Figure 5.2-2. We shall demonstrate how to find the real image locations as an example.

The reconstruction wave illuminating the hologram is, according to Fresnel diffraction,

$$\psi_{pr}(x,y) = \delta(x-b,y) * h(x,y;l_2) = e^{-jk_0 l_2}\frac{jk_0}{2\pi l_2}e^{-j\frac{k_0}{2l_2}\left[(x-b)^2+y^2\right]}$$

$$\propto e^{-j\frac{k_0}{2l_2}\left[(x-b)^2+y^2\right]}. \tag{5.2-9}$$

The complex field after propagation through a distance z behind the hologram is given by multiplying this field with the hologram given by Eq. (5.2-4):

$$\psi_{pr}(x,y)t(x,y)*h(x,y;z),$$

but the term in the hologram responsible for real image reconstruction of point 1 is

$$\psi_{pr}(x,y)t_{rel_1}(x,y)=\psi_{pr}(x,y)\psi_{pR}(x,y)\psi_{p1}^*(x,y)$$

$$=e^{-j\frac{k_0}{2l_2}\left[(x-b)^2+y^2\right]}e^{-j\frac{k_0}{2l_1}\left[(x+a)^2+y^2\right]}e^{j\frac{k_0}{2R}\left[\left(x-\frac{h}{2}\right)^2+y^2\right]}.$$

So we need to find

$$\psi_{p,rel_1}(x,y;z)=\psi_{pr}(x,y)\,t_{rel_1}(x,y)*\,h(x,y;z). \tag{5.2-10}$$

Again, instead of straightforward calculation of the above, we investigate the term $\psi_{pr}(x,y)t_{rel_1}(x,y)$ and try to find its transparency function in terms of an ideal lens.

By expanding the quadratic terms and grouping the coefficients of x^2+y^2 and x, $\psi_{pr}(x,y)t_{rel_1}(x,y)$ becomes

$$\psi_{pr}(x,y)t_{rel_1}(x,y)\propto e^{jk_0\left(-\frac{1}{2l_2}+\frac{1}{2R}-\frac{1}{2l_1}\right)(x^2+y^2)}\times e^{jk_0\left(\frac{b}{l_2}-\frac{h}{2R}-\frac{a}{l_1}\right)x}$$

$$=e^{j\frac{k_0}{2z_{r1}}(x^2+y^2)}e^{jk_0\left(\frac{b}{l_2}-\frac{h}{2R}-\frac{a}{l_1}\right)x},$$

where

$$z_{r1}=\left[\frac{1}{R}-\frac{1}{l_1}-\frac{1}{l_2}\right]^{-1}=\frac{Rl_1l_2}{l_1l_2-(l_1+l_2)R}, \tag{5.2-11}$$

which can be identified as a focal length of an ideal lens. So employing the result shown in Figure 5.1-9, for $\psi_{pr}(x,y)t_{rel_1}(x,y)$, $\psi_{fl}(x,y)=e^{jk_0\left(\frac{b}{l_2}-\frac{h}{2R}-\frac{a}{l_1}\right)x}$ because $e^{j\frac{k_0}{2z_{r1}}(x^2+y^2)}$ serves as a transparency function of a lens. At a distance z_{r1} away from the hologram, that is, equivalently from Eq. (5.2-10), we have

$$\psi_{p,rel_1}(x,y;z_{r1})=\psi_{pr}(x,y)t_{rel_1}(x,y)*\,h(x,y;z_{r1})$$

$$\propto e^{-j\frac{k_0}{2z_r}(x^2+y^2)}\mathcal{F}\{\psi_{fl}(x,y)\}\Big|_{k_x=\frac{k_0x}{z_r},k_y=\frac{k_0y}{z_r}}$$

$$=e^{-j\frac{k_0}{2z_r}(x^2+y^2)}\mathcal{F}\left\{e^{jk_0\left(\frac{b}{l_2}-\frac{h}{2R}-\frac{a}{l_1}\right)x}\right\}\Big|_{k_x=\frac{k_0x}{z_r},k_y=\frac{k_0y}{z_r}}$$

$$=e^{-j\frac{k_0}{2z_r}(x^2+y^2)}4\pi^2\delta\left(k_x+k_0\left(\frac{b}{l_2}-\frac{h}{2R}-\frac{a}{l_1}\right),k_y\right)\Big|_{k_x=\frac{k_0x}{z_r},k_y=\frac{k_0y}{z_r}}$$

$$\propto e^{-j\frac{k_0}{2z_r}(x^2+y^2)}\delta\left(\frac{k_0x}{z_r}+k_0\left(\frac{b}{l_2}-\frac{h}{2R}-\frac{a}{l_1}\right),\frac{k_0y}{z_r}\right)\propto\delta\left(x+z_{r1}\left(\frac{b}{l_2}-\frac{h}{2R}-\frac{a}{l_1}\right),y\right).$$

Note that this result corresponds to the convolution result of Eq. (5.2-10) for $z = z_{r1}$, and this is the real image of the reconstructed point object 1, shifted along the lateral location at

$$x = x_1 = -z_{r1}\left(\frac{b}{l_2} - \frac{h}{2R} - \frac{a}{l_1}\right), \qquad (5.2\text{-}12)$$

and located z_{r1} away from the hologram [see Figure 5.2-2].

While we have just found out that

$$\psi_{pr}(x,y)\, t_{rel_1}(x,y) = \psi_{pr}(x,y)\psi_{pR}(x,y)\psi_{p1}^*(x,y)$$

will give rise to the real image reconstruction of point 1. Similarly, the term

$$\psi_{pr}(x,y)t_{rel_2}(x,y) = \psi_{pr}(x,y)\psi_{pR}(x,y)\psi_{p2}^*(x,y)$$

in the hologram will give rise to the real image reconstruction of point 2. By carrying out the same procedures done for point 1, we state the following result:

$$\psi_{p,rel_2}(x,y;z_{r2}) = \psi_{pr}(x,y)t_{rel_2}(x,y) * h(x,y;z = z_{r2})$$

$$\propto \delta\left[x + z_{r2}\left(\frac{b}{l_2} + \frac{h}{2(R+d)} - \frac{a}{l_1}\right), y\right], \qquad (5.2\text{-}13)$$

where

$$z_{r2} = \left[\frac{1}{R+d} - \frac{1}{l_1} - \frac{1}{l_2}\right]^{-1} = \frac{(R+d)l_1 l_2}{l_1 l_2 - (l_1 + l_2)(R+d)}.$$

z_{r2} is the distance of the real image reconstruction of point 2 behind the hologram and the image point is located laterally at

$$x = x_2 = -z_{r2}\left(\frac{b}{l_2} + \frac{h}{2(R+d)} - \frac{a}{l_1}\right). \qquad (5.2\text{-}14)$$

Figure 5.2-2 shows the real image reconstruction of point 2.

5.2.2 Lateral and Longitudinal Holographic Magnifications

Lateral Reconstruction

From the results of last subsection, we can now evaluate *lateral holographic magnification* of holographic imaging. The lateral distance (along x) between the two real image points 1 and 2 is $x_1 - x_2$ with the original separation between the two points being h. Therefore, the lateral holographic magnification for the real image is

$$M_{\text{Lat}}^r = \frac{x_1 - x_2}{h}$$

$$= \frac{-z_{r1}\left(\frac{b}{l_2} - \frac{h}{2R} - \frac{a}{l_1}\right) + z_{r2}\left(\frac{b}{l_2} + \frac{h}{2(R+d)} - \frac{a}{l_1}\right)}{h}$$

$$\cong \frac{(z_{r2} - z_{r1})\left(\frac{b}{l_2} - \frac{a}{l_1}\right) + (z_{r2} + z_{r1})\frac{h}{2R}}{h} \qquad (5.2\text{-}15)$$

for $R \gg d$. The magnification is a function of h, which is undesirable. To make this magnification independent of the lateral separation, h, we set

$$\frac{b}{l_2} - \frac{a}{l_1} = 0,$$

or

$$\frac{b}{l_2} = \frac{a}{l_1}. \tag{5.2-16}$$

Then, Eq. (5.2-15) becomes

$$M_{\text{Lat}}^r = \frac{(z_{r2} + z_{r1})}{2R} \simeq \frac{l_1 l_2}{l_1 l_2 - (l_1 + l_2)R} \tag{5.2-17}$$

for $R \gg d$.

Lateral Reconstruction

The longitudinal distance (along z) between the two real point images is $z_{r2} - z_{r1}$, so that the *longitudinal holographic magnification* is defined as

$$M_{\text{Long}}^r = \frac{z_{r2} - z_{r1}}{d}. \tag{5.2-18}$$

Using Eqs. (5.2-11) and (5.2-13) and assuming $R \gg d$, the longitudinal magnification becomes

$$M_{\text{Long}}^r \simeq \frac{(l_1 l_2)^2}{(l_1 l_2 - Rl_1 - Rl_2)^2}. \tag{5.2-19}$$

By comparing Eqs. (5.2-17) and (5.2-19), we have the following relationship between the magnifications in 3-D imaging:

$$M_{\text{Long}}^r = \left(M_{\text{Lat}}^r \right)^2. \tag{5.2-20}$$

This result corresponds to the result obtained in 3-D imaging of a single lens [see 3-D imaging using a single lens in Chapter 1], which introduces distortion in volume imaging.

Example: Plane Wave Reference and Plane Wave Reconstruction

When both the reference and the reconstruction waves are plane waves, we let l_1 and l_2 approach infinity in Eqs. (5.2-17) and (5.2-18):

$$M_{\text{Lat}}^r = \lim_{l_1, l_2 \to \infty} \frac{(z_{r2} + z_{r1})}{2R} = \lim_{l_1, l_2 \to \infty} \frac{\left[\frac{1}{R+d} - \frac{1}{l_1} - \frac{1}{l_2} \right]^{-1} + \left[\frac{1}{R} - \frac{1}{l_1} - \frac{1}{l_2} \right]^{-1}}{2R} = \frac{2R+d}{2R} \approx 1,$$

for $R \gg d$, and

$$M_{\text{Long}}^r = \lim_{l_1, l_2 \to \infty} \frac{z_{r2} - z_{r1}}{d} = \lim_{l_1, l_2 \to \infty} \frac{\left[\frac{1}{R+d} - \frac{1}{l_1} - \frac{1}{l_2} \right]^{-1} - \left[\frac{1}{R} - \frac{1}{l_1} - \frac{1}{l_2} \right]^{-1}}{d} = 1.$$

Thus, we see that $M_{\text{Lat}}^r = M_{\text{Long}}^r$ and there is no volume distortion for 3-D imaging. Note that Eq. (5.2-16) is satisfied when $l_1 = l_2 \to \infty$.

Example: Lateral Magnification Using Spherical Wave Reference and Plane Wave Reconstruction

Let us start from the definition of the lateral holographic magnification in Eq. (5.2-15), which is

$$M_{\text{Lat}}^r = \frac{x_1 - x_2}{h} = \frac{-z_{r1}\left(\dfrac{b}{l_2} - \dfrac{h}{2R} - \dfrac{a}{l_1}\right) + z_{r2}\left(\dfrac{b}{l_2} + \dfrac{h}{2(R+d)} - \dfrac{a}{l_1}\right)}{h}.$$

We consider the recording and reconstruction point sources are on the z-axis, that is, $a = b = 0$. Also we take $d = 0$ as we are considering a planar image, M_{Lat}^r then becomes

$$M_{\text{Lat}}^r = \frac{z_{r2} + z_{r1}}{2R}$$

with $z_{r2} = z_{r1} = \left[\dfrac{1}{R} - \dfrac{1}{l_1} - \dfrac{1}{l_2}\right]^{-1}$. For plane wave reconstruction, $l_2 \to \infty$, we have a simple expression given by

$$M_{\text{Lat}}^r = \left[1 - \frac{R}{l_1}\right]^{-1}.$$

For example, taking $l_1 = 2R$, $M_{\text{Lat}}^r = 2$, we have a magnification of a factor of 2. For $l_1 = \left(\dfrac{1}{4}\right)R < R$, $M_{\text{Lat}}^r = -\dfrac{1}{3}$, we have demagnification. Note that if the recording reference wave is also a plane wave, that is, $l_1 \to \infty$, there is no magnification for using plane wave for recording and reconstruction as discussed in our earlier example.

5.3 Types of Holograms

In this section, we will introduce some common types of holograms and their basic principles.

5.3.1 Gabor Hologram and On-axis Hologram

Recording

Let us consider Gabor's original method, which is shown in Figure 5.3-1a. The object is assumed to be rather transparent but with a small variation of $\Delta(x, y)$. Hence, the total transparency of the object $\sigma(x, y)$ can be expressed as

$$\sigma(x, y) = \sigma_0 + \Delta(x, y) \tag{5.3-1}$$

with the condition that $\Delta(x, y) \ll \sigma_0$, where σ_0 is some constant, representing a uniform background of the transparency. Since $\Delta(x, y)$ is small, its scattered wave does not disturb the uniform reference wave.

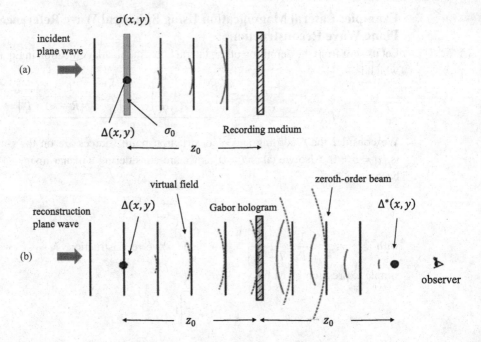

Figure 5.3-1 (a) Recording of a Gabor hologram and (b) reconstruction

According to Fresnel diffraction, the complex field on the hologram plane, at a distance z_0 away from the object, is given by

$$A\left[\sigma_0 + \Delta(x,y)\right] * h(x,y;z_0) = \psi_c + \psi_0(x,y), \tag{5.3-2}$$

where A is the amplitude of the incident plane wave, and $h(x,y;z_0)$ is the spatial impulse response in Fourier optics [see Eq. (3.4-16)]. In Eq. (5.3-2),

$$\psi_c = A\sigma_0 * h(x,y;z_0), \tag{5.3-3}$$

which is basically a plane wave of complex amplitude of $A\sigma_0$, propagating a distance z_0 to reach the recording medium. To show it, let us use the Fourier transform method. Taking the Fourier transform of the equation, we have

$$\mathcal{F}\{\psi_c\} = \mathcal{F}\{A\sigma_0\}\,\mathcal{F}\{h(x,y;z_0)\} = A\sigma_0\,4\pi^2\delta\left(k_x,k_y\right)H\left(k_x,k_y;z_0\right)$$

$$= A\sigma_0\,4\pi^2\delta\left(k_x,k_y\right)e^{-jk_0z_0}e^{j\frac{z_0}{2k_0}\left(k_x^2+k_y^2\right)} = A\sigma_0\,4\pi^2\delta\left(k_x,k_y\right)e^{-jk_0z_0},$$

where we recall that $H\left(k_x,k_y;z\right)$ is the spatial frequency transfer function in Fourier optics [see Eq. (3.4-18)], and we have used the Product Property of the delta function [see Eq. (2.1-4)] to obtain the last step of the above equation. Now, taking the inverse transform, we have

$$\mathcal{F}^{-1}\{\mathcal{F}\{\psi_c\}\} = \psi_c = A\sigma_0 e^{-jk_0z_0},$$

which is the complex amplitude of a plane wave after propagating a distance z_0. This plane wave serves as the reference light in holography. Now from Eq. (5.3-2),

$$\psi_0(x,y) = A\Delta(x,y) * h(x,y;z_0) \qquad (5.3\text{-}4)$$

is the scattered field due to the object $\Delta(x,y)$ and is regarded as the object wave.

As a result, the intensity on the hologram plane or the transparency function $t(x,y)$ of the hologram is given by

$$t(x,y) = |\psi_c + \psi_0(x,y)|^2 = |\psi_c|^2 + |\psi_0(x,y)|^2 + \psi_c\psi_0^*(x,y) + \psi_c^*\psi_0(x,y). \qquad (5.3\text{-}5)$$

Nowadays, this type of hologram is known as a *Gabor hologram*. This type of holography is called *Gabor holography* and has found use in particle sizing as well as distribution analysis.

Reconstruction

In the reconstruction process, the hologram $t(x,y)$ is illuminated by a reconstruction plane wave, that is, $\psi_{\text{rec}} = $ constant. The complex field emerging from the hologram is given by

$$\psi_{\text{rec}} \times t(x,y) * h(x,y,z). \qquad (5.3\text{-}6)$$

According to our previous analysis, let us deal with each term in the hologram separately.

Real Image (Twin Image)

The term $\psi_c\psi_0^*(x,y)$ in the hologram carries $\psi_0^*(x,y)$ and it reconstructs a real image at a distance z_0 from the hologram. Mathematically, we have

$$[\psi_{\text{rec}} \times \psi_c\psi_0^*(x,y)] * h(x,y;z_0) = \psi_{\text{rec}}\psi_c\left[A\Delta(x,y) * h(x,y;z_0)\right]^* * h(x,y;z_0)$$

$$= \psi_{\text{rec}}\psi_c A\left[\Delta^*(x,y) * h^*(x,y;z_0)\right] * h(x,y;z_0), \qquad (5.3\text{-}7)$$

where we have used $\psi_0(x,y)$ from Eq. (5.3-4). As it turns out convolution is *associative*, that is, when convolving three functions together, it does not matter which two functions we perform convolution first. Mathematically, the Associative Property of Convolution is stated as follows. Given three functions $f(x,y)$, $g(x,y)$, and $h(x,y)$,

$$[f(x,y) * g(x,y)] * h(x,y) = f(x,y) * [g(x,y) * h(x,y)]. \qquad (5.3\text{-}8)$$

Using the Associative Property, Eq. (5.3-7) can be rewritten as

$$\psi_{\text{rec}}\psi_c\psi_0^*(x,y) * h(x,y;z_0) = \psi_{\text{rec}}\psi_c A\Delta^*(x,y) * \left[h^*(x,y;z_0) * h(x,y;z_0)\right]. \qquad (5.3\text{-}9)$$

Let us first evaluate $K(x,y) = h^*(x,y;z_0) * h(x,y;z_0)$. We take the Fourier transform of the equation to get

$$\mathcal{F}\{K(x,y)\} = \mathcal{F}\{h^*(x,y;z_0) * h(x,y;z_0)\} = \mathcal{F}\{h^*(x,y;z_0)\}\mathcal{F}\{h(x,y;z_0)\}$$

$$= H^* \left(-k_x, -k_y; z_0 \right) H \left(k_x, k_y; z_0 \right),$$

where we have used Table 2.2 to find $\mathcal{F} \left\{ h^* \left(x, y; z_0 \right) \right\}$.

Now, since $H \left(k_x, k_y; z_0 \right) = e^{-j k_0 z_0} e^{j \frac{z_0}{2 k_0} \left(k_x^2 + k_y^2 \right)}$,

$$\mathcal{F} \left\{ K \left(x, y \right) \right\} = H^* \left(-k_x, -k_y; z_0 \right) H \left(k_x, k_y; z_0 \right) = 1.$$

Therefore,

$$K \left(x, y \right) = h^* \left(x, y; z_0 \right) * h \left(x, y; z_0 \right) = \delta \left(x, y \right). \tag{5.3-10}$$

With this result, Eq. (5.3-9) becomes

$$\psi_{rec} \psi_c \psi_0^* \left(x, y \right) * h \left(x, y; z_0 \right) = \psi_{rec} \psi_c A \Delta^* \left(x, y \right) * \delta \left(x, y \right) \propto \Delta^* \left(x, y \right). \tag{5.3-11}$$

This is the real image reconstructed at a distance z_0 on the right side of the hologram as shown in Figure 5.3-1b.

Virtual Image

Similarly, $\psi_c^* \psi_0 \left(x, y \right)$ from the hologram in Eq. (5.3-5) reconstructs a virtual image of the amplitude fluctuation $\Delta \left(x, y \right)$ at the location of the original object. To show this is the case, let us first discuss how to find a *virtual field* in wave optics.

Let us take a negative lens to illustrate the calculation of a virtual field. The transparency function of the negative lens is

$$t_f \left(x, y \right) = e^{-j \frac{k_0}{2 |f|} \left(x^2 + y^2 \right)} \tag{5.3-12}$$

as $f < 0$ [see Eq. (3.4-24)]. For a plane wave of unit amplitude incident on the lens, the real field z_0 away from the lens, according to Fresnel diffraction, is given by

$$t_f \left(x, y \right) * h \left(x, y; z = z_0 \right) = e^{-j \frac{k_0}{2 |f|} \left(x^2 + y^2 \right)} * h \left(x, y; z = z_0 \right).$$

To find the virtual field z_0 in front of the lens, that is, the field from where it appears to be originating as seen by an observer, we perform the back-propagation calculation as follows:

$$e^{-j \frac{k_0}{2 |f|} \left(x^2 + y^2 \right)} * h \left(x, y; z = -z_0 \right). \tag{5.3-13}$$

The situation is shown in Figure 5.3-2. From the definition of $h \left(x, y; z_0 \right)$ [see Eq. (3.4-16)], we deduce that $h^* \left(x, y; z_0 \right) = h \left(x, y; -z_0 \right)$. With this identity, the virtual field can be calculated according to

$$e^{-j \frac{k_0}{2 |f|} \left(x^2 + y^2 \right)} * h \left(x, y; z = -z_0 \right) = e^{-j \frac{k_0}{2 |f|} \left(x^2 + y^2 \right)} * h^* \left(x, y; z = z_0 \right).$$

The situation is summarized in Figure 5.3-2.

We now return to the term $\psi_c^* \psi_0 \left(x, y \right)$ in the hologram when the hologram is illuminated by plane wave ψ_{rec}. After back propagating a distance of z_0 from the hologram, the virtual field is

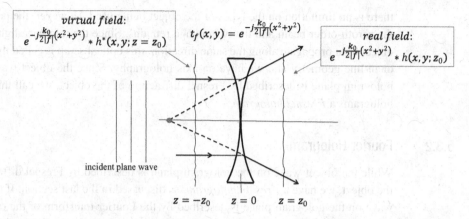

Figure 5.3-2 Calculation of real field and virtual field for an incident plane wave of unit amplitude

$$\psi_{rec}\psi_c^*\psi_0(x,y)*h^*(x,y;z_0) = \psi_{rec}\psi_c^*\left[A\Delta(x,y)*h(x,y;z_0)\right]*h^*(x,y;z_0)$$
$$= \psi_{rec}\psi_c^*\left[A\Delta(x,y)*\delta(x,y)\right]$$
$$\propto \Delta(x,y), \qquad\qquad (5.3\text{-}14)$$

where we have used Eq. (5.3-10) to simplify the equation and substituted the object field $\psi_0(x,y)$ given by Eq. (5.3-4) into Eq. (5.3-14). The result indicates that we have the virtual field $\Delta(x,y)$ at $z=-z_0$ as shown in Figure 5.3-1b.

Zeroth-Order Beam

The first two terms of the hologram equation from Eq. (5.3-5) correspond to the transmitted zeroth-order beam given by

$$\psi_{rec}\left[|\psi_c|^2 + |\psi_0(x,y)|^2\right]*h(x,y;z)$$

$$\approx \psi_{rec}|\psi_c|^2*h(x,y;z) = \psi_{rec}(A\sigma_0)^2*h(x,y;z) \qquad (5.3\text{-}15)$$

as $|\psi_c|^2 = (A\sigma_0)^2 \gg |\psi_0(x,y)|^2 = |A\Delta(x,y)*h(x,y;z_0)|^2$ because $\Delta(x,y) \ll \sigma_0$. The above equation represents a plane wave propagation along the z-direction [see Figure 5.3-1b].

The merit of Gabor holography is that the setup is very simple. However, one of the drawbacks is that the transmitted zeroth-order beam always blurs the reconstructed virtual image. Another drawback is the problem of twin image as the Gabor hologram is an on-axis hologram. In holography, this is known as the "twin-image problem," leading to research known as twin-image elimination. Another problem of the Gabor hologram is that the amplitude variation of the object must be small enough, that is, $\Delta(x,y) \ll \sigma_0$, to make the technique useful. To overcome this shortcoming, an independent light is used as a reference wave, as shown in Figure 5.1-1a, and there is no limitation on the types of the object being used. However, the problem of the zeroth-order beam and the twin image remains. Since the reference light and the object light propagate along the same direction, this optical setup is called the on-axis or in-line geometry and we have on-axis holography. Since the object wave on the hologram plane is described by Fresnel diffraction of the object, we call this type of holograms a *Fresnel hologram*.

5.3.2 Fourier Hologram

While the object wave on the hologram plane is described by Fresnel diffraction of the object, we have a *Fresnel hologram*, as discussed in the last section. If the object wave on the hologram plane is described by the Fourier transform of the object, we have a *Fourier hologram*. A Fourier transform lens can conveniently provide Fourier transformation of the object.

Recording

We use the setup shown in Figure 5.3-3a to record a Fourier hologram. In the setup the input or the object $s(x, y)$ is located at the front focal plane of lens 2. Hence, the complex field at the back focal plane of lens 2 is the Fourier transform of the object light. Meanwhile, a focused light spot next to the object, through the use of lens 1, makes a tilted reference plane wave on the recording medium.

According to the result shown in Figure 3.4-7, we can write the total complex field at the back focal plane of lens 2 as

$$\psi_t(x, y) = \mathcal{F}\{s(x, y) + A\delta(x + x_0, y)\}_{\substack{k_x = k_0 x/f \\ k_y = k_0 y/f}}$$

$$= S\left(\frac{k_0 x}{f}, \frac{k_0 y}{f}\right) + A e^{-j\frac{k_0 x_0 x}{f}}, \tag{5.3-16}$$

where $s(x, y)$ is the transparency function of the object and its Fourier transform is $S(k_x, k_y)$; $\delta(x + x_0, y)$ is the focused spot at $x = -x_0, y = 0$; f is the focal length of lens 2, and A is the amplitude of the reference tilted plane wave.

Consequently, the Fourier hologram can be expressed as

$$t_{\text{FH}}(x, y) = |\psi_t(x, y)|^2$$

$$= \left|S\left(\frac{k_0 x}{f}, \frac{k_0 y}{f}\right)\right|^2 + |A|^2 + S\left(\frac{k_0 x}{f}, \frac{k_0 y}{f}\right) \times A^* e^{jk_0 x_0 x/f} + S^*\left(\frac{k_0 x}{f}, \frac{k_0 y}{f}\right) \times A e^{-jk_0 x_0 x/f}.$$

$$\tag{5.3-17}$$

Reconstruction

In the reconstruction process, the hologram placed at the front focal plane of a lens is illuminated using a normal-incident reconstruction plane wave with unit amplitude, as shown in Figure 5.3-3b. After Fourier transformation by the lens, the total complex field at the back focal plane of the lens contains three terms:

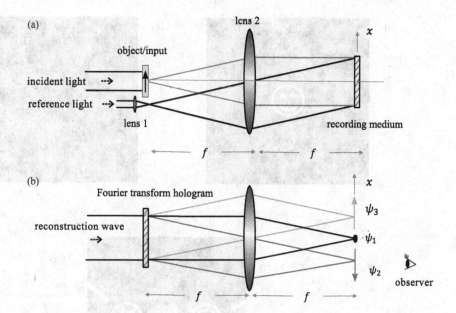

Figure 5.3-3 (a) Recording geometry and (b) reconstruction geometry for Fourier hologram

Zeroth-Order Beam

$$\psi_1(x,y) = \mathcal{F}\left\{\left|S\left(\frac{k_0 x}{f},\frac{k_0 y}{f}\right)\right|^2 + |A|^2\right\}\Bigg|_{\substack{k_x = k_0 x/f \\ k_y = k_0 y/f}}$$

$$= \frac{4\pi^2 f^2}{k_0^2}s(-x,-y) \otimes s(-x,-y) + \frac{4\pi^2 f^2}{k_0^2}|A|^2\,\delta(x,y), \qquad (5.3\text{-}18a)$$

where $\otimes$ denotes correlation. This is the zeroth-order beam located at $x = 0, y = 0$ on the back focal plane.

Inverted Image

$$\psi_2(x,y) = \mathcal{F}\left\{S\left(\frac{k_0 x}{f},\frac{k_0 y}{f}\right)\times A^* e^{j\frac{k_0 x_0 x}{f}}\right\}\Bigg|_{\substack{k_x = k_0 x/f \\ k_y = k_0 y/f}}$$

$$= \frac{4\pi^2 f^2}{k_0^2} A^* s(-x - x_0, -y). \qquad (5.3\text{-}18b)$$

This is an inverted reconstructed image at $x = -x_0, y = 0$.

Conjugate Image

$$\psi_3(x,y) = \mathcal{F}\left\{S^*\left(\frac{k_0 x}{f},\frac{k_0 y}{f}\right)\times A e^{-j\frac{k_0 x_0 x}{f}}\right\}\Bigg|_{\substack{k_x = k_0 x/f \\ k_y = k_0 y/f}}$$

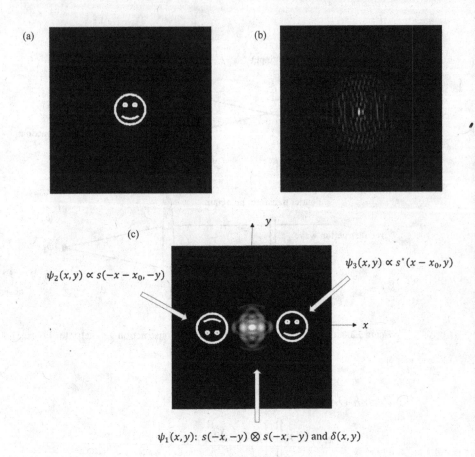

$$\psi_2(x,y) \propto s(-x-x_0,-y)$$

$$\psi_3(x,y) \propto s^*(x-x_0,y)$$

$$\psi_1(x,y): s(-x,-y) \otimes s(-x,-y) \text{ and } \delta(x,y)$$

Figure 5.3-4 (a) Input image, (b) Fourier transform hologram of (a), and (c) reconstruction of the Fourier transform hologram in (b). Note that the image in (b) has been zoomed in from the Matlab's output to show fringe contrast

$$= \frac{4\pi^2 f^2}{k_0^2} A s^*(x-x_0,y). \tag{5.3-18c}$$

This is the conjugate image at $x = x_0, y = 0$.

Figure 5.3-4 illustrates the input image $s(x,y)$, its Fourier transform hologram $t_{FH}(x,y)$, and the reconstruction of the transform hologram.

5.3.3   Image Hologram

Recording

An *image hologram* can be recorded as shown in Figure 5.3-5a, where a real image of the object is formed on the recording medium by an imaging lens.

The real image serves as the object wave. Upon recording, the image hologram is given by

$$t(x,y) = \left| \psi_i(x,y) + \psi_r e^{jk_0 \sin\theta x} \right|^2, \tag{5.3-19}$$

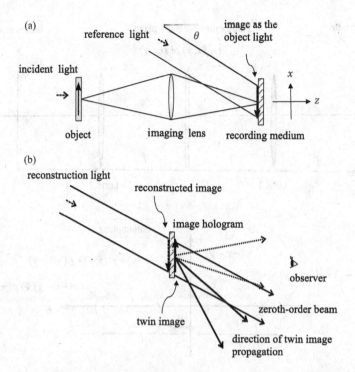

Figure 5.3-5 (a) Recording geometry and (b) reconstruction geometry for image hologram

where $\psi_i(x,y)$ represents the complex field of the real image on the hologram and $\psi_r e^{jk_0\sin\theta x}$ is the off-axis reference plane wave on the hologram.

Reconstruction

Let us take the reconstruction wave ψ_{rec} the same as the reference wave as shown in Figure 5.3-5b. Hence, $\psi_{rec} = \psi_r e^{jk_0\sin\theta x}$, and the complex field just behind the hologram can be expressed as

$$\psi_{rec} t(x,y) = \psi_r e^{jk_0\sin\theta x} \times t(x,y)$$
$$= [|\psi_i|^2 + |\psi_r|^2]e^{jk_0\sin\theta x} + \psi_i(x,y)\psi_r^* + \psi_i^*(x,y)\psi_r e^{j2k_0\sin\theta x}. \qquad (5.3\text{-}20)$$

Zeroth-Order Beam
The first term of the right hand side of Eq. (5.3-20) is the zeroth-order beam propagating along the direction of the reconstruction wave.

Reconstructed Image
The second term contains $\psi_i(x,y)$ and is the reconstructed image. It is right on the surface of the hologram. The image propagates away from the hologram along the z-direction.

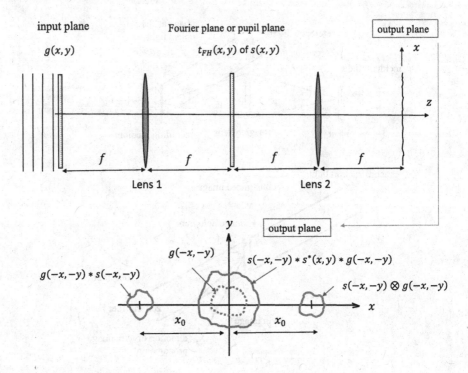

Figure 5.3-6 Complex spatial filtering

Conjugate Image (Twin Image)

The third term is the twin image, which is reconstructed also on the hologram plane but propagating along the direction that is 2θ away from the z-axis. Since the twin image is also reconstructed on the hologram plane as shown in Figure 5.3-5b, off-axis recording is necessary in obtaining the image hologram, and the reconstructed image can be observed normal to the hologram.

5.3.4   Complex Spatial Filtering and Joint-Transform Correlation

The correlation of two images is one of the most important mathematical operations in image processing and pattern recognition. In optical correlation systems, there are two classical techniques available in practice: one is to utilize complex spatial filtering and the other is to utilize joint-transform correlation.

Complex Spatial Filtering

We perform spatial filtering in the 4-f two-lens coherent system by putting a spatial filter in the Fourier plane to perform spatial filtering of the input [see Chapter 4, Section 4.2]. Indeed, we could use holograms as spatial filters. Since a hologram contains the amplitude and phase information of a given object, the notion of *complex spatial filtering* becomes clear as the hologram can modify the amplitude and phase of the input spectrum. For this reason, the hologram placed on the pupil plane is called a *complex filter*, introduced by Vander Lugt in 1963.

Let us insert the Fourier hologram of $s(x,y)$, given by Eq. (5.3-17), on the Fourier plane of the two-lens system shown in Figure 5.3-6.

Now, we let a transparency $g(x,y)$ be placed in the front focal plane of lens 1 and illuminated by a plane wave of unit amplitude. Then the complex field immediately after the complex filter, that is, the hologram, is $G\left(\dfrac{k_0 x}{f},\dfrac{k_0 y}{f}\right)\times t_{FH}(x,y)$, where $G\left(\dfrac{k_0 x}{f},\dfrac{k_0 y}{f}\right)$ is the spectrum of $g(x,y)$ on the Fourier plane of the 4-f system. Hence, on the complex filtered output plane, that is, the back focal plane of lens 2, the complex field is

$$\psi_p(x,y) \propto \mathcal{F}\left\{ G\left(\frac{k_0 x}{f},\frac{k_0 y}{f}\right)\times t_{FH}(x,y)\right\}_{\substack{k_x = k_0 x/f \\ k_y = k_0 y/f}}$$

$$= \mathcal{F}\left\{ G\left(\frac{k_0 x}{f},\frac{k_0 y}{f}\right)\left[\left|S\left(\frac{k_0 x}{f},\frac{k_0 y}{f}\right)\right|^2 + |A|^2 + S\left(\frac{k_0 x}{f},\frac{k_0 y}{f}\right)\times A^* e^{jk_0 x_0 x/f}\right.\right.$$

$$\left.\left. + S^*\left(\frac{k_0 x}{f},\frac{k_0 y}{f}\right)\times A e^{-jk_0 x_0 x/f}\right]\right\}_{\substack{k_x = k_0 x/f \\ k_y = k_0 y/f}}. \tag{5.3-21}$$

Let us look at the contributions due to each term in the hologram separately.

Zeroth-Order Beam
The first two terms contribute to complex fields of

$$\mathcal{F}\left\{ G\left(\frac{k_0 x}{f},\frac{k_0 y}{f}\right)\left|S\left(\frac{k_0 x}{f},\frac{k_0 y}{f}\right)\right|^2\right\}_{\substack{k_x = k_0 x/f \\ k_y = k_0 y/f}} \propto s(-x,-y)*s^*(x,y)*g(-x,-y) \tag{5.3-22a}$$

and

$$\mathcal{F}\left\{ G\left(\frac{k_0 x}{f},\frac{k_0 y}{f}\right)|A|^2\right\}_{\substack{k_x = k_0 x/f \\ k_y = k_0 y/f}} \propto g(-x,-y) \tag{5.3-22b}$$

Both of the fields are centered on the origin of the output plane.

Convolution
The third term corresponds to the convolution of $g(x,y)$ and $s(x,y)$, centered at $(x,y)=(-x_0,0)$:

$$\mathcal{F}\left\{ G\left(\frac{k_0 x}{f},\frac{k_0 y}{f}\right)\left[S\left(\frac{k_0 x}{f},\frac{k_0 y}{f}\right)\times A^* e^{\frac{jk_0 x_0 x}{f}}\right]\right\}_{\substack{k_x = k_0 x/f \\ k_y = k_0 y/f}}$$

$$\propto g(-x,-y)*s(-x,-y)*\delta(x+x_0,y)= g(-x,-y)*s(-x-x_0,-y). \tag{5.3-23}$$

Correlation

The last term corresponds to the cross-correlation of $s(x,y)$ and $g(x,y)$, centered at $(x,y) = (x_0, 0)$.

$$\mathcal{F}\left\{G\left(\frac{k_0 x}{f}, \frac{k_0 y}{f}\right)\left[S^*\left(\frac{k_0 x}{f}, \frac{k_0 y}{f}\right) \times Ae^{-jk_0 x_0 x/f}\right]\right\}_{\substack{k_x = k_0 x/f \\ k_y = k_0 y/f}}$$

$$\propto \left[s(-x,-y) \otimes g(-x,-y)\right] * \delta(x - x_0, y)$$

$$= s(-x,-y) \otimes g(-x + x_0, -y) = C_{sg}(-x + x_0, -y), \qquad (5.3\text{-}24)$$

where, as a reminder, the correlation of $s(x,y)$ and $g(x,y)$ is [see Eq. (2.-3-30)]

$$C_{sg}(x,y) = s(x,y) \otimes g(x,y) = \int\!\!\!\int\limits_{-\infty}^{\infty} s^*(x',y') g(x+x', y+y') dx' dy'.$$

On the bottom of Figure 5.3-6, we summarize the locations of these terms on the output plane. Note that *complex spatial filtering* is also called *matched filtering* introduced in Chapter 2, as correlation of two functions are performed.

MATLAB Example: Complex Filtering

In this example, we illustrate some MATLAB results on complex spatial filtering. The results are shown in Figure 5.3-7. We first obtain the Fourier transform hologram of $s(x,y)$, given by Eq. (5.3-17). Subsequently an input $g(x,y)$ is Fourier transformed for complex filtering, according to Eq. (5.3-21), to provide the output of the $4f$-coherent system. On the output of complex filtering, we expect to have the zeroth-order beam on the origin of the output coordinates. On both side of the zeroth-order beam, there are convolution and correlation of $s(x,y)$ and $g(x,y)$. In Figure 5.3-7a, we show $s(x,y) = g(x,y)$, and we clearly see the convolution result on the left side of the zeroth-order beam with the auto-correlation result on the right side, as shown in the output of complex filtering shown in Figure 5.3-7d. Remember that auto-correlation should give a peak-like output and that is what we see in the result. In this case, convolution also shows some peak-like output, that is because $s(x,y)$ and $g(x,y)$ are very symmetrical, as the outer ring of the two images should give a fairly strong convolution response. On the other hand, Figure 5.3-7e shows no distinct spots for convolution and cross-correlation of $s(x,y)$ and $g(x,y)$ given by Figure 5.3-7b. Finally, Figure 5.3-7c shows identical random patterns for $s(x,y)$ and $g(x,y)$. Clearly, in the output of complex filtering shown in Figure 5.3-7f, the convolution result shows a diffused patch of light as output, while the autocorrelation result shows a distinct spot. Indeed, correlation is useful when comparing the similarity of two images, and it has been implemented through complex spatial filtering and used for *optical pattern recognition* applications.

The results obtained in Figures 5.3-4 and 5.3-7 use the m-file below.

```
% FT_hologram_complex_filtering2
% Adapted from "Introduction to Modern Digital Holography"
% by T.-C. Poon & J.-P. Liu
% Cambridge University Press (2014), Table 3.3.
```

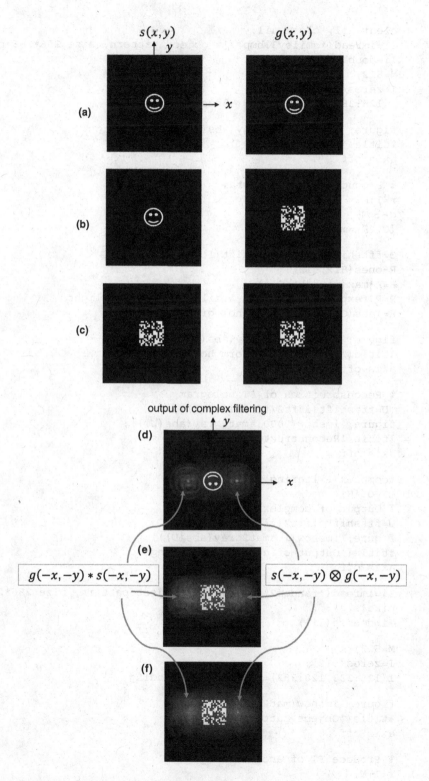

Figure 5.3-7 Complex spatial filtering examples

```
clear all, close all;
Ii=imread('smiley3.bmp');% object pattern, size 256*256 pixels
Ii=double(Ii);
M=512;
I=zeros(512);
I(128:383,128:383)=Ii;  % zero-padding

figure; imshow(mat2gray(abs(I)));
%title('Object pattern')
axis off

% Produce the FT hologram
r=1:M;
c=1:M;
[C, R]=meshgrid(c, r);

O=fftshift(ifft2(fftshift(I)));
R=ones(512,512);
R=R*max(max(abs(O)));
R=R.*exp(-2i*pi.*C/4); % tilted reference light
H=(abs(O+R)).^2; % FT hologram

figure; imshow(mat2gray(abs(H)));
%title('Fourier transform hologram')
axis off

% Reconstruction of FT hologram
 U=fftshift(ifft2(fftshift(H)));
figure; imshow(9000.*mat2gray(abs(U)));
%title('Reconstructed images')
axis off

%complex filtering
CF=O.*H;
% Output of complex filtering
U=fftshift(ifft2(fftshift(CF)));
figure; imshow(1.*mat2gray(abs(U)));
%title('Output of complex filtering')
axis off

Ii=imread('random256_2.bmp');% object pattern, size 256*256
pixels
Ii=double(Ii);

M=512;
I=zeros(512);
I(128:383,128:383)=Ii;  % zero-padding

figure; imshow(mat2gray(abs(I)));
%title('Object pattern2')
axis off

% Produce FT of another image, O2
r=1:M;
```

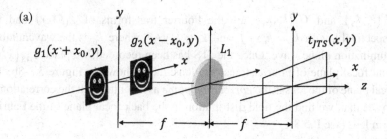

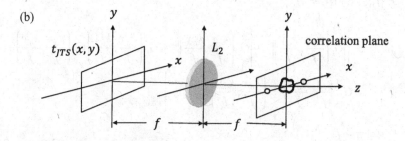

Figure 5.3-8 Joint-transform correlation: (a) recording the joint transform spectrum and (b) display the various correlation outputs

```
c=1:M;
[C, R]=meshgrid(c, r);

O2=fftshift(ifft2(fftshift(I)));
%complex filtering 2
CF=H.*O2;
% Output of complex filtering
U=fftshift(ifft2(fftshift(CF)));
figure; imshow(1.*mat2gray(abs(U)));
%title('Output of complex filtering')
axis off
```

Joint-Transform Correlation (JTC)

The *joint-transform correlation (JTC)* was first proposed by Weaver and Goodman in the 1960s. A typical JTC system is shown in Figure 5.3-8.

$g_1(x,y)$ and $g_2(x,y)$ are the two images to be correlated, and they are in the form of transparencies, which are illuminated by plane waves. The two images are separated by a distance of $2x_0$ in the front focal plane of Fourier transform lens L_1 of focal length f, as shown in Figure 5.3-8a. The spectra of the two images are jointly recorded by a recording medium. The resulting recorded intensity, which is called the *joint-transform spectrum* (JTS), yields a transparency given by

$$t_{JTS}(x,y) = \left| \mathcal{F}\{g_1(x+x_0,y)\} + \mathcal{F}\{g_2(x-x_0,y)\} \right|^2$$

$$= \left| G_1(k_x,k_x) \right|^2 + \left| G_2(k_x,k_x) \right|^2 + G_1^*(k_x,k_x)G_2(k_x,k_x)e^{j2k_xx_0}$$

$$+ G_1(k_x,k_x)G_2^*(k_x,k_x)e^{-j2k_xx_0}. \tag{5.3-25}$$

$G_1(k_x, k_x)$ and $G_2(k_x, k_x)$ are the Fourier transforms of $g_1(x,y)$ and $g_2(x,y)$, respectively, with $k_x = k_0 x / f$ and $k_y = k_0 y / f$, where k_0 is the wavenumber of the illumination plane wave. Once the JTS has been recorded, we place $t_{JTS}(x,y)$ in the front focal plane of Fourier transform lens L_2, as shown in Figure 5.3-8b. The back focal plane of the lens is the *correlation plane* as it contains all the correlation outputs. To see that, we find the field distribution in the back focal plane of the Fourier transform lens [see P.5.8]:

$$\mathcal{F}\{t_{JTS}(x,y)\}_{\substack{k_x=k_0x/f \\ k_y=k_0y/f}}$$

$$= \mathcal{F}\left\{ \left|G_1\left(\frac{k_0 x}{f}, \frac{k_0 y}{f}\right)\right|^2 + \left|G_2\left(\frac{k_0 x}{f}, \frac{k_0 y}{f}\right)\right|^2 + G_1^*\left(\frac{k_0 x}{f}, \frac{k_0 y}{f}\right) G_2\left(\frac{k_0 x}{f}, \frac{k_0 y}{f}\right) e^{j2\frac{k_0 x}{f}x_0} \right.$$

$$\left. + G_1\left(\frac{k_0 x}{f}, \frac{k_0 y}{f}\right) G_2^*\left(\frac{k_0 x}{f}, \frac{k_0 y}{f}\right) e^{-j2\frac{k_0 x}{f}x_0} \right\}_{\substack{k_x=k_0x/f \\ k_y=k_0y/f}}$$

$$= \frac{4\pi^2 f^2}{k_0^2}\left[C_{11}(-x,-y) + C_{22}(-x,-y) + C_{12}(-x-2x_0,-y) + C_{21}(-x+2x_0,-y) \right],$$

$$(5.3\text{-}26)$$

where

$$C_{mn}(x,y) = g_m(x,y) \otimes g_n(x,y),$$

with $m = 1$ or 2, and $n = 1$ or 2. $C_{mn}(x,y)$ is the autocorrelation when $m = n$ and is cross-correlation when $m \neq n$. Hence if the two images (or patterns) are the same, besides a strong peak at the origin of the correlation plane due to the first two terms of Eq. (5.3-26), we have two strong peaks centered at $x = \pm 2x_0$ due to the last two terms of Eq. (5.3-26).

MATLAB Example: Joint-Transform Correlation

In this example, we illustrate some MATLAB results on the joint-transform correlation. The results are shown in Figure 5.3-9. We first obtain the joint-transform spectrum given by Eq. (5.3-25) from the two images shown in Figure 5.3-9a and d. Figure 5.3-9b and e shows the spectrum corresponding to Figure 5.3-9a and d, respectively. Subsequently, the spectrum is Fourier transformed for correlation outputs according to Eq. (5.3-26). Figure 5.3-9c and f shows the correlation outputs corresponding to the image pairs shown in Figure 5.3-9a and d, respectively. For a matched pair, we clearly observe the two strong autocorrelation peaks next to the patch of light at the origin of the correlation plane.

The above results are obtained using the m-file below.

```
% JTC
clear, close all;
Ii=zeros(1024,1024);
A=imread('smiley3.bmp');
A=imresize(A, [1024, 1024]);
A=double(A(:,:,1));
```

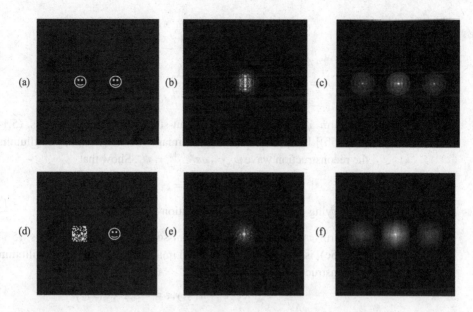

Figure 5.3-9 Joint-transform correlation results: (a) and (d) are the image pairs, (b) and (e) are the joint transform spectrum corresponding to (a) and (d), respectively; (c) and (f) show the correlation outputs on the correlation plane corresponding to the image pairs in (a) and (d), respectively

```
A=imresize(A, 0.25);
A=A./(max(max(A)));
B=imread('random256_2.bmp');
B=imresize(B, [1024, 1024]);
B=double(B);
B=imresize(B, 0.25);
B=B./(max(max(B)));
Ii(385:640,237:492)=A;
Ii(385:640,533:788)=B;
FT=fftshift(fft2(Ii));
CO=abs(FT).*abs(FT);
figure; imshow(mat2gray(Ii)); % two images on front focal plane
of lens
figure; imshow(100.*mat2gray(CO)); % Joint-transform spectrum
(JTS)
IFT=fftshift(fft2(CO));
figure; imshow(2.*mat2gray(3.*abs(IFT))); % FT of JTC
```

Problems

5.1 Show that for an off-axis point source $\delta(x-x_0, y-y_0)$ that is z_0 away from the recording medium as shown in Figure 5.1-1a, the recorded hologram has a transparency function given by

$$A + B \sin\left[\frac{k_0}{2z_0}\left[(x-x_0)^2 + (y-y_0)^2\right]\right],$$

where $A = a^2 + \left(\frac{k_0}{2\pi z_0}\right)^2$ and $B = \frac{ak_0}{\pi z_0}$.

5.2 The term $\psi_r \psi_0^*$ in the off-axis point-source hologram [see Eqs. (5.1-14) and (5.1-15b)] is responsible for a real image reconstruction upon illumination by the reconstruction wave $\psi_{rec} = ae^{jk_0 \sin\theta x} = \psi_r$. Show that

$$[\psi_{rec} \times \psi_r \psi_0^*] * h(x,y;z=z_0) \propto \delta(x + 2z_0\sin\theta, y)$$

by carrying out directly the convolution involved.

5.3 The term $\psi_r^* \psi_0$ in the off-axis point-source hologram [see Eqs. (5.1-14) and (5.1-15c)] is responsible for a virtual image reconstruction upon illumination by a reconstruction wave $\psi_{rec} = ae^{jk_0 \sin\theta x} = \psi_r$. Show that

$$\psi_{rec} \times \psi_r^* \psi_0 * h(x,y;z=-z_0) \propto \delta(x,y)$$

by carrying out directly the convolution involved.

5.4 We have an off-axis point source hologram recorded as shown in Figure 5.1-7 and the recorded hologram of a transparency function is given by Eq. (5.1-16):

$$t(x,y) = A + B \sin\left[\frac{k_0}{2z_0}(x^2 + y^2) + k_0\sin\theta x\right].$$

The hologram is now reconstructed by a plane wave shown in Figure P.5.4.

(a) Find the location of the real and virtual images.

(b) Draw a reconstruction ray diagram similar to that in Figure 5.1-8 for both the real and virtual image.

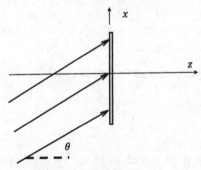

Off-axis point-source hologram

reconstruction plane wave

Figure P.5.4

5.5 We have discussed holographic imaging through the recording and reconstruction of two point sources using spherical waves in Section 5.2.

(a) For real image reconstruction of point 1, we calculate [see Eq. (5.2-10)]

$$\psi_{p,rel_1}(x,y;z) = \psi_{pr}(x,y)t_{rel_1}(x,y)*h(x,y;z).$$

By carrying out directly the above convolution, show that the real image is reconstructed at

$$z = z_{r1} = \left[\frac{1}{R} - \frac{1}{l_1} - \frac{1}{l_2}\right]^{-1},$$

and at this location, $\psi_{p,rel_1}(x,y;z = z_{r1}) \propto \delta\left(x + z_{r1}\left(\frac{b}{l_2} - \frac{h}{2R} - \frac{a}{l_1}\right), y\right)$.

(b) For real image reconstruction of point 2, we calculate [see Eq. (5.2-13)]

$$\psi_{p,rel_2}(x,y;z) = \psi_{pr}(x,y)t_{rel_2}(x,y)*h(x,y;z).$$

By carrying out directly the above convolution, show that the real image is reconstructed at

$$z = z_{r2} = \left[\frac{1}{R+d} - \frac{1}{l_1} - \frac{1}{l_2}\right]^{-1},$$

and at this location, $\psi_{p,rel_2}(x,y;z = z_{r2}) \propto \delta\left(x + z_{r2}\left(\frac{b}{l_2} + \frac{h}{2(R+d)} - \frac{a}{l_1}\right), y\right)$.

5.6 In principle, a hologram can be recorded using a particular color and reconstructed by some other color. This problem explores this aspect. The recording and reconstruction of two point sources are shown in Figures 5.2-1 and 5.2-2, respectively. For recording, let us now record with wavelength $\lambda_1 = 2\pi / k_1$. The complex fields on the recording medium for point 1, point 2, and the spherical reference wave are given by

$$\psi_{p1}(x,y) \propto e^{-j\frac{k_1}{2R}\left[(x-h/2)^2 + y^2\right]},$$

$$\psi_{p2}(x,y) \propto e^{-j\frac{k_1}{2(R+d)}\left[(x+h/2)^2 + y^2\right]},$$

and

$$\psi_{pR}(x,y) \propto e^{-j\frac{k_1}{2l_1}\left[(x+a)^2 + y^2\right]},$$

respectively. On reconstruction, we use wavelength $\lambda_2 = 2\pi / k_2$, and hence the spherical reconstruction wave is

$$\psi_{pr}(x,y) \propto e^{-j\frac{k_2}{2l_2}\left[(x-b)^2 + y^2\right]}.$$

(a) Show that the real image of point 1 is reconstructed at

$$z = z_{r1} = \left[\frac{k_1}{k_2}\left(\frac{1}{R} - \frac{1}{l_1}\right) - \frac{1}{l_2}\right]^{-1},$$

and at this location, $\psi_{p,rel_1}(x,y;z=z_{r1}) \propto \delta\left(x+z_{r1}\left(\dfrac{b}{l_2}-\dfrac{k_1}{k_2}\dfrac{h}{2R}-\dfrac{k_1}{k_2}\dfrac{a}{l_1}\right),y\right)$.

(b) Show that the real image of point 2 is reconstructed at

$$z_{r2} = \left[\frac{k_1}{k_2}\left(\frac{1}{R+d}-\frac{1}{l_1}\right)-\frac{1}{l_2}\right]^{-1},$$

and at this location, $\psi_{p,rel_2}(x,y;z=z_{r2}) \propto \delta\left(x+z_{r2}\left(\dfrac{b}{l_2}+\dfrac{k_1}{k_2}\dfrac{h}{2(R+d)}-\dfrac{k_1}{k_2}\dfrac{a}{l_1}\right),y\right)$.

(c) Show that for magnification that is independent of the lateral separation, h, we must have the following condition:

$$\frac{b}{a} = \frac{\lambda_2}{\lambda_1}\frac{l_2}{l_1},$$

and under this condition,

$$M^r_{\text{Lat}} = \frac{\lambda_2}{\lambda_1}\frac{(z_{r2}+z_{r1})}{2R}.$$

(d) Show that the longitudinal magnification is

$$M^r_{\text{Long}} \cong \frac{\lambda_1\lambda_2\left(l_1 l_2\right)^2}{(\lambda_2 l_1 l_2 - \lambda_1 R l_1 - \lambda_2 R l_2)^2},$$

for $R \gg d$.

(e) Verify that $M^r_{\text{Long}} = \dfrac{\lambda_1}{\lambda_2}\left(M^r_{\text{Lat}}\right)^2$, for $R \gg d$.

(f) What is the advantage of setting $M^r_{\text{Lat}} = \dfrac{\lambda_2}{\lambda_1}$?

5.7 The Fourier hologram of $s(x,y)$ is given by Eq. (5.3-17). Upon reconstruction,

(a) verify that the zeroth-order beam

$$\mathcal{F}\left\{\left|S\left(\frac{k_0 x}{f},\frac{k_0 y}{f}\right)\right|^2 + |A|^2\right\}_{\substack{k_x=k_0 x/f \\ k_y=k_0 y/f}} = \frac{4\pi^2 f^2}{k_0^2}s(-x,-y)\otimes s(-x,-y)+\frac{4\pi^2 f^2}{k_0^2}|A|^2\,\delta(x,y),$$

(b) verify that the inverted image

$$\psi_2(x,y)=\mathcal{F}\left\{S\left(\frac{k_0 x}{f},\frac{k_0 y}{f}\right)\times A^*\exp\left(\frac{jk_0 x_0 x}{f}\right)\right\}_{\substack{k_x=k_0 x/f \\ k_y=k_0 y/f}} = \frac{4\pi^2 f^2}{k_0^2}A^*s(-x-x_0,-y),$$

(c) verify that the conjugate image

$$\psi_3(x,y) = \mathcal{F}\left\{ S^*\left(\frac{k_0 x}{f}, \frac{k_0 y}{f}\right) \times A\exp\left(\frac{-jk_0 x_0 x}{f}\right) \right\}_{\substack{k_x = k_0 x/f \\ k_y = k_0 y/f}} = \frac{4\pi^2 f^2}{k_0^2} As^*(x - x_0, y).$$

5.8 With reference to Figure 5.3-9, the joint-transform spectrum is given by

$$t_{JTS}(x,y) = \left| \mathcal{F}\{g_1(x + x_0, y)\}_{\substack{k_x = k_0 x/f \\ k_y = k_0 y/f}} + \mathcal{F}\{g_2(x + x_0, y)\}_{\substack{k_x = k_0 x/f \\ k_y = k_0 y/f}} \right|^2$$

$$= \left| G_1\left(\frac{k_0 x}{f}, \frac{k_0 y}{f}\right) \right|^2 + \left| G_2\left(\frac{k_0 x}{f}, \frac{k_0 y}{f}\right) \right|^2 + G_1^*\left(\frac{k_0 x}{f}, \frac{k_0 y}{f}\right) G_2\left(\frac{k_0 x}{f}, \frac{k_0 y}{f}\right) e^{j2\frac{k_0 x}{f}x_0}$$

$$+ G_1\left(\frac{k_0 x}{f}, \frac{k_0 y}{f}\right) G_2^*\left(\frac{k_0 x}{f}, \frac{k_0 y}{f}\right) e^{-j2\frac{k_0 x}{f}x_0}.$$

Show that on the correlation plane, the correlation outputs are given by

$$\mathcal{F}\{t_{JTS}(x,y)\}_{\substack{k_x = k_0 x/f \\ k_y = k_0 y/f}}$$

$$= \frac{4\pi^2 f^2}{k_0^2}\left[C_{11}(-x,-y) + C_{22}(-x,-y) + C_{12}(-x - 2x_0, -y) + C_{21}(-x + 2x_0, -y) \right],$$

where

$$C_{mn}(x,y) = g_m(x,y) \otimes g_n(x,y).$$

Bibliography

Gabor, D. (1948). "A new microscopic principle," *Nature* 161, pp. 777–778.

Leith, E. N. and J. Upatnieks (1964). "Wavefront reconstruction with diffused illumination and three-dimensional object," *Journal of the Optical Society of America* 54, pp.1295–1301.

Poon, T.-C. and P. P. Banerjee (2001). *Contemporary Optical Image Processing with MATLAB®*. Elsevier, United Kingdom.

Poon, T.-C. and J.-P. Liu (2014). *Introduction to Modern Digital Holography with MATLAB*. Cambridge University Press, Cambridge, United Kingdom.

Poon, T.-C. (2007). *Optical Scanning Holography with MATLAB®*. Springer, New York.

Poon, T.-C. and T. Kim (2018). *Engineering Optics with MATLAB®*, 2nd ed., World Scientific, New Jersey.

Vander Lugt, A. (1963). "Signal detection by complex spatial filtering," *IEEE Transactions on Information Theory*, IT-10, pp.139–145.

Weaver, C. J. and J. W. Goodman (1966). "A technique for optically convolving two functions," *Applied Optics* 5, pp. 1248–1249

Yu, F. T. S (1982). *Optical Information Processing*. John Wiley & Sons, New York.

6 Digital Holography

Film-based holography employs the use of high-resolution films such as the use of photopolymers or photorefractive materials for recording. Resolution of several thousand-line pairs per millimeter is common. These materials, while having high resolution, have a couple of drawbacks. The film-based techniques are typically slow for real-time applications and difficult to allow direct access to the recorded hologram for manipulation and subsequent processing. With recent advances in high-resolution solid-state 2-D sensors (e.g., the charge-coupled device (CCD) or complementary metal-oxide semiconductor (CMOS) sensor) and the availability of ever-increasing power of computers and digital data storage capabilities, holography coupled with electronic/digital devices has become an emerging technology with an increasing number of applications such as in metrology, nondestructive testing, and three-dimensional (3D) imaging.

While electronic detection of holograms by a TV camera was first performed by Enloe et al. in 1966, hologram numerical reconstruction was initiated by Goodman and Lawrence [1967]. With the advent of high-quality CCDs and faster computing power, digital holography has been widely accepted by the optical metrology community as well as holography community since 2005 [Schnars and Jueptner (2005)]. In *digital holography*, it has meant that holographic information of 3-D objects is captured by a CCD, and reconstruction of holograms is subsequently calculated numerically. Nowadays, digital holography means the following situations as well. Holographic recording is done by an electronic device (such as a CCD, CMOS sensor or a photodetector), and the recorded hologram can be numerically reconstructed or sent to a display device (called a spatial light modulator (SLM), to be discussed in Chapter 7) for optical reconstruction. Or, hologram construction is completely numerically simulated. The resulting hologram is sent subsequently to a display device for optical reconstruction. This aspect of digital holography is often known as *computer-generated holography*.

6.1 Coherent Digital Holography

6.1.1 CCD Limitations

Conventional on-axis and off-axis coherent holography employ an electronic device, such as a CCD or a COMOS sensor, to capture holographic fringes. In this section, we discuss the properties and limitations of these sensor arrays.

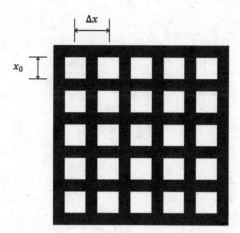

Figure 6.1-1 Structure of a CCD/CMOS 2-D array sensor

A CCD sensor and a CMOS sensor are composed of numerous light sensing units, namely pixels, arranged as a two-dimensional array, shown in Figure 6.1-1. Each pixel of area $x_0 \times x_0$ (assuming a square pixel) is an active area sensitive to light illumination. There is a shielding around the active area of the pixel. The center-to-center distance of the pixels is called the *pixel pitch*, denoted as Δx in Figure 6.1-1. The ratio of the light sensitive area to the whole array area is called the *fill factor, F*. For example, as shown in Figure 6.11, $F = 25\,x_0^2 \big/ (5\Delta x)^2 = x_0^2 \big/ \Delta x^2$. For practical 2-D array sensors, the pixel pitch $\Delta x \approx 4-8\,\mu m$, and the fill factor $F \approx 0.8$ to 0.9. A common CCD sensor has 1024×768 pixels. Neglecting the shielding area, that is, $F = 1$, the pixel pitch has the same linear dimension of the pixel size x_0. The pixel size then determines spatial resolution. With the pixel size of $6\,\mu m$, the chip size is about $6.0 \times 4.5\ mm^2$. A line pair consists of a dark line and a bright line. If one line is 0.5 micron (μm), then a line pair is 1 micron width. Hence, there is one line pair (lp) per micron, that is, 1 lp /micron, which is 1000 lp/mm. Again neglecting the shielding area, Eq. (6.1-1) allows us to approximate the *spatial resolution* in lp/mm for a given pixel size

$$\text{resolution} \left[\frac{\text{lp}}{\text{mm}} \right] = \frac{1}{2 \times \text{pixel size}[\text{in mm}]}. \tag{6.1-1}$$

For a pixel size of $5\,\mu m$, resolution $= 1 \big/ (2 \times 5 \times 10^{-3}) = 100\,\text{lp/mm}$. Note that this resolution is an order of magnitude smaller than that typically available with films. In other words, when we work with digital holography using CCDs, we have to deal with this detrimental issue and care must be taken in performing experiments.

Another important issue that needs to be addressed is the fact that CCD chip is made of discrete elements (pixels) as shown in Figure 6.1-1. When a continuous fringe intensity pattern $g(x, y)$ falls on the 2-D array (grid) formed by a CCD, the pixels capture a spatially sampled version of the pattern.

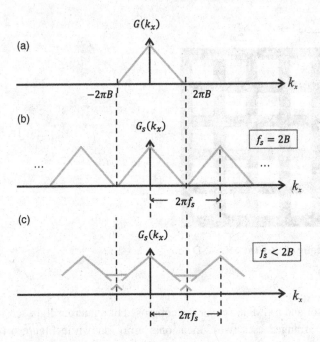

Figure 6.1-2 (a) Spectrum of $g(x)$, (b) sampled spectrum for $f_s = 2B$, and (c) under-sampled spectrum when $f_s < 2B$

Since the pixel is small enough, we assume the sampling pixels are given by $\delta_{\Delta x}(x) = \sum_{n=-\infty}^{\infty} \delta(x - n\Delta x)$. Therefore, the sampled pattern is

$$g_s(x) = g(x) \times \sum_{n=-\infty}^{\infty} \delta(x - n\Delta x), \qquad (6.1\text{-}2)$$

where we have assumed a 1-D analysis for simplicity. Let us consider a signal $g(x)$ that is bandlimited to B [cycle/mm], that is, the signal has no spatial frequencies higher than B; therefore the signal is said to have a *bandwidth* of B. In Figure 6.1-2a, we show the spectrum $G(k_x)$ of $g(x)$. Let us find the spectrum of the sampled pattern $G_s(k_x)$:

$$G_s(k_x) = \mathcal{F}\{g_s(x)\} = \mathcal{F}\left\{g(x) \times \sum_{n=-\infty}^{\infty} \delta(x - n\Delta x)\right\}. \qquad (6.1\text{-}3)$$

Since $\delta_{\Delta x}(x)$ is a periodic function with *sampling period* $T_s = \Delta x$, we can express the function in terms of Fourier series as [see the definition of Fourier series in Eq. (2.2-4a)]

$$\delta_{\Delta x}(x) = \sum_{n=-\infty}^{\infty} d_n e^{jn\frac{2\pi}{\Delta x}x}, \qquad (6.1\text{-}4a)$$

where

$$d_n = \frac{1}{\Delta x} \int_{-\Delta x/2}^{\Delta x/2} \delta_{\Delta x}(x) e^{-jn\frac{2\pi}{\Delta x}x} dx. \tag{6.1-4b}$$

Over the range of integration from $-\Delta x/2$ to $\Delta x/2$, $\delta_{\Delta x}(x) = \delta(x)$. Hence, the integration becomes

$$d_n = \frac{1}{\Delta x} \int_{-\Delta x/2}^{\Delta x/2} \delta(x) e^{-jn\frac{2\pi}{\Delta x}x} dx = \frac{1}{\Delta x}. \tag{6.1-5}$$

Using the result from Eq. (6.1-5), Eq. (6.1-4a) becomes

$$\delta_{\Delta x}(x) = \sum_{n=-\infty}^{\infty} \delta(x - n\Delta x) = \frac{1}{\Delta x} \sum_{n=-\infty}^{\infty} e^{jn\frac{2\pi}{\Delta x}x}. \tag{6.1-6}$$

Substituting Eq. (6.1-6) into Eq. (6.1-3), we have

$$G_s(k_x) = \frac{1}{\Delta x} \mathcal{F} \left\{ g(x) \times \sum_{n=-\infty}^{\infty} e^{jn\frac{2\pi}{\Delta x}x} \right\} = \frac{1}{\Delta x} \mathcal{F} \left\{ \sum_{n=-\infty}^{\infty} g(x) e^{jn\frac{2\pi}{\Delta x}x} \right\}$$

$$= \frac{1}{\Delta x} \sum_{n=-\infty}^{\infty} \mathcal{F} \left\{ g(x) e^{jn\frac{2\pi}{\Delta x}x} \right\}$$

$$= \frac{1}{\Delta x} \sum_{n=-\infty}^{\infty} G\left(k_x + \frac{2\pi n}{\Delta x} \right). \tag{6.1-7}$$

While $T_s = \Delta x$ is the sampling period, its reciprocal is called the *sampling frequency* $f_s = 1/T_s = 1/\Delta x$. With this definition, the spectrum of the sampled function can be written as

$$G_s(k_x) = f_s \sum_{n=-\infty}^{\infty} G(k_x + 2\pi n f_s). \tag{6.1-8}$$

In Figure 6. 1-2b and 6.1-2c, we show the sampled spectra for $f_s = 2B$ and $f_s < 2B$, respectively. Figure 6.1-2b displays the spectrum $G_s(k_x)$ that consists of a periodic replica of $G(k_x)$, the spectrum of the original signal $g(x)$. As long as the sampling frequency f_s is greater than twice the signal bandwidth B, $G_s(k_x)$ consists of nonoverlapping repetitions of $G(k_x)$. The signal can be recovered faithfully without error by passing the sampled signal through an ideal low-pass filter with cutoff frequency B [cycle/mm]. The filtering and recovering processes are summarized in Figure 6.1-3.

The sampling frequency

$$f_s = 2B \tag{6.1-9}$$

is called the *Nyquist sampling frequency* or *Nyquist rate* and it represents a lower bound to the sampling frequency at which $g(x)$ should be sampled in order to faithfully

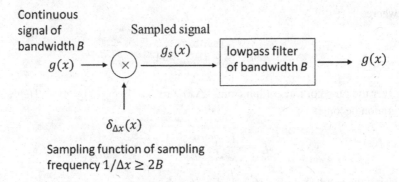

Sampling function of sampling
frequency $1/\Delta x \geq 2B$

Figure 6.1-3 Sampling process and recovery process

recover the original signal. Sampling must be at a rate higher than the Nyquist rate in order to faithfully recover the original signal.

When the signal is under-sampled, that is, the sampling frequency is less than the Nyquist sampling frequency, the retrieved signal is incorrect, resulting in errors known as *aliasing* or *spectral folding*, illustrated in Figure 6.1-2c.

6.1.2   Optical Recording of Digital Holograms

We have found out from the last section that typically CCD arrays have limited resolution of the order of about 100 lp/mm due to the finite size of the pixel. In order to avoid aliasing, we need to sample the fine fringe pattern of the hologram higher than the Nyquist rate, which is dictated by the pixel size Δx [see Eq. (6.1-9)].

Let us take a point source at the furthest extent of the object. For that point source, we have an off-axis point-source hologram [see Eq. (5.1-10)]:

$$FZP(x - x_0, y - y_0; z_0) = A + B \sin\left[\frac{k_0}{2z_0}\left[(x - x_0)^2 + (y - y_0)^2\right]\right].$$

and its local frequency is [see Eq. (5.1-9)]

$$f_{\text{local}} = \frac{1}{2\pi}\frac{d}{dx}\left(\frac{k_0}{2z_0}(x - x_0)^2\right) = \frac{1}{\lambda_0 z_0}(x - x_0). \tag{6.1-10}$$

For $x < 0$ on the hologram plane [see Figure 5.1-3], we see that the local fringe frequency $|f_{\text{local}}|$ is getting larger. However, we can lower the local fringe frequency by limiting the extent of the object, that is, letting x_0 smaller. Therefore, we see that the larger the object in lateral dimension, the finer the local fringe frequency is for holographic recording. Hence, the CCD could only capture objects with small lateral dimension if we do not want to record with aliasing. Alternatively, we can avoid aliasing by placing the object further away from the CCD as z_0 is getting larger and hence, the local frequency gets smaller.

What happens when we want to employ off-axis holographic recording? For an off-axis point-source hologram, we have [see Eq. (5.1-16)]

$$t(x,y) = A + B \sin\left[\frac{k_0}{2z_0}(x^2 + y^2) + k_0\sin\theta x\right].$$

Let use rewrite the above equation, and we have

$$t(x,y) = A + B \sin\left[\frac{k_0}{2z_0}(x^2 + y^2) + 2\pi f_c x\right], \qquad (6.1\text{-}11)$$

where $f_c = k_0\sin\theta / 2\pi = \sin\theta / \lambda_0$ is called the *spatial carrier*. The terminology of carrier is actually borrowed from communication theory. In the context of our situation, the high-frequency carrier signal $\sin(2\pi f_c x)$ carries the holographic information $\frac{k_0}{2z_0}(x^2 + y^2)$. For a recording angle $\theta = 45^0$ and $\lambda_0 = 0.6\,\mu m$ for a red laser, $f_c \sim 1,000$ cycle/mm. This high spatial frequency translates into a recording medium of resolution at least 1,000 lp/mm in order to employ this technique for holographic recording. Indeed, the local frequency of the hologram is, according to Eq. (5.1-9),

$$f_{\text{local}} = \frac{1}{2\pi}\frac{d}{dx}\left(\frac{k_0}{2z_0}x^2 + 2\pi f_c x\right) = \frac{k_0 x}{2\pi z_0} + f_c = \frac{x}{\lambda_0 z_0} + f_c. \qquad (6.1\text{-}12)$$

From the above discussion, we can conclude that when capturing holographic information with CCD arrays, we must guard against aliasing errors. To do this, we can limit the lateral dimension of the object, place the object farther away from the CCD (making z_0 larger), or make the recording angle small (making f_c smaller) if off-axis recording is employed.

On-axis Fresnel Hologram

For the recording of on-axis Fresnel holograms, the setup is shown in Figure 6.1-4. In the analysis presented below, we involve the x–z plane only. Assume that the CCD contains $M \times M$ pixels, so its size is $M\Delta x \times M\Delta x$. The size of the object is D and centered on the optical axis. The distance between the object and the CCD sensor is z, and the plane-wave reference wave is coupled to the object light via a beamsplitter (BS). At the CCD plane, each object point together with the reference wave forms a FZP.

The ray of an object point emitting from the object's lower edge to the upper edge of the CCD sensor (indicating by a dashed line in Figure 6.1-4) indicates a maximum lateral extent of the Fresnel zone plate due to the object point on the CCD sensor. Thus, the maximum lateral dimension of the on-axis Fresnel zone plate is $(D/2 + (M\Delta x)/2)$, and the corresponding maximum local fringe frequency [see Eq. (6.1-10) for $x_0 = 0$] is

$$f_{\text{local}} = \frac{x}{\lambda_0 z}\bigg|_{x=D/2+(M\Delta x)/2} = \frac{1}{2z\lambda_0}(D + M\Delta x).$$

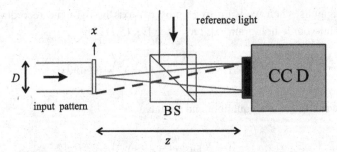

Figure 6.1-4 Typical setup of on-axis Fresnel holography

To avoid under-sampling of the holographic fringes, the sampling frequency of the CCD must be at least twice the highest local frequency of the incoming fringe pattern, that is, $2B$ [see Eq. (6.1-9)]:

$$f_s = \frac{1}{\Delta x} \geq 2B = 2\frac{1}{2z\lambda_0}(D + M\Delta x), \tag{6.1-13}$$

giving

$$z \geq \frac{\Delta x}{\lambda_0}(D + M\Delta x). \tag{6.1-14}$$

Therefore, there is a minimum distance between the object and the CCD senor if aliasing is to be avoided.

Off-axis Fresnel Hologram

For recording of off-axis Fresnel holograms, the setup is shown in Figure 6.1-5. The analysis is similar to that in the last section for on-axis recording. Again, for an object point emitting from the object's lower edge, the maximum lateral dimension of the Fresnel zone plate is $(D/2 + (M\Delta x)/2)$. The corresponding maximum local fringe frequency [see Eq. (6.1-12)] is

$$f_{\text{local}} = \frac{x}{\lambda_0 z}\bigg|_{x=D/2+(M\Delta x)/2} + f_c = \frac{D/2 + (M\Delta x)/2}{\lambda_0 z} + \frac{\sin\theta}{\lambda_0}. \tag{6.1-15}$$

To avoid aliasing, we need

$$f_s = \frac{1}{\Delta x} \geq 2B, \tag{6.1-16}$$

where $B = f_{\text{local}}$ is given by Eq. (6.1-15). Combining Eqs. (6.1-15) and (6.1-16), we derive

$$z \geq \frac{\Delta x(D + M\Delta x)}{(\lambda_0 - 2\sin\theta\Delta x)},$$

which is the minimum distance the object must be positioned away from the CCD sensor.

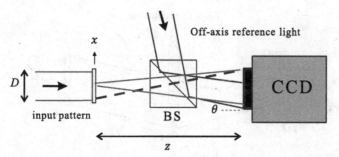

Figure 6.1-5 Typical setup of off-axis Fresnel holography. θ is the recording angle

Fourier Hologram

For recording of *Fourier transforms*, the setup is shown in Figure 6.1-6 [also see Figure 5.3-3a]. The input of size D is centered on the optical axis and is on the front focal plane of the Fourier transform lens. A CCD is at the back focal plane of the Fourier lens. A single point of the input will produce a plane wave on the CCD, and at the same time, a reference point source $\delta(x+x_0)$ produced by the reference light provides a tilted plane wave $e^{-jk_0x_0x/f}$ on the CCD plane [see Section 5.3.2]. The two plane waves result in a uniform sinusoidal fringe on the CCD. However, the finest fringe is contributed from the point source at $\delta(x-D/2)$, which emits at the edge of the input at $x=D/2$. This point source produces a plane wave $e^{jk_0Dx/2f}$ on the CCD. Hence, the finest fringe on the CCD plane is given by

$$\left|e^{jk_0Dx/2f}+e^{-jk_0x_0x/f}\right|^2=2+2\cos\left[\pi\frac{(D+2x_0)x}{\lambda_0f}\right]. \qquad (6.1\text{-}17)$$

In order to resolve the finest fringe, the Nyquist frequency associated with the CCD must be at least twice the local frequency, which is

$$f_s=\frac{1}{\Delta x}\geq 2B, \qquad (6.1\text{-}18)$$

where $B=f_{\text{local}}=\dfrac{1}{2\pi}\dfrac{d}{dx}\left(\pi\dfrac{(D+2x_0)x}{\lambda_0f}\right)=\dfrac{(D+2x_0)}{2\lambda_0f}$ is obtained from Eq. (6.1-17). Therefore, we have

$$\frac{1}{\Delta x}\geq\frac{(D+2x_0)}{\lambda_0f},$$

or

$$D\leq\frac{\lambda_0f}{\Delta x}-2x_0. \qquad (6.1\text{-}19)$$

To avoid aliasing during recording, the size of the input, that is, D must be less than $\dfrac{\lambda_0f}{\Delta x}-2x_0$. To maximize the recording size, x_0 can be set to zero. However, this is not

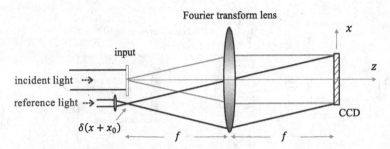

Figure 6.1-6 Typical setup of Fourier holography

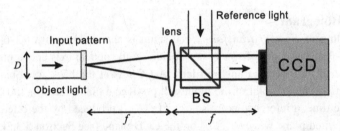

Figure 6.1-7 Alternative setup of Fourier holography to maximize the recording size of the object

possible in practice using the recoding geometry shown in Figure 6.1-6. An alternative and practical way to record a Fourier hologram is shown in Figure 6.1-7, and in this case, x_0 is effective zero, giving

$$D \leq \frac{\lambda_0 f}{\Delta x}. \tag{6.1-20}$$

6.2   Modern Digital Holographic Techniques

We have seen that off-axis holography is an effective technique to separate the zeroth-order light and the twin image. However, because of the limited resolution of the CCD, it is difficult to record a high-quality off-axis digital Fresnel hologram. Digital phase-shifting holography is a modern-day technique to fully utilize the limited resolution of the CCD. In this section, we also cover optical scanning holography that is a single-pixel holographic recording technique, completely bypassing the use of a CCD for recording.

6.2.1   Phase-Shifting Holography

Gabor and Goss pioneered the phase-shifting technique in holography in 1966. A few years later, Burckhardt and Enloe [1969] proposed the television transmission of a hologram using phase shifting and finally the first experiments were demonstrated by Berrang [1970]. The central theme of phase shifting was to reduce the resolution requirements on the recording medium.

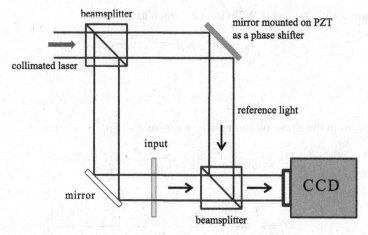

beamsplitter

mirror mounted on PZT
as a phase shifter

collimated laser

reference light

input

mirror

CCD

beamsplitter

Figure 6.2-1 Typical setup of phase-shifting digital holography

A typical setup for digital phase-shifting holography (PSH) is shown in Figure 6.2-1. The setup is basically identical to that of conventional on-axis Fresnel digital holography with the exception that on the reference arm of the optical system, a mirror is mounted on a piezoelectric transducer (PZT) as a phase shifter to effectuate phase shifting. As the PZT is moved, it creates an optical path difference between the object wave and the reference wave.

In PSH, we need to take multiple holograms corresponding to the various phase differences between the object wave and the reference wave. We can express the phase-shifted hologram as

$$I_\delta = \left| \psi_0 + \psi_r e^{-j\delta} \right|^2 \tag{6.2-1}$$
$$= \left| \psi_0 \right|^2 + \left| \psi_r \right|^2 + \psi_0 \psi_r^* e^{j\delta} + \psi_0^* \psi_r e^{-j\delta},$$

where ψ_0 and ψ_r are the complex amplitude of the object wave and the reference wave, respectively on the CCD; δ stands for the phase induced by the phase shifter.

Two-Step Phase-Shifting Holography (Quadrature Phase-Shifting Holography)

In this case, only two holograms with a zero phase shift and a $\pi/2$ phase shift are acquired sequentially. According to Eq. (6.2-1), the two holograms for $\delta = 0$ and $\delta = \pi/2$ are

$$I_0 = \left| \psi_0 \right|^2 + \left| \psi_r \right|^2 + \psi_0 \psi_r^* + \psi_0^* \psi_r, \tag{6.2-2a}$$

and

$$I_{\pi/2} = \left| \psi_0 + \psi_r e^{-j\pi/2} \right|^2 = \left| \psi_0 \right|^2 + \left| \psi_r \right|^2 + j\psi_0 \psi_r^* - j\psi_0^* \psi_r \tag{6.2-2b}$$

Equations (6.2-2a) and (6.2-2b) can be written as

$$I_0 - |\psi_0|^2 - |\psi_r|^2 = \psi_0\psi_r^* + \psi_0^*\psi_r,$$

and

$$I_{\pi/2} - |\psi_0|^2 - |\psi_r|^2 = j\psi_0\psi_r^* - j\psi_0^*\psi_r.$$

Hence, from the above two equations, we have

$$I_0 - |\psi_0|^2 - |\psi_r|^2 - j\left(I_{\pi/2} - |\psi_0|^2 + |\psi_r|^2\right) = 2\psi_0\psi_r^*,$$

or

$$\psi_0 = \frac{\left(I_0 - |\psi_0|^2 - |\psi_r|^2\right) - j\left(I_{\pi/2} - |\psi_0|^2 - |\psi_r|^2\right)}{2\psi_r^*}. \tag{6.2-3}$$

This is the complex amplitude of the object at the hologram plane, that is, the object wave on the hologram plane. In digital holography, it is often referred to as the *complex hologram* of the object as we can retrieve the amplitude distribution of the object in the object plane by performing back propagation. In Eq. (6.2-3), ψ_r is usually known as the reference wave, which is either a plane wave or a spherical wave in practice. However, we still need two quantities, that is, the intensities of the object wave, $|\psi_0|^2$, and the reference wave, $|\psi_r|^2$, in order to completely solve for ψ_0. These two intensities can be measured easily. We, therefore, need a total of four measurements (two holograms and two intensity patterns) in order to find the complex hologram of the object from Eq. (6.2-3) [Guo and Devaney (2004), Wang et al. (2004)]. Since the two holograms are with a zero phase shift and a $\pi/2$ phase shift, I_0 and $I_{\pi/2}$ are called the *in-phase hologram* and the *quadrature-phase hologram*, respectively. Hence, the term quadrature phase shifting holography is often used. Figure 6.2-2 shows the simulation results for two-step PSH.

The images in Figure 6.2-2 are generated using the m-file shown below.

```
% simulation_of_two_step_PSH.m
% Adapted from "Introduction to Modern Digital Holography"
% by T.-C. Poon & J.-P. Liu
% Cambridge University Press(2014), Tables 5.2
clear all; close all;

%%Reading input bitmap file
I=imread('smiley3.bmp','bmp');    % original object, 256x256
I=double(I);

% parameter setup
M=256;
deltax=0.001; % pixel pitch 0.001 cm (10 um)
w=633*10^-8;  % wavelength 633 nm
z=25; % 25 cm, propagation distance
delta=pi/2; % phase step
```

In-phase hologram

(a)

Q-phase hologram

(b)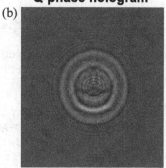

intensity pattern of object wave

(c)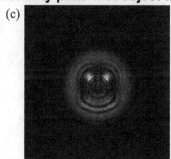

intensity pattern of reference wave

(d)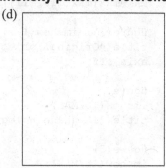

Reconstructed image

(e)

Figure 6.2-2 Simulation of two-step PSH. (a) In-phase hologram, (b) quadrature-phase hologram, (c) intensity pattern of object wave, (d) intensity pattern of reference wave (all-white pattern denoting uniform illumination), and (e) reconstructed image from the complex hologram, Eq. (6.2-3).

```
%Step 1: simulation of propagation
r=1:M;
c=1:M;
[C, R]=meshgrid(c, r);
A0=fftshift(ifft2(fftshift(I)));
deltaf=1/M/deltax;
p=exp(-2i*pi*z.*((1/w)^2-((R-M/2-1).*deltaf).^2-...
```

```
      ((C-M/2-1).*deltaf).^2).^0.5);
Az=A0.*p;
EO=fftshift(fft2(fftshift(Az)));

%Step 2: Interference at the hologram plane
AV=(min(min(abs(EO)))+max(max(abs(EO))));
% the amplitude of reference light

% Recording of two phase-shifting holograms
I0=(EO+AV).*conj(EO+AV);   %zero phase shift
I1=(EO+AV*exp(-1j*delta)).*conj(EO+AV*exp(-1j*delta));      %90
  degrees phase shift
I2=(EO).*conj(EO); %intensity of object wave
I3=(AV).*conj(AV);% intensity of reference wave
MAX=max(max([I0, I1]));

figure(1); imshow(I);
title('Original object')
axis off

figure(2)
imshow(I0/MAX);
title('In-phase hologram')

figure(3)
imshow(I1/MAX);
axis off
title('Q-phase hologram')

I2=I2./max(max(I2));
figure (4)
imshow(I2);
axis off
title('intensity pattern of object wave')

I3=I3./max(max(I3));
figure (5)
imshow(I3);
axis off
title('intensity pattern of reference wave')

%Step 3: Reconstruction
CH=(I0-I2-I3)-1j*(I1-I2-I3); % the complex hologram (2-step PSH)
A1=fftshift(ifft2(fftshift(CH)));
Az1=A1.*conj(p);
EI=fftshift(fft2(fftshift(Az1)));
EI=(EI.*conj(EI));
EI=EI/max(max(EI));

figure(6);
imshow(EI);
title('Reconstructed image')
axis off
```

Three-Step Phase-Shifting Holography

In three-step PSH, three holograms are acquired sequentially. The phase differences for the three holograms are $\delta = 0$, $\delta = \pi/2$, and $\delta = \pi$, and the three holograms are

$$I_0 = |\psi_0|^2 + |\psi_r|^2 + \psi_0\psi_r^* + \psi_0^*\psi_r, \tag{6.2-4a}$$

$$I_{\pi/2} = |\psi_0 + \psi_r e^{-j\pi/2}|^2 = |\psi_0|^2 + |\psi_r|^2 + j\psi_0\psi_r^* - j\psi_0^*\psi_r, \tag{6.2-4b}$$

and

$$I_\pi = |\psi_0 + \psi_r e^{-j\pi}|^2 = |\psi_0|^2 + |\psi_r|^2 - \psi_0\psi_r^* - \psi_0^*\psi_r. \tag{6.2-4c}$$

Similar to some mathematical manipulations in the last section, we can find that the complex hologram of the object is given by

$$\psi_0 = \frac{(1+j)(I_0 - I_{\pi/2}) + (j-1)(I_\pi - I_{\pi/2})}{4\psi_r^*}. \tag{6.2-5}$$

Four-Step Phase-Shifting Holography

In four-step PSH, four holograms are acquired sequentially. The phase differences for the four holograms are $\delta = 0$, $\delta = \pi/2$, $\delta = \pi$, and $\delta = 3\pi/2$. Therefore, according to Eq. (6.2-1), the four holograms are

$$I_0 = |\psi_0|^2 + |\psi_r|^2 + \psi_0\psi_r^* + \psi_0^*\psi_r, \tag{6.2-6a}$$

$$I_{\pi/2} = |\psi_0|^2 + |\psi_r|^2 + j\psi_0\psi_r^* - j\psi_0^*\psi_r, \tag{6.2-6b}$$

$$I_\pi = |\psi_0|^2 + |\psi_r|^2 - \psi_0\psi_r^* - \psi_0^*\psi_r, \tag{6.2-6c}$$

$$I_{3\pi/2} = |\psi_0|^2 + |\psi_r|^2 - j\psi_0\psi_r^* + j\psi_0^*\psi_r. \tag{6.2-6d}$$

From the four holograms, we can derive the complex hologram of the object and it is given by

$$\psi_0 = \frac{(I_0 - I_\pi) - j(I_{\pi/2} - I_{3\pi/2})}{4\psi_r^*}. \tag{6.2-7}$$

In standard quadrature PSH, two holograms and two intensity patterns are needed to completely calculate the complex hologram for zeroth-order-free and twin image-free reconstruction. In four-step PSH, also four measurements are needed. The measurement of the intensity pattern of the object wave can be omitted in three-step PSH, thus reducing the overall holographic recording process. Recent advances in PSH research lead to the use of only two holograms and a reference intensity to extract the complex hologram [Meng et al. (2006)]. While it also takes three measurements, it is advantageous as compared to those in three-step PSH because the recording of the intensity pattern is not sensitive to any vibration that causes phase variation between the object wave and the reference wave. Most recent development in two-step PSH has led to what is known as *two-step-only quadrature PSH* in which only two holograms are needed to retrieve the complex hologram of the object [Liu and Poon (2009)]. Subsequently, Liu et al. (2011) have provided a report on the comparison of two-, three- and four-exposure quadrature PSH. Interested readers are referred to the references in the biography list at the end of this chapter.

6.2.2   Optical Scanning Holography in Coherent Mode

Two-Pupil Heterodyning Image Processor
Optical scanning holography (OSH) is a radically different approach to holographic recording as it bypasses the use of a CCD for recording. Indeed, it is a single-pixel holographic recording technique where a photodetector such as a photodiode is used for recording. The original idea of OSH was dated back to the late 1970s when Poon and Korpel investigated bipolar image processing with their two-pupil heterodyning image processor [1979]. We have briefly encountered a *two-pupil system* in Section 4.5. In this section, we will first describe a generalized version of the two-pupil system, shown in Figure 6.2-3. We will then show how the system is configured to become a holographic recording system and such a technique is known as optical scanning holography. The overall system is hybrid (optical/electrical). We shall describe the optical system first before explaining the electrical processing part of the system.

We essentially have a 4-*f* system with lenses L1 and L2 when z_0 is zero in the optical system. z_0 is a distance measured away from the back focal plane of lens L1. We model the 3-D object as a stack of transverse slices where each slide is represented by a transparency function $T(x, y; z)$, which is thin and weakly scattering. $p_1(x, y)$ and $p_2(x, y)$ are the two pupils located at the front focal plane of lens L1, and they are illuminated by plane waves at frequencies ω_0 and $\omega_0 + \Omega$, respectively, where ω_0 is the frequency of a laser. The frequency shift of one of the plane waves from ω_0 to $\omega_0 + \Omega$ can be conveniently done by the use of a frequency shifter, such as an acousto-optic or electro-optic modulator (to be covered in Chapter 7). The beam splitter is used to combine the two beams exiting from the two pupils and the combined beam is projected onto the 3-D object through the xy-scanner for a raster scan as shown in Figure 6.2-3.

The complex field oscillating at frequency ω_0 from $p_1(x, y)$ just before the object slice at position $z_0 + z$ from the focal plane of lens L1 is given by

$$s_1(x, y; z + z_0) e^{j\omega_0 t} = \left[\mathcal{F}\{p_1(x, y)\}\Big|_{k_x = \frac{k_0 x}{f}, k_y = \frac{k_0 y}{f}} * h(x, y; z + z_0) \right] e^{j\omega_0 t} \quad (6.2\text{-}8)$$

as the pupil has been Fourier transformed by lens L1 and then propagated for a distance $z + z_0$. Similarly, the complex field exiting from $p_2(x, y)$ with a time oscillation $\omega_0 + \Omega$ at the object slice is

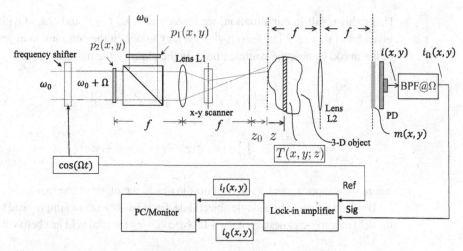

Figure 6.2-3 Typical setup of two-pupil heterodyning image processing BPF @Ω: bandpass filter tuned at frequency Ω, PD: photodetector. Adapted from Zhang et al. [2019].

$$s_2\left(x,y;z+z_0\right)e^{j(\omega_0+\Omega)t} = \left[\mathcal{F}\left\{p_2\left(x,y\right)\right\}\Big|_{k_x=\frac{k_0 x}{f},k_y=\frac{k_0 y}{f}} * h\left(x,y;z+z_0\right)\right]e^{j(\omega_0+\Omega)t}. \quad (6.2\text{-}9)$$

Hence, the total complex field at the object slide is

$$s\left(x,y;z\right) = s_1\left(x,y;z+z_0\right)e^{j\omega_0 t} + s_2\left(x,y;z+z_0\right)e^{j(\omega_0+\Omega)t}. \quad (6.2\text{-}10)$$

This is the total scanning field, which consists of two scanning beams s_1 and s_2. According to the principle of optical scanning developed in Section 4.4, the field just after the object slide, $T\left(x,y;z\right)$, is given by $s\left(x''-x(t),y''-y(t);z\right)T\left(x'',y'';z\right)$ or alternatively given by

$$s\left(x',y';z\right)\,T\left(x'+x(t),y'+y(t);z\right)$$

if $x''-x=x'$ and $y''-y=y'$ have been used. Again, $x(t)$ and $y(t)$ represent the instantaneous 2-D position of the object with respect to the scanning field. This field now propagates through the Fourier transform lens L2 and reaches a mask $m(x,y)$ placed in front of the photodetector, PD.

According to Figure 3.4-6, the complex field, $t(x,y)$, d_0 in front of the Fourier transform lens of focal length f, gives the field distribution on the focal plane proportional to [see Eq. (3.4-29)]

$$e^{-j\frac{k_0}{2f}\left[1-\frac{d_0}{f}\right]\left(x^2+y^2\right)}\,\mathcal{F}\left\{t(x,y)\right\}\Big|_{k_x=\frac{k_0 x}{f},\,k_y=\frac{k_0 y}{f}} = e^{-j\frac{k_0}{2f}\left[1-\frac{d_0}{f}\right]\left(x^2+y^2\right)}\int\!\!\!\int_{-\infty}^{\infty} t\left(x',y'\right)e^{\frac{jk_0}{f}(xx'+yy')}\,dx'dy'$$

Using this result in our situation, we replace d_0 with $f - z$, and $t(x, y)$ with the field given by $s(x', y'; z)T(x' + x(t), y' + y(t); z)$ to obtain the complex field just in front of the mask. Hence, the complex field on the plane of the mask is

$$e^{-j\frac{k_0}{2f}\left[1-\frac{f-z}{f}\right]\left(x_m^2 + y_m^2\right)} \mathcal{F}\left\{s(x', y'; z)\ T(x' + x, y' + y; z)\right\}\Big|_{k_x = -\frac{k_0 x_m}{f}, k_y = -\frac{k_0 y_m}{f}}$$

$$= e^{-j\frac{k_0 z}{2f^2}\left(x_m^2 + y_m^2\right)} \int\!\!\int_{-\infty}^{\infty} s(x', y'; z)\ T(x' + x, y' + y; z) e^{\frac{jk_0}{f}(x_m x' + y_m y')}\, dx'dy',$$

where we denote x_m and y_m as the coordinates in the plane of the mask.

This field is caused by a single object slide. For a 3-D object, we simply need to integrate the field over the thickness, z, of the 3-D object, giving the total field just before the mask as

$$\int e^{-j\frac{k_0 z}{2f^2}\left(x_m^2 + y_m^2\right)} \int\!\!\int_{-\infty}^{\infty} s(x', y'; z)\ T(x' + x, y' + y; z) e^{\frac{jk_0}{f}(x_m x' + y_m y')}\, dx'dy'dz\ .$$

Finally, the complex field after the mask is

$$\psi(x, y; x_m, y_m) =$$

$$\left\{\int e^{-j\frac{k_0 z}{2f^2}\left(x_m^2 + y_m^2\right)} \int\!\!\int_{-\infty}^{\infty} s(x', y'; z)\ T(x' + x, y' + y; z)\ e^{\frac{jk_0}{f}(x_m x' + y_m y')}\, dx'dy'dz\right\} m(x_m, y_m).$$

$$(6.2\text{-}11)$$

The photodetector, PD, responding to intensity gives out a current $i(x, y)$ as an output by spatially integrating over the active area of the detector [see Eq. (4.4-3)]:

$$i(x, y) \propto \int\!\!\int |\psi(x, y; x_m, y_m)|^2\, dx_m dy_m. \qquad (6.2\text{-}12)$$

Coherent Processing Using Point Detector

Let us examine a simple case that when the mask is a pinhole, that is, $m(x_m, y_m) = \delta(x_m, y_m)$. In this case, we have a *point detector*. In other words, the detector only receives light at a single point location within its active area. When this happens, the complex field after the mask is, according to Eq. (6.2-11) [also see the product property of the delta function, Eq. (2.1-4)],

$$\psi(x, y; x_m, x_m) = \left\{\int\!\!\int\!\!\int_{-\infty}^{\infty} s(x', y'; z)\ T(x' + x, y' + y; z)\, dx'dy'dz\right\} \delta(x_m, y_m).$$

Putting this field into Eq. (6.4-12) and after integrating over x_m and y_m, Eq. (6.2-12) becomes

$$i(x, y) \propto \left\{\int\!\!\int\!\!\int_{-\infty}^{\infty} s(x', y'; z')\ T(x' + x, y' + y; z')\, dx'dy'dz'\right\}$$

$$\times \left\{\int\!\!\int\!\!\int_{-\infty}^{\infty} s^*(x'', y''; z'')\ T^*(x'' + x, y'' + y; z'')\, dx''dy''dz''\right\}.$$

Substituting the scanning field, $s(x, y; z)$, from Eq. (6.2-10) into the above equation, we have

$$i(x, y) \propto$$

$$\left\{ \iiint_{-\infty}^{\infty} \left[s_1(x', y'; z' + z_0) e^{j\omega_0 t} + s_2(x', y'; z' + z_0) e^{j(\omega_0 + \Omega)t} \right] T(x' + x, y' + y; z') \, dx' dy' dz' \right\}$$

$$\left\{ \iiint_{-\infty}^{\infty} \left[s_1^*(x'', y''; z'' + z_0) e^{-j\omega_0 t} + s_2^*(x'', y''; z'' + z_0) e^{-j(\omega_0 + \Omega)t} \right] T^*(x'' + x, y'' + y; z'') \, dx'' dy'' dz'' \right\}$$

Rearranging the above equation, we have

$$\int \left[s_1(x', y'; z' + z_0) e^{j\omega_0 t} + s_2(x', y'; z' + z_0) e^{j(\omega_0 + \Omega)t} \right] \left[s_1^*(x'', y''; z'' + z_0) e^{-j\omega_0 t} \right.$$
$$\left. + s_2^*(x'', y''; z'' + z_0) e^{-j(\omega_0 + \Omega)t} \right] T(x' + x, y' + y; z')$$
$$\times T^*(x'' + x, y'' + y; z'') \, dx' dy' dx'' dy'' dz' dz''$$

The terms involving $s_1 s_1^*$ and $s_2 s_2^*$ give a baseband current $i_{base}(t)$ as

$$i_{base}(x, y) =$$
$$\int (s_1 s_1^* + s_2 s_2^*) T(x' + x, y' + y; z') T^*(x'' + x, y'' + y; z'') \, dx' dy' dx'' dy'' dz' dz'',$$

whereas the terms involving frequency Ω give a *heterodyne current* $i_\Omega(t)$ as

$$i_\Omega(x, y) =$$
$$\int \left(s_2 s_1^* e^{j\Omega t} + s_1 s_2^* e^{-j\Omega t} \right) T(x' + x, y' + y; z') T^*(x'' + x, y'' + y; z'') \, dx' dy' dx'' dy'' dz' dz''.$$

$$(6.2\text{-}13a)$$

Grouping the primed and double primed coordinates, and rearranging, we have

$$i_\Omega(x, y) = \left[\int s_2(x', y'; z' + z_0) T(x' + x, y' + y; z') \, dx' dy' dz' \right.$$
$$\left. \int s_1^*(x'', y''; z'' + z_0) T^*(x'' + x, y'' + y; z'') \, dx'' dy'' dz'' \right] e^{j\Omega t}$$
$$+ \left[\int s_2^*(x'', y''; z'' + z_0) T^*(x'' + x, y'' + y; z'') \, dx'' dy'' dz'' \right.$$
$$\left. \int s_1(x', y'; z' + z_0) T(x' + x, y' + y; z') \, dx' dy' dz' \right] e^{-j\Omega t}. \qquad (6.2\text{-}13b)$$

We now recognize that the second term is simply the complex conjugate of the first term. We know $2\text{Re}\,[e^{j\theta}] = e^{j\theta} + e^{-j\theta}$, where Re [.] denotes taking the real part of the quantity being bracketed. By neglecting a constant factor, we can write $i_\Omega(t)$ as

$$i_\Omega(x, y) = \text{Re}\left[\int s_2(x', y'; z' + z_0) T(x' + x, y' + y; z') \, dx' dy' dz' \right.$$
$$\left. \int s_1^*(x'', y''; z'' + z_0) T^*(x'' + x, y'' + y; z'') \, dx'' dy'' dz'' e^{j\Omega t} \right]. \quad (6.2\text{-}14)$$

Equation (6.2-14) is fairly complicated, albeit quite general. We observe that the transparency function of the object slide $T(x, y; z)$ and $T^*(x, y; z)$, which are complex in general, are processed by s_2 and s_1^*, respectively. The system therefore performs *coherent processing*, that is, the manipulation and processing of the complex amplitude of the object.

Specific Case with One of the Pupils Being a Delta Function

Recording   For a specific case, let us specify some pupil functions. We leave $p_2(x, y)$ as is but let $p_1(x, y) = \delta(x, y)$, that is, a point source, and this will give a uniform plane wave illuminating the object by inspecting Figure 6.2-3. Mathematically, the point source pupil, according to Eq. (6.2-8), gives a uniform plane wave illumination:

$$s_1(x, y; z + z_0) = \mathcal{F}\{p_1(x, y)\}\Big|_{k_x = \frac{k_0 x}{f}, k_y = \frac{k_0 y}{f}} * h(x, y; z + z_0)$$

$$= \mathcal{F}\{\delta(x, y)\}\Big|_{k_x = \frac{k_0 x}{f}, k_y = \frac{k_0 y}{f}} * h(x, y; z + z_0) \propto \text{constant},$$

and this subsequently gives the following term in Eq. (6.2-14) to become a constant, that is,

$$\int s_1^*(x'', y''; z'' + z_0) T^*(x'' + x, y'' + y; z'') dx'' dy'' dz'' \propto \text{constant},$$

as the integral is a correlation integral where the correlation of a function with a constant will give a constant. With the above result, Eq. (6.2-14) becomes

$$i_\Omega(x, y) \propto \text{Re}\left[i_p^\delta(x, y) e^{j\Omega t}\right], \tag{6.2-15}$$

where

$$i_p^\delta(x, y) = \int s_2(x', y'; z + z_0) T(x' + x, y' + y; z) dx' dy' dz$$

is a complex quantity [phasor in electrical engineering]. Indeed, the amplitude and phase information of i_Ω contains the scanned and processed [by $p_2(x, y)$] information of the 3-D object given by the complex function $T(x, y; z)$.

Electronic Demodulation   Remember that $i(x, y)$ contains a baseband signal $i_{\text{base}}(x, y)$ and a heterodyned signal $i_\Omega(x, y)$. We can reject the baseband signal by passing $i(x, y)$ through an electronic bandpass filter tuned at the heterodyne frequency, Ω, as shown in Figure 6.2-3. The heterodyne signal, $i_\Omega(x, y)$, can now be processed according to the lock-in demodulation scheme shown in Figure 6.2-4, where "Sig" and "Ref" are the input signal and reference signal terminals of the amplifier, respectively. The reference signal $\cos(\Omega t)$ is typically provided from an external source. Let

$$i_p^\delta(x, y) = \left|i_p^\delta(x, y)\right| e^{j\theta(x, y)},$$

then

$$i_\Omega(x, y) = \left|i_p^\delta(x, y)\right| \cos(\Omega t + \theta(x, y)). \tag{6.2-16}$$

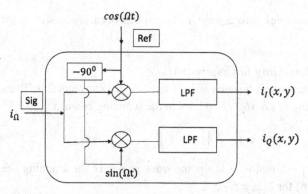

Figure 6.2-4 Lock-in amplifier, illustrating demodulation scheme. $\otimes$: electronic multiplier, LPF: low-pass filter. The block showing $-90°$ represents a phase shifter of $-90°$ of the incoming sinusoidal signal

This input signal splits into two channels to obtain two outputs, i_I and i_Q, as shown in Figure 6.2-4. Basically, each channel performs lock-in detection, which consists of electronically multiplying the incoming signal with the sine or cosine of the heterodyne frequency, Ω, and then performing low-pass filtering to extract the amplitude and phase of the incoming electrical signal.

Consider the upper channel where the electronic multiplier gives

$$i_\Omega(x,y) \times \cos\Omega = \left|i_p^\delta(x,y)\right| \cos(\Omega t + \theta(x,y)) \times \cos\Omega t$$
$$= \left|i_p^\delta(x,y)\right| \frac{1}{2}\left[\cos\theta(x,y) + \cos(2\Omega t + \theta(x,y))\right].$$

After low-pass filtering (which means we are rejecting the frequency of 2Ω), we have, apart from some constant, the so-called *in-phase component* of the incoming heterodyne signal $i_\Omega(x,y)$:

$$i_I(x,y) = \left|i_p^\delta(x,y)\right| \cos\theta(x,y). \tag{6.2-17}$$

Similarly, the lower channel gives the *quadrature component* of the heterodyne incoming signal, which is given by

$$i_Q(x,y) = \left|i_p^\delta(x,y)\right| \sin\theta(x,y). \tag{6.2-18}$$

Once $i_I(x,y)$ and $i_Q(x,y)$ have been extracted, they can be stored in a computer as two digital records, and the computer can perform the following complex addition to get a complex record $i_C(x,y)$:

$$i_C(x,y) = i_I(x,y) + j\, i_Q(x,y) = \left|i_p^\delta(x,y)\right| e^{j\theta(x,y)} = i_p^\delta(x,y). \tag{6.2-19}$$

Note that we have extracted the full complex information from the heterodyne current, that is, $i_p^\delta(x,y)$ given by Eq. (6.2-15) as follows:

$$i_C(x,y) = i_p^\delta(x,y) = i_I(x,y) + j\, i_Q(x,y)$$
$$= \int s_2(x',y';z+z_0)T(x'+x,y'+y;z)\,dx'dy'dz, \tag{6.2-20}$$

where the 3-D complex information $T(x,y;z)$ has been processed by the scanning beam s_2.

Holographic Recording in Coherent Mode

Let us see how the overall system can perform holographic recording. Remember that $p_2(x,y)$, according to Eq. (6.2-9), relates to the scanning beam, s_2, as

$$s_2(x,y;z+z_0) = \mathcal{F}\{p_2(x,y)\}\Big|_{k_x=\frac{k_0 x}{f}, k_y=\frac{k_0 y}{f}} * h(x,y;z+z_0).$$

We take $p_2(x,y) = 1$, that is, it is a plane wave as one of the scanning beams. Then, using Eq. (3.4-16) for $h(x,y;z+z_0)$,

$$s_2(x,y;z+z_0) = \mathcal{F}\{1\}\Big|_{k_x=\frac{k_0 x}{f}, k_y=\frac{k_0 y}{f}} * h(x,y;z+z_0)$$

$$\propto h(x,y;z+z_0) = \frac{jk_0}{2\pi(z+z_0)} e^{-jk_0(z+z_0)} e^{\frac{-jk_0}{2(z+z_0)}(x^2+y^2)}.$$

With this result, Eq. (6.2-20) becomes

$$i_C(x,y) = \int \frac{jk_0}{2\pi(z+z_0)} e^{-jk_0(z+z_0)} e^{\frac{-jk_0}{2(z+z_0)}(x'^2+y'^2)} T(x'+x,y'+y;z)\,dx'dy'dz.$$

$$= \int \frac{jk_0}{2\pi(z+z_0)} e^{-jk_0(z+z_0)} \left[e^{\frac{-jk_0}{2(z+z_0)}(x^2+y^2)} * T(x,y;z) \right] dz, \qquad (6.2\text{-}21)$$

where we have expressed the above result conveniently using the convolution operation. This is a *complex Fresnel hologram* of the 3-D object $T(x,y;z)$. This single-pixel digital holographic recording technique is known as *Optical Scanning Holography* in a coherent mode [Poon (2007)].

Example: Complex Fresnel Hologram of a Point Source

Consider $T(x,y;z)$ as a point object, located z_1 away from the back focal plane of lens L1 in the two-pupil system shown in Figure 6.2-3. For the point object, we model $T(x,y;z) = \delta(x,y)\delta(z-z_1)$. Let us find its complex hologram. According to Eq. (6.2-21), we have

$$i_C(x,y) = \int \frac{jk_0}{2\pi(z+z_0)} e^{-jk_0(z+z_0)} e^{\frac{-jk_0}{2(z+z_0)}(x'^2+y'^2)} \delta(x'+x,y'+y)\delta(z-z_1)\,dx'dy'dz.$$

After integrating over the depth of the object, we obtain

$$i_C(x,y) = \frac{jk_0}{2\pi(z_1+z_0)} e^{-jk_0(z_1+z_0)} \int e^{\frac{-jk_0}{2(z_1+z_0)}(x'^2+y'^2)} \delta(x'+x,y'+y)\,dx'dy'$$

$$\propto e^{\frac{-jk_0}{2(z_1+z_0)}(x^2+y^2)} = H^c(x,y). \qquad (6.2\text{-}22)$$

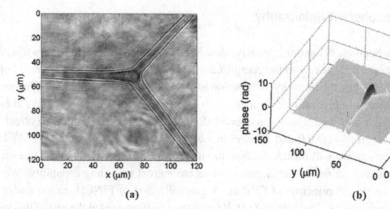

(a)   (b)

Figure 6.2-5 Reconstruction of a complex Fresnel hologram obtained by optical scanning holography operating in a coherent mode. The object is a three-pronged spongilla spicule. (a) Absolute value of the reconstruction amplitude and (b) 3-D phase profile. Reprinted from T.-C. Poon, *Journal of the Optical Society of Korea* 13, pp. 406–415 (2009), with permission. Original work is from Guy Indebetouw, Yoshitaka Tada, and John Leacock, Quantitative phase imaging with scanning holographic microscopy: an experimental assessment, BioMedical Engineering OnLine, 5:60 (2006).

This is a complex Fresnel hologram of the point source. The complex hologram does not contain the information of the zeroth-order beam and the twin image [see Figure 5.1-5 where the FZP reconstructs the zeroth-order beam, the virtual image as well as the twin image]. We obtain the virtual image reconstruction by $H^c(x,y) * h^*(x,y; z_1 + z_0) \propto \delta(x,y)$ at $z = -z_1 - z_0$ [see Figure 5.3-2]. Note that if we want to reconstruct a real image, we could take a complex conjugate of $H^c(x,y)$ to obtain $H^{c^*}(x,y)$, and that leads to real image reconstruction as $H^{c^*}(x,y) * h(x,y; z_1 + z_0) \propto \delta(x,y)$ at $z = z_1 + z_0$.

Figure 6.2-5 is an example of *quantitative phase imaging (QPI)* achievable with optical scanning holography operating in a coherent mode. In contrast to conventional phase imaging methods, such as Zernike phase contrast imaging [see P4.5], which provide visualization of the phase of biological structures in a qualitative way, QPI is a collective name for a group of microscopy methods that involves the study of weakly scattering and absorbing specimens [Popescu (2011)] and provides quantitative measurements. Figure 6.2-5a shows the absolute value of the reconstruction amplitude of a three-pronged spongilla spicule, and Figure 6.2-5b is the 3-D phase profile of the object. The specimen is almost a pure phase object and its amplitude image is rendered visible due to diffraction from the specimen's edges. To calculate the 3-D phase profile in Figure 6.2-5b, we simply first reconstruct the complex field $\psi_p(x,y)$ from the complex hologram [see Eq. (6.2-21)] by the Fresnel diffraction formula [see Eq. (3.4-17)] or the angular spectrum method [see Eq. (3.4-8)] and then calculate the phase $\theta(x,y)$ according to

$$\theta(x,y) = \tan^{-1}\left\{\frac{\mathrm{Im}[\psi_p(x,y)]}{\mathrm{Re}[\psi_p(x,y)]}\right\}. \tag{6.2-23}$$

6.3   Incoherent Holography

In Sections 6.1 and 6.2, we have discussed coherent holography. The word "coherent" means that we record the complex amplitude of the object. In *incoherent holography*, we capture the intensity distribution of a 3-D object holographically, allowing a higher signal-to-noise ratio as compared to its coherent counterpart. Nowadays, there are two major and well-known incoherent digital holographic techniques: *optical scanning holography (OSH)* and *Fresnel incoherent correlation holography (FINCH)*. In this section, we will further develop the theory of OSH discussed in the last section, and concentrate on the development of its incoherent recording capability. We will then explain the principle of FINCH. A generalization of FINCH, called *coded aperture correlation holography (COACH)*, will then be discussed at the end of this subsection.

6.3.1   Optical Scanning Holography in Incoherent Mode

We have discussed that with one of the pupils being a pinhole, that is, $p_1(x,y) = \delta(x,y)$ and the other pupil being open, that is, $p_2(x,y) = 1$, in the two-pupil system with a point detector, that is, $m(x,y) = \delta(x,y)$ [see Figure 6.2-3], we have coherent holographic recording in that the complex amplitude of the object is being recorded. In this section, we first leave the mask as is and develop a more general situation to some extent. We then let the mask $m(x,y) = 1$ to investigate incoherent processing.

According to Eq. (6.2-11), the complex field after the mask is

$$\psi(x,y;x_m,x_m) =$$

$$\left\{ \int e^{-j\frac{k_0 z}{2f^2}\left(x_m^2+y_m^2\right)} \int\limits_{-\infty}^{\infty}\!\!\int s(x',y';z)\, T(x'+x,y'+y;z) e^{\frac{jk_0}{f}\left(x_m x'+y_m y'\right)} dx'dy'dz \right\} m(x_m,y_m),$$

and the current from the photodetector is

$$i(x,y) \propto \iint \left|\psi(x,y;x_m,x_m)\right|^2 dx_m dy_m$$

$$= \int \left\{ \int e^{-j\frac{k_0 z'}{2f^2}\left(x_m^2+y_m^2\right)} \int\limits_{-\infty}^{\infty}\!\!\int s(x',y';z')\, T(x'+x,y'+y;z') e^{\frac{jk_0}{f}\left(x_m x'+y_m y'\right)} dx'dy'dz' \right\}$$

$$\times m(x_m,y_m)$$

$$\times \left\{ \int e^{j\frac{k_0 z''}{2f^2}\left(x_m^2+y_m^2\right)} \int\limits_{-\infty}^{\infty}\!\!\int s^*(x'',y'';z'')\, T^*(x''+x,y''+y;z'') e^{\frac{-jk_0}{f}\left(x_m x''+y_m y''\right)} dx''dy''dz'' \right\}$$

$$\times m^*(x_m,y_m)\, dx_m dy_m.$$

$$(6.3\text{-}1)$$

According to Eq. (6.2-10), the scanning beam is

$$s(x,y;z) = s_1(x,y;z+z_0) e^{j\omega_0 t} + s_2(x,y;z+z_0) e^{j(\omega_0+\Omega)t},$$

and with this quantity substituted into Eq. (6.3-1), we will have a baseband current $i_{base}(x,y)$ as well as a heterodyne current, $i_\Omega(x,y)$, as demonstrated earlier in the last section. Since the baseband current will be rejected by the bandpass filter placed after the photodetector, we only need to concentrate on the heterodyne term coming out from Eq. (6.3-1). By collecting the heterodyne terms, we have, similar to what has been shown earlier in the last section [see Eq. (6.2-13)], the heterodyne current is

$$i_\Omega(x,y) =$$
$$\int \left(s_2 s_1^* e^{j\Omega t} + s_1 s_2^* e^{-j\Omega t}\right) e^{j\frac{k_0}{f}[x_m(x'-x'')+y_m(y'-y'')]} e^{-j\frac{k_0(z'-z'')}{2f^2}(x_m^2+y_m^2)} \times T(x'+x,y'+y;z')$$
$$T^*(x''+x,y''+y;z'') dx'dy'dx''dy''dz'dz'' |m(x_m,y_m)|^2 dx_m dy_m. \tag{6.3-2}$$

Let us group all the x_m and y_m variables together and the resulting integral is called the coherence function of the scanning system:

$$\Gamma(x'-x'', y'-y''; z'-z'')$$
$$= \int |m(x_m,y_m)|^2 e^{j\frac{k_0}{f}[x_m(x'-x'')+y_m(y'-y'')]} e^{-j\frac{k_0(z'-z'')}{2f^2}(x_m^2+y_m^2)} dx_m dy_m. \tag{6.3-3}$$

With definition of the coherence function, Eq. (6.3-2) becomes

$$i_\Omega(x,y) = \int \left(s_2 s_1^* e^{j\Omega t} + s_1 s_2^* e^{-j\Omega t}\right) \Gamma(x'-x'', y'-y''; z'-z'')$$
$$\times T(x'+x,y'+y;z') T^*(x''+x,y''+y;z'') dx'dy'dx''dy''dz'dz''. \tag{6.3-4}$$

Indeed, the coherence function controls the coherency of the optical scanning system. For example, when $|m(x,y)|^2 = \delta(x,y)$,

$$\Gamma(x'-x'', y'-y''; z'-z'') = \int \delta(x_m,y_m) e^{j\frac{k_0}{f}[x_m(x'-x'')+y_m(y'-y'')]} e^{-j\frac{k_0(z'-z'')}{2f^2}(x_m^2+y_m^2)} dx_m dy_m$$
$$= 1,$$

and Eq. (6.3-4) becomes

$$i_\Omega(x,y) = \int \left(s_2 s_1^* e^{j\Omega t} + s_1 s_2^* e^{-j\Omega t}\right) T(x'+x,y'+y;z')$$
$$\times T^*(x''+x,y''+y;z'') dx'dy'dx''dy''dz'dz'',$$

which is identical to Eq. (6.2-13a), bringing us to the use of a point detector for coherent processing. So, for a point detector, that is, $|m(x,y)|^2 = \delta(x,y)$, we have a coherent processing system. Now, let us look at another extreme case for the mask, which is when $|m(x,y)|^2 = 1$. According to Eq. (6.3-3),

$$\Gamma(x'-x'', y'-y''; z'-z'') = \int e^{j\frac{k_0}{f}[x_m(x'-x'')+y_m(y'-y'')]} e^{-j\frac{k_0(z'-z'')}{2f^2}(x_m^2+y_m^2)} dx_m dy_m$$

$$= \mathcal{F}\left\{e^{-j\frac{k_0}{2f^2/(z'-z'')}(x_m^2+y_m^2)}\right\}\Bigg|_{k_x=k_0(x'-x'')/f, k_y=k_0(y'-y'')/f}$$

$$\sim \frac{1}{z'-z''} e^{j\frac{k_0}{2(z'-z'')}\left[(x'-x'')^2+(y'-y'')^2\right]}$$

$$\sim \delta(x'-x'', y'-y''; z'-z''), \tag{6.3-5}$$

where we recognize that the quadratic term in the above integral represents a spherical wave with a radius of curvature $R = f^2/(z'-z'')$, which can be made arbitrarily large [see Section 3.5 on radius of curvature]. Therefore, we have ended up taking the Fourier transform of 1 in the limit that $z' \to z''$, giving us a 3-D delta function. With this result, Eq. (6.3-2) is simplified to

$$i_\Omega(x,y) = \int \left(s_2 s_1^* e^{j\Omega t} + s_1 s_2^* e^{-j\Omega t}\right) \delta(x'-x'', y'-y''; z'-z'')$$
$$\times T(x'+x, y'+y; z') \, T^*(x''+x, y''+y; z'') \, dx'dy'dx''dy''dz'dz''.$$

We can now write out the terms $s_2 s_1^*$ and $s_1 s_2^*$, where they are given by Eqs. (6.2-8) and (6.2-9), respectively. Hence, the above equation becomes

$$i_\Omega(x,y) = \int \left[s_2(x',y'; z'+z_0) s_1^*(x'',y''; z''+z_0) e^{j\Omega t} \right.$$
$$\left. + s_1(x',y'; z'+z_0) s_2^*(x'',y''; z''+z_0) e^{-j\Omega t} \right]$$
$$\times \delta(x'-x'', y'-y''; z'-z'') T(x'+x, y'+y; z')$$
$$\times T^*(x''+x, y''+y; z'') \, dx'dy'dx''dy''dz'dz''.$$

After evaluating the integral involving the delta function, we replace all the double primed coordinates by primed coordinates, giving

$$i_\Omega(x,y) = \int \left[s_2(x',y'; z'+z_0) s_1^*(x',y'; z'+z_0) e^{j\Omega t} \right.$$
$$\left. + s_1(x',y'; z'+z_0) s_2^*(x',y'; z'+z_0) e^{-j\Omega t} \right]$$
$$\times T(x'+x, y'+y; z') \, T^*(x'+x, y'+y; z') \, dx'dy'dz'$$

$$= \int \left[s_2(x',y'; z'+z_0) s_1^*(x',y'; z'+z_0) e^{j\Omega t} \right.$$
$$\left. + s_1(x',y'; z'+z_0) s_2^*(x',y'; z'+z_0) e^{-j\Omega t} \right]$$
$$\times \left| T(x'+x, y'+y; z') \right|^2 dx'dy'dz'.$$

Finally, we write the heterodyne current as

$$i_\Omega(x,y) \propto \mathrm{Re}\left[i_p(x,y) e^{j\Omega t} \right],$$

where

$$i_p(x,y) = \int s_2(x',y'; z'+z_0) s_1^*(x',y'; z'+z_0) \left| T(x'+x, y'+y; z') \right|^2 dx'dy'dz'. \tag{6.3-6}$$

This equation is fairly general and it represents the intensity distribution of the 3D object being processing by the two scanning beams, s_1 and s_2.

Similar to the coherent case [see Eq. (6.2-20)], the heterodyne current is demodulated electronically and finally gives a complex record as follows:

$$i_C(x,y) = i_p(x,y) = i_I(x,y) + j\, i_Q(x,y)$$

$$= \int s_2(x',y';z+z_0)s_1^*(x',y';z+z_0)|T(x'+x,y'+y;z)|^2\, dx'dy'dz, \quad (6.3\text{-}7)$$

where $i_I(x,y)$ and $i_Q(x,y)$ are the in-phase and the quadrature-phase outputs from the lock-in amplifier, and the 3-D intensity information, $|T(x,y;z)|^2$, has been processed by the scanning beams s_1 and s_2. The system performs *incoherent processing* as it manipulates and processes the intensity distribution of the 3D object.

Holographic Recording

Let us specifically choose $p_1(x,y) = \delta(x,y)$ and $p_2(x,y) = 1$. According to Eqs. (6.2-8) and (6.2-9), s_1 and s_2 become

$$s_1(x,y;z+z_0) = \mathcal{F}\{\delta(x,y)\}\big|_{k_x=\frac{k_0 x}{f},k_y=\frac{k_0 y}{f}} * h(x,y;z+z_0) \propto e^{-jk_0(z+z_0)} \quad (6.3\text{-}8)$$

and

$$s_2(x,y;z+z_0) = \mathcal{F}\{1\}\big|_{k_x=\frac{k_0 x}{f},k_y=\frac{k_0 y}{f}} * h(x,y;z+z_0)$$

$$\propto e^{-jk_0(z+z_0)}\frac{jk_0}{2\pi(z+z_0)}e^{-j\frac{k_0}{2(z+z_0)}(x^2+y^2)}, \quad (6.3\text{-}9)$$

and the complex record from Eq. (6.3-7) becomes

$$i_C(x,y) = i_I(x,y) + j\, i_Q(x,y)$$

$$= \int \frac{jk_0}{2\pi(z+z_0)}e^{\frac{-jk_0}{2(z+z_0)}(x'^2+y'^2)}|T(x'+x,y'+y;z)|^2\, dx'dy'dz$$

$$= \int \frac{jk_0}{2\pi(z+z_0)}\left[e^{\frac{-jk_0}{2(z+z_0)}(x^2+y^2)} * |T(x,y;z)|^2\right]dz. \quad (6.3\text{-}10)$$

This is a complex holographic record of the 3-D object $|T(x,y;z)|^2$. This is optical scanning holography operating in the incoherent mode.

Example: Cosine and Sine Holograms
From Eq.(6.3-10), we have

$$i_C(x,y) = \int \frac{jk_0}{2\pi(z+z_0)}e^{\frac{-jk_0}{2(z+z_0)}(x^2+y^2)} * |T(x,y;z)|^2\, dz = i_I(x,y) + j\, i_Q(x,y). \quad (6.3\text{-}11a)$$

From Eq. (6.3-11a), we see that

$$i_I(x,y) = \text{Re}[i_C(x,y)] = \int \frac{k_0}{2\pi(z+z_0)}\sin\left(\frac{k_0(x^2+y^2)}{2(z+z_0)}\right) * |T(x,y;z)|^2\, dz$$

$$= H_{\sin}(x,y), \quad (6.3\text{-}11b)$$

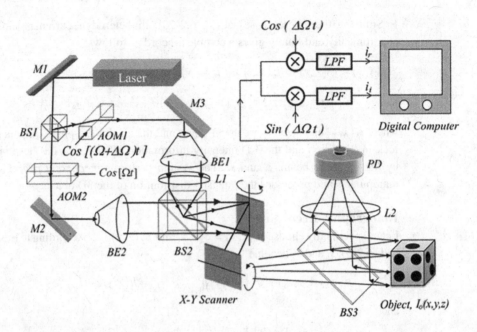

Figure 6.3-1 Typical setup of optical scanning holography operating in incoherent mode. BE : beam expander, BS: beam splitter, AOM: acousto-optic modulator, PD: photodetector. Reprinted from Kim et al., "Speckle-free digital holographic recording of a diffusely reflecting object," *Optics Express* 21, pp. 8183–8189 (2013), with permission. © OSA.

and

$$i_Q(x,y) = \mathrm{Im}\left[i_C(x,y)\right] = \int \frac{k_0}{2\pi(z+z_0)} \cos\left(\frac{k_0(x^2+y^2)}{2(z+z_0)}\right) * \left|T(x,y;z)\right|^2 dz = \mathrm{H}_{\cos}(x,y).$$

$$(6.3\text{-}11c)$$

$\mathrm{H}_{\sin}(x,y)$ and $\mathrm{H}_{\cos}(x,y)$ are called the *sine and cosine holograms*, respectively as these holograms are obtained through the convolution of the sine and cosine zone plates with the intensity distribution of the object. Note that the sine and cosine holograms are obtained from the two outputs of the lock-in amplifier. A *complex Fresnel zone plate hologram* $\mathrm{H}^c(x,y)$ can be constructed according to

$$\mathrm{H}^c_\pm(x,y) = \mathrm{H}_{\cos}(x,y) \pm j\mathrm{H}_{\sin}(x,y)$$

$$= \int \frac{k_0}{2\pi(z+z_0)} e^{\pm j\frac{k_0}{2(z+z_0)}(x^2+y^2)} * \left|T(x,y;z)\right|^2 dz. \quad (6.3\text{-}11d)$$

Similar to Eq. (6.2-22) in the coherent case, we can reconstruct a virtual image or real image by selecting the sign of the exponent in Eq. (6.3-11d) as $\mathrm{H}^c_+(x,y) = \left[\mathrm{H}^c_-(x,y)\right]^*$.

Figure 6.3-1 shows a typical setup of optical scanning holography (OSH) operating in the incoherent mode, where we see that we have an open mask, that is, $\left|m(x,y)\right|^2 = 1$ in front of the photodetector PD.

From the general theory developed for the two-pupil system in Figure 6.2-3, we see that optical scanning holography (OSH) is achieved when $p_1(x,y) = \delta(x,y)$, at light frequency ω_0, sends a plane wave illuminating the object, while $p_2(x,y) = 1$ oscillating at $\omega_0 + \Omega$ illuminates the object with a spherical wave. Therefore, in practice for OSH, we have the interference of a plane wave and a spherical wave of different temporal frequencies to scan the object to acquire the holographic information of a 3-D object. Coherency of OSH depends on the mask in front of the photodetector. *Heterodyning* is performed at the photodetector to give out a heterodyne signal at heterodyne frequency Ω [being the frequency difference between the two scanning light beams]. In Figure 6.3-1, a typical experimental setup of OSH is shown. The laser is at frequency ω_0. A plane wave at frequency $\omega_0 + \Omega$ exiting from beam expander 2 (BE2) illuminates the object. The upshifting of the light frequency from ω_0 to $\omega_0 + \Omega$ is with *acousto-optic modulator* 2 (AOM2) operating at frequency Ω. We will cover frequency-shifting capabilities of AOM in Chapter 7. From the other arm of the interferometer, through the use of beam expander 1 (BE1) and lens L1 we generate a spherical wave at frequency $\omega_0 + \Omega + \Delta\Omega$ toward the object, where $\Omega \gg \Delta\Omega$. The upshifting of the light frequency is with AOM1 operating at frequency $\Omega + \Delta\Omega$. So in the experiment, a heterodyne frequency of $\Delta\Omega$ is generated and inputted to the lock-in detection system. In the experiment, two AOMs operating at $\Omega / 2\pi = 40$ MHz and $(\Omega + \Delta\Omega)/2\pi = 40.01$ MHz are used so as to achieve a lower heterodyne frequency at 0.01 MHz for ease of subsequent electronic processing.

At the outputs of the lock-in system, i_r and i_i are the in-phase and quadrature-phase outputs, giving the sine and cosine holograms, respectively. Figure 6.3-2a and b shows the cosine and sine holograms of the dice. From the two holograms, a complex hologram is constructed according to $H_{\cos}(x,y) \pm jH_{\sin}(x,y)$ for either a real image reconstruction or virtual image reconstruction. Figure 6.3-2c shows a reconstruction of a complex hologram without any coherent speckle noise. Finally, for comparison, in Figure 6.3-2d, an image taken with CCD imaging of a coherently illuminated dice is shown, illustrating speckle noise due to coherent illumination.

Based on the principle of optical scanning holography operating in the incoherent mode, holograms of fluorescent specimens were recorded by a digital holographic technique for the first time since 1997 [Schilling et al. (1997)]. Holographic imaging of fluorescent inhomogeneities embedded in a turbid medium was subsequently demonstrated [Indebetouw et al. (1998)]. Three-dimensional reconstructions of a complex hologram of fluorescent *Oscillatoria* strands of resolution better than 1 μm were also reported [Indebetouw and Zhong (2006)].

As with scanning image processing studied in Section 4.4, the coherence properties of the acquired digital holograms in optical scanning holography also depend on the size of the mask in front of the active area of the photodetector. For a point detector, that is, $|m(x,y)|^2 = \delta(x,y)$, we have coherent digital holographic recording. For another extreme case when we use an open mask, that is, $|m(x,y)|^2 = 1$, we have incoherent digital holographic processing. Clearly, we can envision *partial coherent digital holographic recording* when the mask employed is finite in size. Recently partial coherent digital holographic imaging has been demonstrated [Liu et al. (2015)].

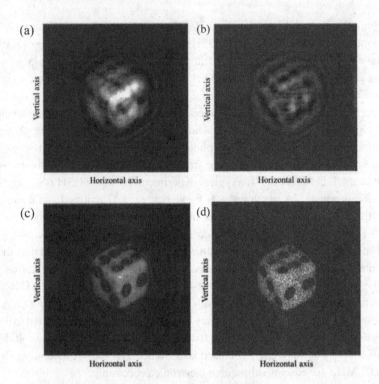

Figure 6.3-2 (a) Cosine hologram, (b) sine hologram, (c) reconstruction of a complex hologram, (d) CCD imaging of a coherently illuminated dice, illustrating speckle noise due to coherent illumination. Reprinted from Kim et al., "Speckle-free digital holographic recording of a diffusely reflecting object," *Optics Express* 21, pp. 8183–8189 (2013), with permission. © OSA.

To conclude this subsection, we want to point out that recent approaches to the implementation of optical scanning holography include motionless optical scanning holography using SLM technology [Yoneda et al. (2020), Yoneda, Saita and Nomura (2020)] and coaxial scanning holography using a geometric phase lens [Kim and Kim (2020), Tsai et al. (2021)]. A holographic system for recording a curved digital hologram has also been proposed and demonstrated [Liu et al. (2020)].

6.3.2   Fresnel Incoherent Correlation Holography

Fresnel incoherent correction holography (FINCH) is the second incoherent digital holographic technique widely researched nowadays. The principle of FINCH is based on the concept that each object point generates its own Fresnel zone plate on the hologram plane [Mertz and Young (1962)]. The idea was extended through the concept of interference by Lohmann [1965] and Cochran reported the first experimental result [1966]. This principle of incoherent holography is now known as *self-interference* [Poon (2008)]. Self-interference automatically creates enormous bias buildup for complicated objects, that is, object being made up of many points, and limits the dynamic

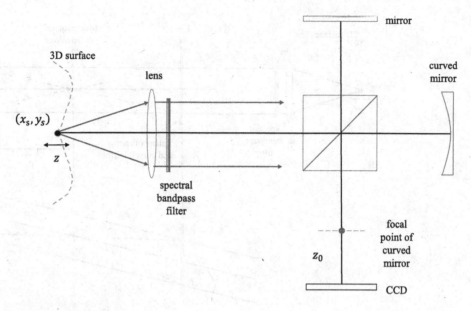

Figure 6.3-3 Self-interference

range of the recording medium. Incoherent holography based on self-interference had little success in terms of practical applications until Rosen and Brooker [2007] demonstrated its usefulness through their digital holographic technique called *Fresnel incoherent correlation holography* (FINCH). They have employed modern spatial-light-modulator technology as well as phase-shifting technique to alleviate the issue of bias buildup. In passing, we mention that optical scanning holography has avoided the bias buildup using optical heterodyning [see Eq. (6.3-4)].

Self-Interference

The principle of self-interference is based on the idea that each point object's wave front is split into two parts and then re-combined to form an inference pattern. Figure 6.3-3 shows a version of the setup for self-interference. A single point object in the 3-D surface on the focal plane of the lens generates a spherical wave toward the lens, which then sends a plane wave toward a modified Michelson interferometer configuration with one of the paths ending up with a curved mirror. After the lens, a spectral bandpass filter is placed to increase the coherence length of the light so as to make the system more efficient to produce high-contrast interference patterns. Recall that as we make the spectral width of the source narrower, the coherence length of the source increases [see Eq. (4.1-1)]. The mirror and the curved mirror reflect a plane wave and a spherical wave (assuming the focal point of the curved mirror is at a distance z_0 in front of the CCD), respectively toward the CCD plane for recording the interference pattern, which is a Fresnel zone plate given by [see Figure 5.1-1a and Eq. (5.1-7)]

$$\text{FZP}(x, y; z_0) = A + B \sin\left[\frac{k_0}{2z_0}(x^2 + y^2)\right].$$

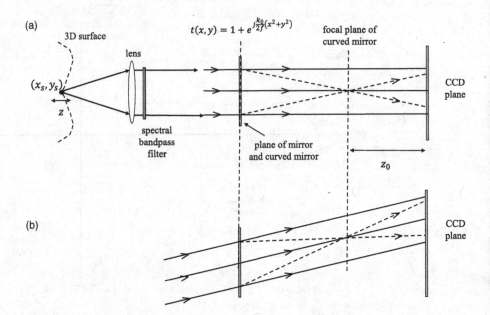

Figure 6.3-4 "Unfolded" interferometry to explain the generation of a Fresnel zone plate under self-interference (a) when the object point is on the optical axis and (b) when the object point is off in the transverse direction

The situation, shown in Figure 6.3-3, is easier to analyze using Figure 6.3-4a if the interferometer is "unfolded." The word "unfolded" means that on the plane of the mirror and the curved mirror, instead of allowing reflections, we simply consider that the light waves continue propagating forward. Hence, we see that the optical element serving to replace the effect of the plane and curved mirrors has a transparency function of the form

$$t(x,y) = 1 + e^{j\frac{k_0}{2f}(x^2+y^2)}, \qquad (6.3\text{-}12)$$

where f is the focal length of the curved mirror.

If the object point is off-center transversely, it will send inclined waves as shown in Figure 6.3-4b, giving an off-axis Fresnel zone plate on the hologram recording plane:

$$\mathrm{FZP}(x,y;x_s,y_s,z_0) = A' + B'\sin\left[\frac{k_0}{2z_0}\left[(x-x_s)^2 + (y-y_s)^2\right]\right].$$

For an object point that is located away from the focal plane of the lens but on the optical axis, that is, $(x_s,y_s) = (0,0)$, it will record interference of two spherical waves with different radii of curvature, still giving a Fresnel zone plate but with a different depth parameter $\gamma(z)$ that is a function of z:

$$\mathrm{FZP}(x,y;z) = A'' + B''\sin\left[\frac{k_0}{2\gamma(z)}(x^2+y^2)\right].$$

The actual functional form of $\gamma(z)$ depends on the optical geometry under consideration. Since the source is incoherent, which means there is no interference between

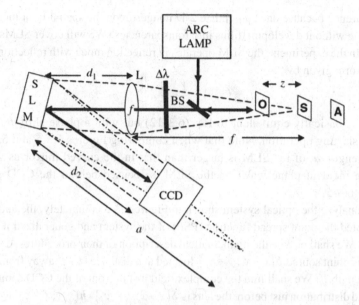

Figure 6.3-5 Schematic of FINCH. Reprinted from Rosen and Brooker, "Digital spatially incoherent Fresnel holography," *Optics Letters* 32, pp. 912–914 (2007), with permission. © OSA.

object points, we simply add up all the intensity patterns [see Eq. (4.1-6)] from the FZPs on the hologram plane due to all the points from the 3D object. Hence, for an object that consists of three points given above, we simply add all the FZPs on the hologram plane. Clearly, we can see the constant bias terms, A, A', and A'' being added up. In short, the useful signals, that is, the sine variations, are being buried within a huge constant and that huge constant will lower the dynamic range of the recording device. In general, for a 3-D object with intensity distribution $I(x_s, y_s, z)$, a self-interference hologram will have a form

$$\mathrm{H}(x,y) \cong C + \iiint I(x_s, y_s, z) \sin\left[\frac{k_0}{2\gamma(z)}\left[(x-x_s)^2 + (y-y_s)^2\right]\right] dx_s dy_s dz,$$

where C is some constant bias.

FINCH Description

Rosen and Brooker employed modern spatial-light-modulator technology as well as phase-shifting technique to alleviate the issue of bias buildup with the technique called FINCH and demonstrated the practical use of digital incoherent holography based on self-reference. Their system is shown in Figure 6.3-5. Note that in principle the optical configuration is identical to that in Figure 6.3-4a.

An arc lamp illuminates a 3-D object, and a CCD camera captures the reflected light from the object after passing through a lens L and a spatial light modulator SLM. A 2-D spatial light modulator is a device with which one can imprint a 2-D pattern on a coherent light beam by passing the beam through it (or by reflecting the beam off the device). In fact, we can think of a 2-D spatial light modulator as a real-time transparency because one can update 2-D images upon the spatial light modulator in real time without developing films into transparencies (We will cover SLMs in Chapter 7). In the experiment, the SLM operates in reflection mode with reflection function of the form given by

$$t_{SLM}(x,y) = \frac{1}{2} + \frac{1}{2}e^{j\left[\frac{k_0}{2a}(x^2+y^2)-\theta\right]},$$

which is basically equivalent to Eq. (6.3-12) but with a phase shift θ, allowing phase-shifting operation. Note that when comparing Figures 6.3-5 and 6.3-4a, the focal length, a, of the SLM is larger than that in the curved mirror as we notice that an incident plane wave on the SLM is focused behind the CCD shown in Figure 6.3-5.

To analyze the optical system shown in Figure 6.3-5 completely, the authors have calculated the point spread function (PSF) of the system and generalized it for a 3-D object. We shall outline the mathematical description of their procedures. Consider an off-axis point source $\delta(x-x_s,y-y_s)$ located at a distance $f-z$ away from lens L of focal length f. We shall find the complex field just in front of the CCD camera. Now, the field distribution just before the lens is $\delta(x-x_s,y-y_s)*h(x,y;f-z)$ according to Fresnel diffraction. After the lens, we have $[\delta(x-x_s,y-y_s)*h(x,y;f-z)]e^{j\frac{k_0}{2f}(x^2+y^2)}$. After propagating a distance of d_1 to the SLM plane, the complex field is

$$\psi_{front,SLM}(x,y) = \left\{ [\delta(x-x_s,y-y_s)*h(x,y;f-z)]e^{j\frac{k_0}{2f}(x^2+y^2)} \right\} * h(x,y;d_1).$$

After the SLM, we have $\psi_{front,SLM}(x,y)t_{SLM}(x,y)$. Finally, in the CCD plane, a distance d_2 from the SLM, the PSF of the system is

$$I_p(x,y) = |\psi_{front,SLM}(x,y)t_{SLM}(x,y)*h(x,y;d_2)|^2,$$

which has been evaluated by the authors to be, for $f \gg z$,

$$I_p(x,y) \propto 2 + e^{j\frac{k_0}{2\gamma(z)}\left[\left(x-\frac{ax_s}{f}\right)^2+\left(y-\frac{ay_s}{f}\right)^2\right]+j\theta} + c.c.,$$

where c.c. denotes the complex conjugate, and

$$\gamma(z) = \frac{\left[d_2 - a - z(d_1a + d_2f - af + d_2a - d_2d_1)f^{-2}\right]}{1 - z(a+f-d_1)f^{-2}}.$$

For a general 3-D object, the record hologram is simply given by integrating over the object intensity distribution $g(x_s,y_s,z)$:

$$H_\theta(x,y) \propto D + \iiint g(x_s,y_s,z)e^{j\left\{\frac{k_0}{2\gamma(z)}\left[\left(x-\frac{ax_s}{f}\right)^2+\left(y-\frac{ay_s}{f}\right)^2\right]+\theta\right\}}dx_s dy_s dz$$

$$+ \iiint g(x_s,y_s,z)e^{-j\left\{\frac{k_0}{2\gamma(z)}\left[\left(x-\frac{ax_s}{f}\right)^2+\left(y-\frac{ay_s}{f}\right)^2\right]+\theta\right\}}dx_s dy_s dz,$$

where D is some constant bias.

To extract the complex hologram $H_F(x,y)$ of the object, three phase-shifted holograms are obtained with $\theta = (\theta_1,\theta_2,\theta_3) = (0, 2\pi/3, 4\pi/3)$:

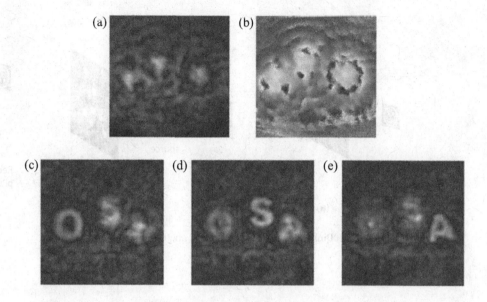

Figure 6.3-6 FINCH results: (a) magnitude of $H_F(x, y)$, that is, $|H_F(x, y)|$, (b) phase of $H_F(x, y)$, (c) reconstruction of $H_F(x, y)$ at the best focus distance for "O," (d) reconstruction at the best focus distance for "S," and (e) reconstruction at the best focus distance for "A." Reprinted from Rosen and Brooker, "Digital spatially incoherent Fresnel holography," *Optics Letters* 32, pp. 912–914 (2007), with permission. © OSA.

$$H_F(x, y) = \iiint g(x_s, y_s, z) e^{j\frac{k_0}{2\gamma(z)}\left[\left(x - \frac{ax_s}{f}\right)^2 + \left(y - \frac{ay_s}{f}\right)^2\right]} dx_s dy_s dz$$

$$= H_{\theta_1}(x, y)[e^{-j\theta_3} - e^{-j\theta_2}] + H_{\theta_2}(x, y)[e^{-j\theta_1} - e^{-j\theta_3}] + H_{\theta_3}(x, y)[e^{-j\theta_2} - e^{-j\theta_1}]. \quad (6.3\text{-}13)$$

This is a Fresnel hologram and reconstruction of it is done by performing the Fresnel diffraction integral:

$$H_F(x, y) * h(x, y; z).$$

Figure 6.3-6 shows the FINCH results. Figure 6.3-6a and b shows the magnitude and the phase of $H_F(x, y)$, respectively. Figure 6.3-6 c, d, and f are the reconstructions at the best focus distances. Since the introduction of FINCH, there has been a plethora of reports in *digital incoherent holography* using self-interference. Most recently, a review comparing the differences between OSH and FINCH has been reported [Liu et al. (2018)].

6.3.3   Coded Aperture Imaging and Coded Aperture Correlation Holography (COACH)

Coded Aperture Imaging

FINCH and coded aperture correlation holography (COACH) have their roots in coded aperture imaging dated back to the 1960s [Mertz and Young (1961)]. A coded aperture camera is similar to a pinhole camera. In coded aperture camera, the single

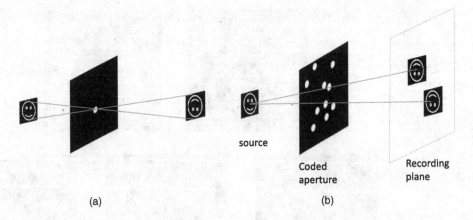

Figure 6.3-7 (a) Pinhole imaging and (b) coded aperture imaging

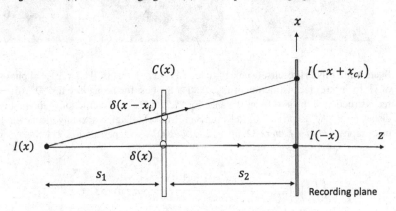

Figure 6.3-8 Coded aperture imaging (illustrated with two pinholes)

opening of the pinhole camera is replaced with a pattern of pinholes (called the coded aperture CA). The coded image is then the superposition of many pinhole images. A multiple-pinhole aperture was first proposed in coded aperture imaging [Dicke (1968)]. Figure 6.3-7a and b shows pinhole imaging and coded aperture imaging, respectively. In Figure 6-3-7b, we just show two images due to the two pinholes under consideration. Let us formulate coded aperture imaging in the $x–z$ plane shown in Figure 6.3-8. $C(x)$ is the coded aperture. We show two light rays from a single point of $I(x)$ going through the two pinholes to give two images on the recording plane. Hence, we can write the coded image as the sum of the two images given by

$$I(-x)+I(-x+x_{c,i}),$$

where $I(-x)$ is the intensity image due to the pinhole $\delta(x)$ on the z-axis, and $I(-x+x_{c,i})$ is the image due to the pinhole $\delta(x-x_i)$ that is x_i away from the z-axis. From the geometry of the figure, we can relate $x_{c,i}$ to x_i as follows:

$$x_{c,i} = \frac{s_1 + s_2}{s_1} x_i = M(z) x_i. \tag{6.3-14}$$

$M(z) = (s_1 + s_2)/s_1$ is the magnification factor, which depends on the axial distance of the source from the aperture as s_1 can be written as $s_1 + z$, where z is the depth of the 3-D object.

For an n-pinhole coded aperture $C(x) = \sum_{i=1}^{n} \delta(x - x_i)$, the coded image $I_{coded}(x)$ is

$$I_{coded}(x) = \sum_i I(-x + x_{c,i}) = \sum_i I(-x + Mx_i),$$

$$= I(-x) * \sum_i \delta(x - Mx_i), \tag{6.3-15a}$$

which can be written in terms of correlation as

$$I_{coded}(x) = I(x) \otimes \sum_i \delta(x - Mx_i). \tag{6.3-15b}$$

Note that the magnification M of $C(x)$ is

$$C\left(\frac{x}{M}\right) = \sum_{i=1}^{n} \delta\left(\frac{x}{M} - x_i\right) = \sum_{i=1}^{n} \delta\left(\frac{1}{M}(x - Mx_i)\right) = M \sum_{i=1}^{n} \delta(x - Mx_i).$$

Using this result, Eq. (6.3-15b) becomes

$$I_{coded}(x) = \frac{1}{M} I(x) \otimes C\left(\frac{x}{M}\right). \tag{6.3-16}$$

We see that the coded image, therefore, is expressed as the correlation of the source intensity $I(x)$ with the coded aperture $C(x)$ magnified by a factor M.

The reconstruction (or decoding) of the coded image is typically done by a computer and may be achieved by correlation with a carefully chosen decoding aperture $D(x)$. The decoded image is given by

$$I_{decoded}(x) = I_{coded}(x) \otimes D(x) = \left[\frac{1}{M} I(x) \otimes C\left(\frac{x}{M}\right)\right] \otimes D(x). \tag{6.3-17}$$

The simplest and most commonly used systems employ the same type of coded aperture and decoding aperture, that is, $D(x) = C\left(\frac{x}{M}\right)$, with the condition that

$$C(x) \otimes D(x) = C(x) \otimes C(x) \sim \delta(x). \tag{6.3-18}$$

In other words, the autocorrelation of the coded aperture is approximately equal to the delta function. With Eq. (6.3-18), let us evaluate the decoded image from Eq. (6.3-17). From the definition of correlation [see Eq. (2.3-30)], we write

$$I_{decoded}(x) = \left[\frac{1}{M} I(x) \otimes C\left(\frac{x}{M}\right)\right] \otimes C\left(\frac{x}{M}\right)$$

$$= \frac{1}{M}\left[\int_{-\infty}^{\infty} I^*(x') C\left(\frac{x + x'}{M}\right) dx'\right] \otimes C\left(\frac{x}{M}\right)$$

$$= \frac{1}{M} \int_{-\infty}^{\infty} \left[\int_{-\infty}^{\infty} I^*(x') C\left(\frac{x'' + x'}{M}\right) dx'\right]^* C\left(\frac{x + x''}{M}\right) dx''$$

$$= \frac{1}{M} \int_{-\infty}^{\infty} \int_{-\infty}^{\infty} I(x') C^* \left(\frac{x'' + x'}{M} \right) C \left(\frac{x + x''}{M} \right) dx' dx''. \qquad (6.3\text{-}19)$$

From Eq. (6.3-18), we establish the following relationship:

$$C(x) \otimes C(x) = \int_{-\infty}^{\infty} C^*(x') C(x + x') dx' = \delta(x).$$

Using the above relationship and by grouping the double-primed coordinates in Eq. (6.3-19), we obtain [see P. 6.6]

$$\int_{-\infty}^{\infty} C^* \left(\frac{x'' + x'}{M} \right) C \left(\frac{x + x''}{M} \right) dx'' = \delta \left(\frac{x - x'}{M} \right). \qquad (6.3\text{-}20)$$

With the above result, we evaluate Eq. (6.1-19) to obtain

$$I_{\text{decoded}}(x) = \frac{1}{M} \int_{-\infty}^{\infty} I(x') \delta \left(\frac{x - x'}{M} \right) dx' = I(x),$$

and we have recovered the original image $I(x)$ from the coded image $I_{\text{coded}}(x)$. Therefore, one needs to design $C \left(\dfrac{x}{M} \right)$ such that Eq. (6.3-18) is satisfied. An infinite random pinhole array is in principle a perfect choice as its autocorrelation is a delta function [Cannon and Fenimore (1980)]. Some of the popular employed coded apertures are Fresnel zone plates [Mertz and Young (1961)], random pinhole arrays [Dicke (1968)], and annular apertures [Simpson et al. (1975)]. In one of the most recent coded aperture imaging techniques, Wu et al. [2020] have employed a Fresnel zone aperture for *lensless imaging* under incoherent illumination.

Coded Aperture Correlation Holography (COACH)

Recently, a self-interference incoherent digital holography technique using a coded aperture has been developed. The technique is known as COACH and it is a generalized version of the self-interference incoherent digital holography technique of FINCH.

Instead of using a quadratic phase mask, that is, $t_{\text{SLM}}(x, y) = \dfrac{1}{2} + \dfrac{1}{2} e^{j \left[\frac{k_0}{2a}(x^2 + y^2) - \theta \right]}$, in FINCH as shown in Figure 6.3-5, COACH is operational with any random phase mask in principle as long as the phase mask has the property that its auto-correlation approximates a delta function. In COACH, it requires to record two complex holograms, one from a point source as the PSF hologram and the other as the object hologram. Reconstruction of the object is to perform correlation of the two complex holograms. The holograms recorded by COACH, therefore, cannot be classified as Fourier or Fresnel holograms as neither Fourier transform nor Fresnel back propagation can reconstruct the image. With reference to Figure 6.3-9, we discuss the basic principle of COACH.

Apart from the two additional polarizers, the optical system is basically identical to that of FINCH shown in Figure 6.3-5. The illumination is focused by lens L1 directly into the plane of the specimen. This type of illumination is called *source-focused or critical illumination* in microscopy. A *coded phase mask (CPM)* is displayed on the

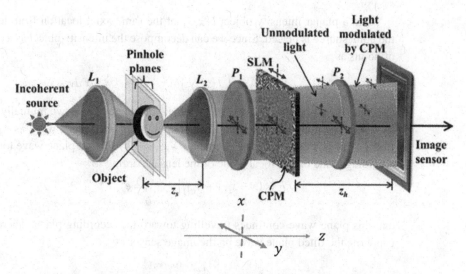

Figure 6.3-9 Optical configuration of COACH. CPM – Coded phase mask; L1, L2 – Refractive lenses; P1, P2 –Polarizers; SLM – Spatial light modulator; blue arrows indicate polarization orientations. The optical system is reprinted from Rosen et al., *Applied Sciences* 9, 605 (2019). https://doi.org/10.3390/app9030605

phase-only SLM. Phase modulation occurs along the y-axis [see Figure 7.3-6 on phase modulation by a spatial light modulator]. The polarization axes of polarizers P1 and P1 are oriented $45°$ with respect to the y-axis, as shown in the figure. The light emitted from each object point is polarized along the polarization axis of P1. The polarized wave is decomposed into the two orthogonal directions, one along the x-axis and the other along the y-axis. The y-polarized wave is phase-modulated but the x-polarized wave is unmodulated. Polarized P2 allows the two waves with the same polarization at $45°$ away from the y-axis to interfere on the image sensor, thereby accomplishing self-interference.

Let us now develop the mathematical formulation. As mentioned before, COACH needs to record two complex holograms, the point-source hologram and the object hologram. For a point object located at the front focal plane of lens L2 of focal length f, that is, $z_s = f$, the intensity distribution $I_\delta(x,y)$ is formed between the unmodulated plane wave with amplitude A and the modulated diffracted pattern of CPM, $G(x,y)$, on the image sensor:

$$I_\delta(x,y) = \left| A + G(x,y)e^{j\delta} \right|^2,$$ (6.3-21)

where δ is the phase shift. A possible three-step algorithm has $\delta = 0, \pi/2,$ and π [see Eqs. (6.2-4) and (6.2-5)] and, according to Eq. (6.2-5), we obtain the point source hologram $H_{\mathrm{PSH}}(x,y)$, also called the complex point spread hologram:

$$H_{\mathrm{PSH}}(x,y) = A^* G(x,y).$$ (6.3-22)

Next, a planar intensity object $I(x,y)$ at the same axial location from lens L2 of the point source is placed. Since we can decompose the intensity object as a collection of points as

$$I(x,y) = \int I(x_i, y_i)\delta(x-x_i, y-y_i)\,dx_i dy_i,$$

each ith object point, at (x_i, y_i), on the object plane, generates two mutually coherent beams on the image sensor. One is an unmodulated tilted plane wave as we can see that the off-axis object point $I(x_i, y_i)\delta(x-x_i, y-y_i)$ sends a plane wave towards the recording plane on the focal plane of the lens according to

$$\mathcal{F}\{\delta(x-x_i, y-y_i)\}\Big|_{\substack{k_x = k_0 x/f \\ k_y = k_0 y/f}} = e^{j2\pi(xx_i + yy_i)/\lambda_0 f},$$

and this plane wave continues travelling toward the recording plane. Therefore, we can write the tilted plane wave on the image sensor as

$$A_i e^{j2\pi(xx_i + yy_i)/\lambda_0 f} \tag{6.3-23}$$

with A_i being the amplitude of the plane wave. Now, the other beam is the modulated field given by

$$B_i e^{j2\pi(xx_i + yy_i)/\lambda_0 f} G\left(x + \frac{x_i z_h}{f}, y + \frac{y_i z_h}{f}\right), \tag{6.3-24}$$

where we have a shifted version of $G(x,y)$ multiplied by a tilted plane wave as the modulated beam is also propagating along the same direction of the unmodulated plane wave. The shifted version of $G(x,y)$ can be recognized by inspecting Figure 6.3-10 as we can envision that the point source $I(x_i, y_i)\delta(x-x_i, y-y_i)$ projects a shifted image through the pinhole on the z-axis.

The total intensity on the recording sensor from the point source $I(x_i, y_i)\delta(x-x_i, y-y_i)$ is then given by the sum of the two fields from Eqs. (6.3-23) and (6.3-24)

$$\left| A_i e^{j2\pi(xx_i + yy_i)/\lambda_0 f} + B_i e^{j2\pi(xx_i + yy_i)/\lambda_0 f} G\left(x + \frac{x_i z_h}{f}, y + \frac{y_i z_h}{f}\right) e^{j\delta} \right|^2.$$

For a 2-D object with a collection of N object points, we have

$$\sum_i^N \left| A_i e^{j2\pi(xx_i + yy_i)/\lambda_0 f} + B_i e^{j2\pi(xx_i + yy_i)/\lambda_0 f} G\left(x + \frac{x_i z_h}{f}, y + \frac{y_i z_h}{f}\right) e^{j\delta} \right|^2.$$

By the phase shifting procedure again, similar to the method we obtain the point source hologram $H_{\text{PSH}}(x,y)$ in Eq. (6.3-22), a complex object hologram $H_{\text{OBJ}}(x,y)$ is given by

$$H_{\text{OBJ}}(x,y) = \sum_i A_i^* B_i G\left(x + \frac{x_i z_h}{f}, y + \frac{y_i z_h}{f}\right)$$

$$\propto \sum_i I(x_i, y_i) G\left(x + \frac{x_i z_h}{f}, y + \frac{y_i z_h}{f}\right), \tag{6.3-25}$$

where we recognize that the amplitudes A_i and B_i basically are derived from the point source $I(x_i, y_i)\delta(x-x_i, y-y_i)$.

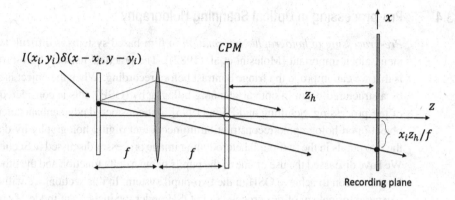

Figure 6.3-10 Shifted $G(x,y)$ due to an off-axis point source from the object

Now that the complex point spread hologram $H_{\mathrm{PSH}}(x,y)$ and the complex object hologram $H_{\mathrm{OBJ}}(x,y)$ have been recorded, we simply perform the correlation of the two holograms to reconstruct the object:

$$H_{\mathrm{OBJ}}(x,y) \otimes H_{\mathrm{PSH}}(x,y)$$

$$\propto \sum_i I(x_i,y_i) G\left(x+\frac{x_i z_h}{f}, y+\frac{y_i z_h}{f}\right) \otimes G(x,y). \tag{6.3-26}$$

In COACH, CPM is designed such that

$$G(x,y) \otimes G(x,y) \sim \delta(x,y). \tag{6.3-27}$$

The CPM function is a pure phase function that is synthesized with the following constraints: (1) its magnitude spectrum is uniform, thereby giving a delta function in the spatial domain and (2) its diffracted pattern on the SLM plane, that is, $G(x,y)$, is to have a pure phase function [Vijayakumar et al. (2016)]. With Eq. (6.3-27), it can be shown that [see P.6.7]

$$G\left(x+\frac{x_i z_h}{f}, y+\frac{y_i z_h}{f}\right) \otimes G(x,y) = \delta\left(x-\frac{x_i z_h}{f}, y-\frac{y_i z_h}{f}\right).$$

This result allows us to write Eq. (6.3-26) as

$$H_{\mathrm{OBJ}}(x,y) \otimes H_{\mathrm{PSH}}(x,y)$$

$$\propto \sum_i I(x_i,y_i) \delta\left(x-\frac{x_i z_h}{f}, y-\frac{y_i z_h}{f}\right)$$

$$\sim \int I(x_i,y_i) \delta\left(x-\frac{x_i z_h}{f}, y-\frac{y_i z_h}{f}\right) dx_i dy_i \propto I\left(\frac{x}{M}, \frac{y}{M}\right),$$

where $M = z_h/f$ is the magnification of the optical system.

6.3.4   Pre-processing in Optical Scanning Holography

Pre-processing of holographic information in film-based systems is difficult to implement, albeit important [Molesini et al. (1982)]. The practical reason of pre-processing is that we can improve the fringe contrast before recording. When the object is scanned by a structured beam in optical scanning holography (OSH), this is considered a form of pre-processing. Schilling and Poon [1995] first considered edge enhancement of an object upon holographic reconstruction in incoherent digital holography by designing the two pupils in the two-pupil heterodyning image processor discussed in Section 6.2.2. We have discussed the use of one of the pupils being a delta function and the other a uniform function to achieve OSH in the two-pupil system. In this section, we will develop a general formalism of pre-processing in OSH under the incoherent mode of operation. This is an important modern development for incoherent digital holography as we have pointed out in Chapter 4 that conventional incoherent optical imaging systems always exhibit low-pass filtering characteristics and many important processing operations, such as edge extraction, are not possible until the introduction of two-pupil systems.

Let us start with a concise version of Figure 6.2-3 in that we take $z_0 = 0$ and the detailed electronics subsystem is replaced simply by a block diagram called "electronic processing." We also set $m(t) = 1$ to reflect on the fact that we perform incoherent optical processing, that is, only the intensity distribution of the object $|T(x,y;z)|^2$ is being processed. The final complex record on the PC is given by [see Eq. (6.3-7) with $z_0 = 0$]

$$i_C(x,y) = i_I(x,y) + j\, i_Q(x,y)$$

$$= \int s_2(x',y';z) s_1^*(x',y';z) |T(x'+x, y'+y; z)|^2\, dx'dy'dz, \qquad (6.3\text{-}28a)$$

where, according to Eqs. (6.2-8) and (6.2-9), the two scanning beams are

$$s_1(x,y;z) = \mathcal{F}\{p_1(x,y)\}\Big|_{k_x = \frac{k_0 x}{f}, k_y = \frac{k_0 y}{f}} * h(x,y;z) \qquad (6.3\text{-}28b)$$

and

$$s_2(x,y;z) = \mathcal{F}\{p_2(x,y)\}\Big|_{k_x = \frac{k_0 x}{f}, k_y = \frac{k_0 y}{f}} * h(x,y;z). \qquad (6.3\text{-}28c)$$

We can re-write Eq. (6.3-28a) in terms of correlation involving x and y coordinates:

$$i_C(x,y) = \int s_1(x,y;z) s_2^*(x,y;z) \otimes |T(x,y;z)|^2\, dz. \qquad (6.3\text{-}29)$$

By taking the Fourier transform of Eq. (6.3-29), we have

$$\mathcal{F}\{i_C(x,y)\} = \int \mathcal{F}^*\{s_1(x,y;z) s_2^*(x,y;z)\} \mathcal{F}\{|T(x,y;z)|^2\} dz$$

$$= \int OTF_\Omega(k_x, k_y; z) \mathcal{F}\{|T(x,y;z)|^2\} dz, \qquad (6.3\text{-}30a)$$

where $OTF_\Omega(k_y, k_y; z)$ is the optical transfer function of the two-pupil heterodyning system:

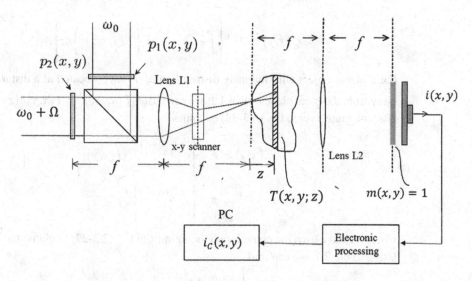

Figure 6.3-11 Optical scanning holography in incoherent mode

$$OTF_\Omega\left(k_x,k_y;z\right) = \mathcal{F}^*\left\{s_1\left(x,y;z\right)s_2^*\left(x,y;z\right)\right\}. \tag{6.3-30b}$$

The complex record can be written in terms of the optical transfer function as

$$i_c\left(x,y\right) = \mathcal{F}^{-1}\left\{\int OTF_\Omega\left(k_x,k_y;z\right)\mathcal{F}\left\{\left|T\left(x,y;z\right)\right|^2\right\}dz\right\}. \tag{6.3-31a}$$

By substituting Eqs. (6.3-28b) and (6.3-28c) into Eq. (6.3-30b), Poon [1985] has expressed the *OTF* in terms of the two pupils p_1 and p_2 [see Prob. 6.8]:

$$OTF_\Omega\left(k_x,k_y;z\right) = e^{j\frac{z}{2k_0}\left(k_x^2+k_y^2\right)}\int p_1^*\left(x',y'\right)p_2\left(x'+\frac{f}{k_0}k_x,y'+\frac{f}{k_0}k_y\right)e^{j\frac{z}{f}\left(x'k_x+y'k_y\right)}dx'dy'.$$

$$\tag{6.3-31b}$$

By manipulating the two pupils, we can design various *OTFs*. For example, Hilbert transformation and bandpass filtering of intensity objects have been investigated [Zhang et al. (2019)].

Example: Optical Scanning Holography in Incoherent Mode
Case A $p_1\left(x\right) = \delta\left(x,y\right)$ and $p_2\left(x\right) = 1$.

This corresponds to the most extreme case in terms of the functional form of the two pupils as one of the pupils is infinitely small and the other is infinitely large. The *OTF*, according to Eq. (6.3-31b), becomes

$$OTF_\Omega\left(k_x,k_y;z\right) = e^{j\frac{z}{2k_0}\left(k_x^2+k_y^2\right)}. \tag{6.3-32}$$

The resulting *OTF* is called the *holographic OTF*, denoted by $OTF_{\mathrm{OSH}}\left(k_x,k_y;z\right)$, and the complex record, according to Eq. (6.3-31a), is

$$i_C(x,y) = \mathcal{F}^{-1}\left\{ \int e^{j\frac{z}{2k_0}(k_x^2+k_y^2)} \mathcal{F}\left\{ |T(x,y;z)|^2 \right\} dz \right\}. \tag{6.3-33}$$

For a planar object with intensity distribution of $I(x,y)$, located at a distance of z_1 away from the focal plane of lens L1, we can write $|T(x,y;z)|^2 = I(x,y)\delta(z-z_1)$ and after integrate over z, Eq. (6.3-33) becomes

$$i_C(x,y) = \mathcal{F}^{-1}\left\{ e^{j\frac{z_1}{2k_0}(k_x^2+k_y^2)} \mathcal{F}\left\{ I(x,y) \right\} \right\}$$

$$= \mathcal{F}^{-1}\left\{ e^{j\frac{z_1}{2k_0}(k_x^2+k_y^2)} \right\} * I(x,y), \tag{6.3-34}$$

where we have used the convolution theorem from Eq. (2.3-27) to obtain the last step. Using Table 2.1, we can find

$$\mathcal{F}^{-1}\left\{ e^{j\frac{z_1}{2k_0}(k_x^2+k_y^2)} \right\} = \frac{jk_0}{2\pi z_1} e^{\frac{-jk_0}{2z_1}(x^2+y^2)}.$$

Equation (6.3-34) can then be written as

$$i_C(x,y) = \frac{jk_0}{2\pi z_1} e^{\frac{-jk_0}{2z_1}(x^2+y^2)} * I(x,y),$$

which is a complex hologram of the 2-D object located z_1 away from the focal plane of lens L1. This is basically the previous result in Eq. (6.3-10). The virtual image reconstruction of the complex hologram is given by

$$i_C(x,y) * h^*(x,y;z_1) = \frac{jk_0}{2\pi z_1} e^{\frac{-jk_0}{2z_1}(x^2+y^2)} * I(x,y) * h^*(x,y;z_1) \propto I(x,y). \tag{6.3-35}$$

Case B $p_1(x)=1$ and $p_2(x)=\delta(x,y)$.

This case corresponds to another extreme case as we simply interchange the functional form of $p_1(x)$ and $p_2(x)$ from Case A. The *OTF*, according to Eq. (6.3-31b), becomes

$$OTF_\Omega(k_x,k_y;z) = e^{-j\frac{z}{2k_0}(k_x^2+k_y^2)}. \tag{6.3-36}$$

In this case, the complex record of $I(x,y)$ becomes

$$i_C(x,y) = \mathcal{F}^{-1}\left\{ e^{-j\frac{z_1}{2k_0}(k_x^2+k_y^2)} \mathcal{F}\left\{ I(x,y) \right\} \right\}$$

$$= \mathcal{F}^{-1}\left\{ e^{-j\frac{z_1}{2k_0}(k_x^2+k_y^2)} \right\} * I(x,y) = \frac{-jk_0}{2\pi z_1} e^{\frac{jk_0}{2z_1}(x^2+y^2)} * I(x,y).$$

This complex hologram gives a real image reconstruction as

$$i_C(x,y) * h(x,y;z_1) = \frac{-jk_0}{2\pi z_1} e^{\frac{jk_0}{2z_1}(x^2+y^2)} * I(x,y) * h(x,y;z_1) \propto I(x,y). \tag{6.3-37}$$

It is clear at this point that the *OTF* either of the form of Eq. (6.3-32) or Eq. (6.3-36) gives rise to holographic recording. Therefore, we can define the *holographic OTF* in optical scanning holography, denoted by $OTF_{OSH}(k_x,k_y;z)$, as

$$OTF_{OSH\pm}(k_x,k_y;z) = e^{\pm j\frac{z}{2k_0}(k_x^2+k_y^2)}. \tag{6.3-38}$$

With the introduction of the holographic *OTF*, we can now in a position to explain pre-processing of holographic information. A general formalism starts from Eq. (6.3-31). However, for the sake of simplicity, we will assume a planar object $|T(x,y;z)|^2 = I(x,y)\delta(z-z_1)$. The complex record, from Eq. (6.3-31), then becomes

$$i_C(x,y) = \mathcal{F}^{-1}\left\{e^{j\frac{z_1}{2k_0}(k_x^2+k_y^2)} \times \left[\int p_1^*(x',y')p_2\left(x'+\frac{f}{k_0}k_x,y'+\frac{f}{k_0}k_y\right)e^{j\frac{z_1}{f}(x'k_x+y'k_y)}dx'dy'\right]\mathcal{F}\{I(x,y)\}\right\}$$

$$= \mathcal{F}^{-1}\left\{OTF_{OSH+}(k_x,k_y;z_1)\left[\int p_1^*(x',y')p_2\left(x'+\frac{f}{k_0}k_x,y'+\frac{f}{k_0}k_y\right)e^{j\frac{z_1}{f}(x'k_x+y'k_y)}dx'dy'\right]\mathcal{F}\{I(x,y)\}\right\}$$

$$= \mathcal{F}^{-1}\left\{OTF_{HFE}(k_x,k_y;z_1)\mathcal{F}\{I(x,y)\}\right\}, \tag{6.3-39a}$$

where $OTF_{HFE}(k_x,k_y;z)$ is the holographic feature extraction OTF:

$$OTF_{HFE}(k_x,k_y;z)$$

$$= OTF_{OSH+}(k_x,k_y;z)\left[\int p_1^*(x',y')p_2\left(x'+\frac{f}{k_0}k_x,y'+\frac{f}{k_0}k_y\right)e^{j\frac{z}{f}(x'k_x+y'k_y)}dx'dy'\right]. \tag{6.3-39b}$$

The holographic feature extraction *OTF* can perform holographic recording through $OTF_{OSH+}(k_x,k_y;z_1)$ and at the same time it can process the spectrum of the object and the processing is controlled by the selection of $p_1(x)$ and $p_2(x)$. To see it clearly, let us reconstruct the complex record by taking the Fourier transform of Eq. (6.3-39a), multiplying by the complex conjugate of $OTF_{OSH+}(k_x,k_y;z_1)$, that is, $e^{-j\frac{z_1}{2k_0}(k_x^2+k_y^2)}$, and finally taking the inverse transform:

$$\mathcal{F}^{-1}\left\{\mathcal{F}\{i_C(x,y)\}OTF_{OSH+}^*\right\} = \mathcal{F}^{-1}\left\{OTF_{HFE}(k_x,k_y;z_1)\mathcal{F}\{I(x,y)\}OTF_{OSH+}^*(k_x,k_y;z_1)\right\}$$

$$= \mathcal{F}^{-1}\left\{\left[\int p_1^*(x',y')p_2\left(x'+\frac{f}{k_0}k_x,y'+\frac{f}{k_0}k_y\right)e^{j\frac{z_1}{f}(x'k_x+y'k_y)}dx'dy'\right]\mathcal{F}\{I(x,y)\}\right\}. \tag{6.3-40}$$

The above result represents that we have the processed version of the original intensity distribution upon holographic reconstruction.

Example: Edge Extraction Using a Gaussian-Ring Pupil

A Gaussian-ring pupil was first considered by Schilling and Poon [1995] to perform edge enhancement of an incoherent object upon holographic reconstruction. We shall use this pupil as an example. The other pupil used is a delta function in the two-pupil system. Hence, we have $p_1(x,y) = e^{-b\left(\sqrt{x^2+y^2}-r_0\right)}$ and $p_2(x,y) = \delta(x,y)$, where b specifies the Gaussian falloff, and r_0 is the radius of the ring. With $p_2(x) = \delta(x,y)$, the holographic feature extraction OTF becomes

$$OTF_{\text{HFE}}(k_x, k_y; z) = e^{-j\frac{z}{2k_0}\left(k_x^2+k_y^2\right)} p_1^*\left(-\frac{f}{k_0}k_x, -\frac{f}{k_0}k_y\right)$$

$$= e^{-j\frac{z}{2k_0}\left(k_x^2+k_y^2\right)} e^{-b\left|\sqrt{\left(\frac{f}{k_0}k_x\right)^2+\left(\frac{f}{k_0}k_y\right)^2}-r_0\right|}. \qquad (6.3\text{-}41)$$

Using this OTF and from Eq. (6.3-39a), the edge-preserved complex hologram is

$$i_C(x,y) = \mathcal{F}^{-1}\left\{OTF_{\text{HFE}}(k_x, k_y; z_1)\mathcal{F}\{I(x,y)\}\right\}$$

$$= \mathcal{F}^{-1}\left\{e^{-j\frac{z_1}{2k_0}\left(k_x^2+k_y^2\right)} e^{-b\left|\sqrt{\left(\frac{f}{k_0}k_x\right)^2+\left(\frac{f}{k_0}k_y\right)^2}-r_0\right|}\mathcal{F}\{I(x,y)\}\right\}. \qquad (6.3\text{-}42)$$

Clearly, the spectrum of the intensity destitution is being bandpass filtered. Figure 6.3-12a and d shows the original binary image and its spectrum, respectively. Figure 6.3-12b shows the magnitude of $p_1^*\left(-fk_x/k_0, -fk_y/k_0\right)$, where the pupil is a Gaussian ring $p_1(x,y) = e^{-b\left(\sqrt{x^2+y^2}-r_0\right)}$. In Figure 6.3-12e, we show the original spectrum of the object multiplied by the Gaussian ring shown in Figure 6.3-12b. Figure 6.3-12c shows the magnitude of the complex hologram given by Eq. (6.3-42) and Figure 6.3-12f shows the real image reconstruction of the complex hologram:

$$i_C(x,y) * h(x,y;z_1) = \mathcal{F}^{-1}\left\{e^{-j\frac{z_1}{2k_0}\left(k_x^2+k_y^2\right)} e^{-b\left|\sqrt{\left(\frac{f}{k_0}k_x\right)^2+\left(\frac{f}{k_0}k_y\right)^2}-r_0\right|}\mathcal{F}\{I(x,y)\}\right\} * h(x,y;z_1)$$

$$= \mathcal{F}^{-1}\left\{e^{-j\frac{z_1}{2k_0}\left(k_x^2+k_y^2\right)}\right\} * \mathcal{F}^{-1}\left\{e^{-b\left|\sqrt{\left(\frac{f}{k_0}k_x\right)^2+\left(\frac{f}{k_0}k_y\right)^2}-r_0\right|}\mathcal{F}\{I(x,y)\}\right\} * h(x,y;z_1)$$

$$= \frac{-jk_0}{2\pi z_1}e^{jk_0\left(x^2+y^2\right)/2z_1} * \mathcal{F}^{-1}\left\{e^{-b\left|\sqrt{\left(\frac{f}{k_0}k_x\right)^2+\left(\frac{f}{k_0}k_y\right)^2}-r_0\right|}\mathcal{F}\{I(x,y)\}\right\} * h(x,y;z_1)$$

$$\propto \mathcal{F}^{-1}\left\{e^{-b\left|\sqrt{\left(\frac{f}{k_0}k_x\right)^2+\left(\frac{f}{k_0}k_y\right)^2}-r_0\right|}\mathcal{F}\{I(x,y)\}\right\}, \qquad (6.3\text{-}43)$$

as $e^{j\frac{k_0}{2z_1}\left(x^2+y^2\right)} * h(x,y;z_1) \propto e^{j\frac{k_0}{2z_1}\left(x^2+y^2\right)} * e^{-j\frac{k_0}{2z_1}\left(x^2+y^2\right)} \propto \delta(x,y)$. The result clearly indicates bandpass filtering of the intensity object $I(x,y)$.

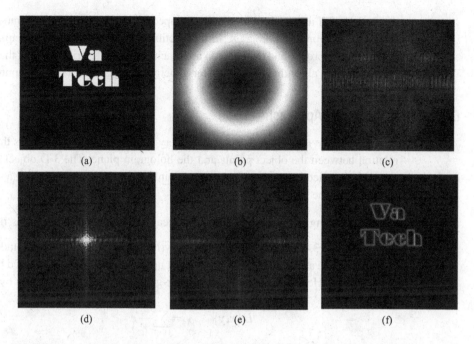

Figure 6.3-12 (a) Original binary image, (b) plot of Gaussian ring in the frequency domain, that is, $p_1^*\left(-fk_x/k_0, -fk_y/k_0\right)$, where $p_1(x) = e^{-b\left(\sqrt{x^2+y^2}-r_0\right)}$, (c) magnitude plot of the edge-preserved complex hologram from Eq.(6.3-42), (d) magnitude spectrum of (a), (e) original spectrum multiplied by the Gaussian ring shown in (b), (f) holographic reconstruction of the complex hologram from Figure (c). Reprinted from Zhang et al., "Review on feature extraction for 3-D incoherent image processing using optical scanning holography," *IEEE Transactions on Industrial Informatics* 15, pp. 6146–6154 (2019), with permission. © IEEE

6.4 Computer-Generated Holography

Computer-generated holography deals with the methods for digitally generating holographic interference patterns. The resulting interference patterns are called *computer-generated holograms* (CGHs). The CGH can then be subsequently printed on a film or inputted to a spatial light modulator (SLM) for optical reconstruction. There are two popular approaches to generate CGHs: point-based approach and polygon-based approach. In the point-based approach, the 3-D object is considered to be made up of a collection of point sources. Each point source generates a spherical wave on the hologram plane and the total complex amplitude on the hologram plane is the object wave calculated by adding up the spherical waves emitted by all the point sources. In the polygon approach, the 3-D object is expressed as a collection of planar polygons. The object wave on the hologram plane is the summation of all the diffracted fields from each polygon. The central idea of using polygons to represent a 3-D object is that the number of polygons representing an object is much smaller than that of points in the point-based approach, thus drastically speeding up the calculation time required for the generation of a hologram. In this section, we cover the basic principles of the point-based approach and the polygon approach. We then describe some of the modern fast calculation algorithms for computer-generated holograms in the two approaches.

6.4.1   Point-Based Approach

In principle, the *point-based approach* is very simple. Figure 6.4-1 shows the spatial relation between the object points and the hologram plane. The 3-D object is represented by a collection of self-illuminating points and each point is emitting a spherical wave toward the hologram plane.

Remembering that $\dfrac{a_i}{r_i} e^{-j\frac{2\pi}{\lambda_0} r_i}$ is a spherical wave of amplitude a_i, as discussed earlier in Eq. (3.4-13), we, therefore, can write the total complex amplitude on the hologram plane as the object wave by adding up the spherical waves emitted by all the point sources. Hence, the complex computer-generated hologram is given by

$$H(x,y)\big|_{\substack{0\leq x < X \\ 0\leq y < Y}} = \sum_{i=0}^{N-1} \frac{a_i}{r_i} e^{-j\frac{2\pi}{\lambda_0} r_i}, \tag{6.4-1}$$

where X and Y are the horizontal and vertical extents of the complex hologram, respectively. λ_0 is the wavelength of the optical beam that is used to generate the hologram. a_i is the light intensity of the ith point source. The term $r_i = \sqrt{(x-x_i)^2 + (y-y_i)^2 + z_i^2}$ is the distance of an ith object point at position (x_i, y_i) to a point at (x, y) on the hologram, and z_i is the distance from the ith object point to the hologram plane.

Since conventional SLMs cannot display a complex hologram, we can display the real part of the complex hologram if an amplitude-type SLM is used:

$$b_0 + \mathrm{Re}[H(x,y)] = \sum_{i=0}^{N-1} \frac{a_i}{r_i} \cos\left(\frac{2\pi}{\lambda_0} r_i\right), \tag{6.4-2}$$

where b_0 is a DC bias added to make the real hologram become a positive real value. For this display, a twin image of the object is also constructed.

If a phase-type SLM is used, we can display

$$\theta(x,y) = \tan^{-1}\left[\frac{\mathrm{Im}[H(x,y)]}{\mathrm{Re}[H(x,y)]}\right]. \tag{6.4-3}$$

We can also convert the complex hologram to an off-axis hologram if we multiply a spatial sinusoidal carrier to the complex hologram and take the real part of the resulting product:

$$H_{\text{off-axis}}(x,y) = b_0 + \mathrm{Re}\left[H(x,y) e^{j\frac{2\pi}{\lambda_0}\sin\theta x}\right], \tag{6.4-4}$$

where θ is the off-axis angle that separates the desired 3D image from the twin image upon holographic reconstruction [see Eq. (5.1-16) of an off-axis point-source hologram and its reconstruction in Figure 5.1-18].

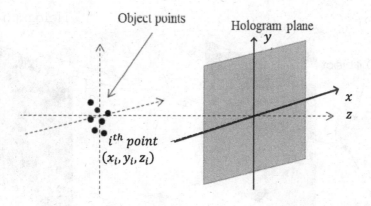

Figure 6.4-1 Spatial relation between object points and the hologram plane

Computer-generated holography is likely to become one of the promising solutions in the next generation of 3-D display. At the same time, such optimism is hampered by a number of practical problems which are difficult to solve in the foreseeable future. Many of these problems could be traced back to the fine pixel size of a hologram. For example, a small 10 mm × 10 mm hologram with a square pixel size of 5 μm × 5 μm, comprises over $2000 \times 2000 = 4 \times 10^6$ points which is about half of the number of pixels in a 4K Ultra HD TV, which has $3840 \times 2160 \approx 8.3$ million pixels. One can easily imagine the formidable amount of computation associated with the generation and processing of a large hologram digitally.

From Eq. (6.4-1), we can see that each object point is contributing to the entire hologram, and the evaluation of each hologram pixel involves the complicated expression enclosed in the summation operation. The calculation time required is proportional to the number of point sources representing the 3-D object. Indeed, one of the major bottlenecks is the intensive computation that is involved in the generation and processing of a large hologram digitally in point-based computer-generated holography. In this section, we discuss a modern framework of computer-generated holography known as the *wavefront recording plane* (WRP).

The concept of the wavefront recording plane (WRP), inspired by the image hologram method [Yoshikawa et al. (2009)], is suggested by Shimobaba et al. [2009]. Traditional approaches aimed at enhancing the speed of generating a hologram directly from a 3-D object. In the WRP approach, a 2-D WRP is a hypothetical plane that is parallel to the hologram plane, and placed at close proximity to the 3D object. The situation is shown in Figure 6.4-2. The self-illuminating object point will generate a spherical waveform to cover the entire hologram plane, and intercept the WRP in its path. If the WRP is near to the object point, the coverage of the object wave front on the WRP plane is limited to a small virtual window that is bounded by a small support. Instead of calculating the wave front on every pixel of the hologram, only wave front on the pixels within the virtual window is calculated. Denoting the spherical wave front and the support of the pth object point on the WRP by $w_p(x, y)$ and S_p, respectively, we therefore have

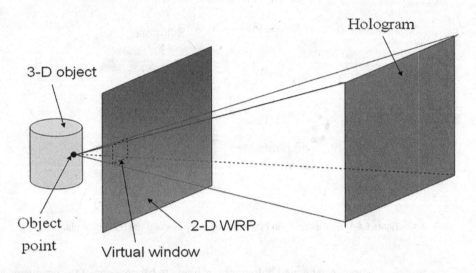

Figure 6.4-2 Spatial relation between object points, the 2-D WRP, and the hologram plane. Adapted from Tsang and Poon, Chinese Optics Letters 11 (1), 2013

$$w_p(x,y)\Big|_{(x,y)\in S_p} = \frac{a_p}{r_p}e^{-j\frac{2\pi}{\lambda_0}r_p}, \tag{6.4-5}$$

where $r_p = \sqrt{(x-x_p)^2 + (y-y_p)^2 + z_p^2}$ is the distance of a pth object point at position (x_p, y_p) to a point at (x,y) on the WRP, and z_p is the distance from the pth object point to the WRP. a_p is the intensity of the pth point. The hologram on the WRP is the sum of the spherical wave fronts from all the object points and is given by

$$W(x,y) = \sum_{p=0}^{N-1} w_p(x,y)\Big|_{(x,y)\in S_p} = \sum_{p=0}^{N-1}\frac{a_p}{r_p}e^{-j\frac{2\pi}{\lambda_0}r_p}. \tag{6.4-6}$$

After the WRP pattern is generated, the field distribution is expanded to the hologram plane given by

$$u(x,y) = W(x,y) * h(x,y;z_w), \tag{6.4-7}$$

where z_w is the separation between the WRP and the hologram plane. As the support area, that is, S_p, is much smaller than X × Y from Eq. (6.4-1), the computation load is significantly reduced compared with that in Eq. (6.4-1). To reconstruct the object from the complex hologram $u(x,y)$, we perform

$$u(x,y) * h^*(x,y;z+z_w), \tag{6.4-8}$$

where z is the distance from the WRP to the object. This step corresponds to the virtual field calculation illustrated in Figure 5.3-2. Figure 6.4-3 shows the simulation results of the WRP method.

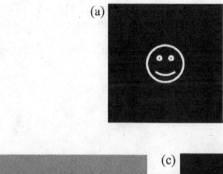

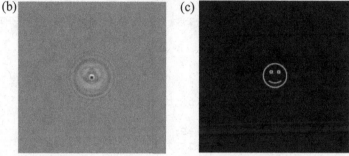

Figure 6.4-3 WRP simulation results. (a) Original object, (b) hologram on WRP, and (c) reconstruction of the complex hologram $u(x, y)$

The images in Figure 6.4-3 are generated using the m-file shown below, where Eqs. (6.4-7) and (6.4-8) have been implemented using the *angular spectrum method (ASM)* discussed in Eq. (3.4-8) in Chapter 3.

```
% Simulation_of_simple_WRP
% All length units are in mm
% Adapted from the one initially developed
% by J.-P. Liu of Feng Chia Univ., Taiwan
close all;clear
lambda = 0.532*10^-3; % wavelength
dx = 6.4*10^-3; % pixel size
dz = 1;  % distance between object plane and WRP plane
z = 150; %distance between WRP to hologram plane
k = 2*pi/lambda; % wavenumber
W = 5; % size of support on WRP, #of pixel=odd integer
w=floor(W/2);

% generation of the tables
[x, y]=meshgrid(-w:w,-w:w);
r=sqrt(x.^2+y.^2+(dz)^2);
WRPT = zeros(W,W,255); % the WRP tables
for g =1:255
    WRPT(:,:,g)=sqrt(g).*exp(-1j*k.*r)./r;
end
```

```
O = imread('front.jpg'); % input object, 512 by 512
figure;imshow(O);

% calculate the field on WRP
[M, N] = size(O);

P=zeros(M+2*w,N+2*w);

for a = 1:M
    for b =1:N
        s = O(a,b);
        if s>0
            P(a:W+a-1,b:b+W-1) = P(a:W+a-1,b:b+W-1)+WRPT(:,:,s);
        end

    end
end

% Propagation to the hologram plane
P2 = zeros (1024);
P2(257-w:768+w,257-w:768+w)=P;
[k,l]=meshgrid(-512:511,-512:511);
TF = exp(-1i*(z)*2*pi/lambda.*sqrt(1-(lambda*k/1024/dx).^2-
(lambda*l/1024/dx).^2));
TF=fftshift(TF);
hologram=fftshift(ifft2(fft2(fftshift(P2)).*TF));
figure; imshow (real(hologram),[]); %title('Hologram')

% Reconstruction using ASM TFr = exp(1i*(z+dz)*2*pi/
lambda.*sqrt(1-(lambda*k/1024/dx).^2-(lambda*l/1024/dx).^2));
TFr=fftshift(TFr);
Er =fftshift(ifft2(fft2(fftshift(hologram)).*TFr));

AFD=(abs(Er)).^2;
AFD=AFD/max(max(AFD));
figure; imshow (AFD); %title('Reconstructed image')
```

The calculation time using the concept of WRP can be further reduced if we employ a *look-up-table (LUT) method*. In a look-up table, we pre-compute Eq. (6.4-5) for all combinations of (x_p, y_p, z_p). As a result, in the generation of $W(x, y)$ from Eq. (6.4-6), each of its constituting virtual window $w_p(x, y)\big|_{(x,y)\in S_p}$ can be retrieved from the corresponding entries in the LUT, giving the process being computational free. A video sequence of digital holograms, each comprising 2048 × 2048 pixels and representing 3×10^4 object points, have been generated at a rate of over 10 frames per second [Shimobaba et al. (2010)]. Although the computation time is decreased, the memory requirement for storing the LUT is increased as the number of object points get larger.

The WRP approach has one major drawback. The WRP must be close to the object space to ensure a small support for all the object points and, therefore, the method is suitable for 3-D objects with a narrow depth range. The situation is illustrated in

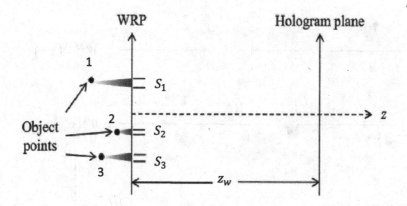

Figure 6.4-4 Example of three object points casting three wavefronts on a WRP with support S_p, where $p = 1, 2, 3$. Note that the nearer the object point to the WRP, the smaller is the support

Figure 6.4-4. Note that the support S_p ($p = 1, 2, 3$) is larger when the object point is further away from the WRP. In order to maintain the computation efficiency of the WRP method, the depth range of the object must not be too large. To address this issue, double WRPs and multiple WRPs have been investigated [Phan et al. (2014a, 2014b)]. Figure 6.4-5 illustrates the multiple WRP approach. The long-depth object is divided into different cross sections along the depth direction, and a unique WRP is generated for each section. The diffracted spherical wave front of each point (dot in the figure) is projected onto the nearest WRP. Suppose there are M sections, and the hologram on the WRP of the mth section and its distance from the hologram are denoted by $W^m(x, y)$ and z^m, respectively. The complex hologram is, therefore, generated for the collective contributions of all the WRPs, and Eq. (6.4-7) can be re-generalized for multiple WRPs as

$$u(x, y) = \sum_{m=0}^{M-1} \mathcal{F}^{-1} \left\{ \mathcal{F} \left\{ W^m(x, y) \right\} \mathcal{F} \left\{ h(x, y; z^m) \right\} \right\}. \tag{6.4-9}$$

Inherent in the point-based method, the computation time of hologram generation in the WRP approach also suffers from the number of object points used in the calculation. In light of this, Tsang et al. [2011] have developed the *interpolated WRP (IWRP) method* to alleviate this problem. In the approach, it is assumed that the scene image is considerably smaller in resolution than that of the hologram, which is in general true in practice. As such, object points that are clustered around a small neighborhood will share similar optical and spatial properties. On this basis, the object intensity profile is evenly partitioned into nonoverlapping square supports with each support uniformly filled with the mean intensity and mean depth values of its constituting pixels. The situation is illustrated in Figure 6.4-6.

The object point at the center of the support is taken as a sample point and within the square support on the object plane, all the pixels have identical intensity and depth

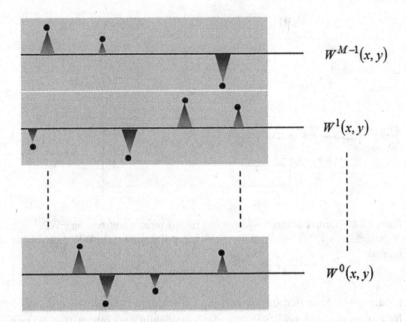

$W^{M-1}(x, y)$

$W^1(x, y)$

$W^0(x, y)$

Figure 6.4-5 Multiple WRP configuration: Dots in the figure are object points. $W^m(x, y)$ is the hologram on the mth section WRP. Reprinted from Tsang et al., "Review on the state-of-the-art technologies for acquisition and display of digital holograms," *IEEE Transactions on Industrial Informatics,* 12, pp. 886–901 (2016), with permission. © IEEE

values as the sample point in the center of the square. In other words, the object point is duplicated to all the pixels within each square support. The overall process is equivalent to subsampling the object scene along with interpolation. Now, due to the close proximity of the object scene and the WRP, the wave front projected from a square support of pixels on the WRP will only cover a small virtual window, which is practically similar in size and location as the object's square support. In other words, both the object space and the WRP are evenly portioned into nonoverlapping square blocks. The WRP is now called the interpolated WRP or IWRP. We denote $w_n(x, y)$ as the fringe pattern on the virtual window on the IWRP with n denoting the nth square block. After the hologram fringe patterns of all the nonoverlapping virtual window have been calculated, the overall hologram on the IWRP is simply given by the sum of the fringe pattern of the nonoverlapping individual virtual window. Since we have nonoverlapping windows, summation can be done by union as

$$W(x, y) = \bigcup_{n=0}^{N-1} w_n(x, y). \tag{6.4-10}$$

It is important to point out that summation can be replaced by the union operation. The amount of computation is negligible with union as it is only a memory copying operation from the LUT [Peter Tsang, City University of Hong Kong, (private communications)]. Video frame rate is thus possible for large holograms. With the

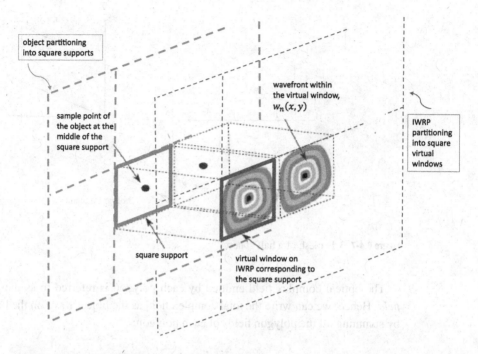

object partitioning into square supports

sample point of the object at the middle of the square support

wavefront within the virtual window, $w_n(x, y)$

IWRP partitioning into square virtual windows

square support

virtual window on IWRP corresponding to the square support

Figure 6.4-6 Spatial relation between the square support on the object space and its corresponding virtual window on the IWRP

IWRP approach, experimental result demonstrating the generation of a 2048 × 2048 hologram of around 4 million object points at 40 frames per second has been demonstrated [Tsang et al. (2011)]. To end this subsection, we want to point out that there are a number of recent review papers on the point-based approach [Tsang and Poon (2013), Tsang et al. (2018)].

6.4.2 Polygon-Based Approach

The *polygon-based method* reduces a vast amount of sampling units as compared with the point-based method. The method divides a 3-D object into a collection of 2-D *polygons* (a triangle is typically used for the method). The polygon method is also partly motivated by the availability of visualization tools or rendering software such as *3ds Max*. 3ds Max is a professional 3-D computer graphics program for making 3-D models and images. A *mesh* is a geometric model of a 3D object in which the basic shape is made up of vertices connected by edges. In 3ds Max, we can edit a mesh by adding or deleting the various polygons. The feature of the method, therefore, makes it possible to apply computer graphics to computer-generated holograms. In the development of this subsection, the polygon employed is a triangle. Figure 6.4-7 illustrates a typical 3-D mesh of a half sphere consisting of triangles and the hologram is in the x–y plane.

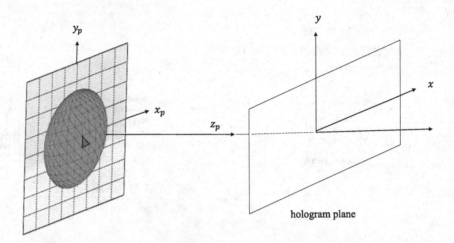

Figure 6.4-7 3-D mesh of a half sphere

The optical complex field emitted by each polygon is referred to as the *polygon field*. Hence, we can write the total complex field as the object wave on the hologram by summing all the polygon fields of each polygon:

$$u(x,y) = \sum_{i=1}^{N} u_i(x,y), \qquad (6.4\text{-}11)$$

where N is the number of polygons and $u_i(x,y)$ is the polygon field on the hologram plane from the ith polygon. Equation (6.4-11) is reminiscent of the complex hologram given by Eq. (6.4-1) in the point-based method. In this subsection, we discuss two classical methods in polygon-based computer-generated holography [Zhang et al. (2022)]: the traditional method [Matsushima (2020), Shimobaba and Ito (2019)] and the analytical method [Ahrenberg et al. (2008)].

Traditional Method

Diffraction involving two parallel planes is well established. For example, we can relate the spectrum on one plane to that on the other plane by the spatial transfer function of propagation [see Eq. (3.4-7)], or we can relate the complex fields between the two parallel planes through the Fresnel diffraction formula under the small-angle approximations [see Eq. (3.4-17)]. However, in a general mesh, most polygons do not lie parallel to the hologram plane.

The main objective first is to find the polygon field $u_i(x,y)$. Figure 6.4-8 shows a general single tilted polygon. We represent the *surface function of an arbitrary polygon* on a tilted plane as $u_s(x_s,y_s) = A(x_s,y_s)e^{j\phi(x_s,x_s)}$, where $A(x_s,y_s)$ and $\phi(x_s,x_s)$ are amplitude and phase distributions. The amplitude distribution gives the shape, texture, and brightness of the polygon, and the phase distribution is taken to be randomized to simulate a light diffuser. We also assume that the amount of tilt is small enough so that the hologram captures all the diffracted light from all the polygons

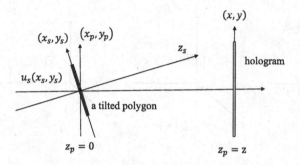

Figure 6.4-8 Coordinate systems: source coordinates (tilted local coordinates) and parallel local coordinates

involved. Coordinate system (x_s, y_s, z_s) is referred to as a *tilted local coordinate system* or a *source coordinate system*. We also define a *parallel local coordinate system* (x_p, y_p, z_p), which shares the origin of the tilted local coordinates. The parallel local coordinate system is parallel to the hologram plane (x, y). Given the complex field $u_s(x_s, y_s)$ in the tilted local coordinate system, we want to find the polygon field $u_i(x, y)$ on the hologram plane. The approach is to first relate the field distribution in the tilted local coordinate system to that in the parallel local coordinate system through rotation transformation. Subsequent diffraction between the two parallel planes is then performed to obtain the polygon field on the hologram.

The two local coordinates can be mutually transformed by coordinate rotation using *transformation matrix T* as follows:

$$r_p^t = \begin{pmatrix} x_p \\ y_p \\ z_p \end{pmatrix} = \begin{pmatrix} a_1 & a_4 & a_7 \\ a_2 & a_5 & a_8 \\ a_3 & a_6 & a_9 \end{pmatrix} \begin{pmatrix} x_s \\ y_s \\ z_s \end{pmatrix} = T r_s^t, \tag{6.4-12a}$$

and

$$r_s^t = T^{-1} r_p^t, \tag{6.4-12b}$$

where $r_p = (x_p, y_p, z_p)$ and $r_s = (x_s, y_s, z_s)$ are position vectors defined by row matrices. The transpose and inverse of a matrix $\mathcal{M}$ are denoted by $\mathcal{M}^t$ and $\mathcal{M}^{-1}$, respectively. In general, the matrix T is a rotation matrix $R(\theta)$ or the product of rotation matrices such as $R_x(\theta), R_y(\theta)$, and $R_z(\theta)$, where the subscripts x, y, and z denote the axes of rotation and θ is the angle of rotation around the axes. Figure 6.4-9 gives the three rotation matrices. The sense of the angle θ is defined by the right-hand rule. A positive rotation, that is, $\theta > 0$, means that if the thumb of the right hand is pointed along the positive direction of the rotation axis, say, the x-axis, then the fingers curl in the positive direction. Therefore, we have a counter-clockwise rotation about the axis. It can be verified readily that any transformation matrix T is an *orthogonal matrix*, that is, $T^t = T^{-1}$, and the determinant of T is unity, that is, $|T| = 1$.

Let us now establish the polygon field due to a single polygon on the hologram plane. With reference to Figure 6.4-8, the complex polygon field with angular plane

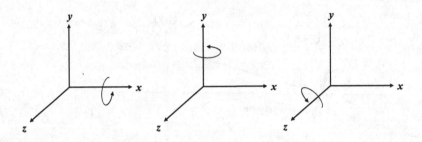

$$R_x(\theta) = \begin{pmatrix} 1 & 0 & 0 \\ 0 & \cos\theta & \sin\theta \\ 0 & -\sin\theta & \cos\theta \end{pmatrix} \quad R_y(\theta) = \begin{pmatrix} \cos\theta & 0 & -\sin\theta \\ 0 & 1 & 0 \\ \sin\theta & 0 & \cos\theta \end{pmatrix} \quad R_z(\theta) = \begin{pmatrix} \cos\theta & \sin\theta & 0 \\ -\sin\theta & \cos\theta & 0 \\ 0 & 0 & 1 \end{pmatrix}$$

Figure 6.4-9 Rotation matrices

wave spectrum $U_s\left(k_{sx},k_{sy}\right) = U_s\left(k_{sx},k_{sy};z_s=0\right)$ propagating along z_s is [see Eq. (3.4-8)]

$$u_s(x_s,y_s,z_s) = \frac{1}{4\pi^2} \iint_{-\infty}^{\infty} U_s\left(k_{sx},k_{sy};0\right) e^{-j\left(k_{sx}x_s + k_{sy}y_s + k_{sz}z_s\right)} dk_{sx}dk_{sy}$$

$$= \frac{1}{4\pi^2} \iint_{-\infty}^{\infty} U_s\left(k_{sx},k_{sy};0\right) e^{-jz_s\sqrt{k_0^2 - k_{sx}^2 - k_{sy}^2}} e^{-j\left(k_{sx}x_s + k_{sy}y_s\right)} dk_{sx}dk_{sy}$$

$$= \mathcal{F}^{-1}\left\{ U_s\left(k_{sx},k_{sy};0\right) e^{-jz_s\sqrt{k_0^2 - k_{sx}^2 - k_{sy}^2}} \right\} \tag{6.4-13}$$

as $k_{sz} = \sqrt{k_0^2 - k_{sx}^2 - k_{sy}^2}$. In vector notations and using the dot product, we have

$$u_s(\boldsymbol{r}_s) = \frac{1}{4\pi^2} \iint_{-\infty}^{\infty} U_s\left(k_{sx},k_{sy};0\right) e^{-j\boldsymbol{k}_s \cdot \boldsymbol{r}_s} dk_{sx}dk_{sy}, \tag{6.4-14}$$

where $\boldsymbol{k}_s = \left(k_{sx},k_{sy},k_{sz}\right)$. Upon rotation transformation, the complex field in Eq. (6.4-14) in the parallel local coordinates becomes

$$u_p\left(\boldsymbol{r}_p\right) = u_s\left(\boldsymbol{r}_s\right)\Big|_{\boldsymbol{r}_s = \boldsymbol{r}_p T} = \frac{1}{4\pi^2} \iint_{-\infty}^{\infty} U_s\left(k_{sx},k_{sy};0\right) e^{-j\boldsymbol{k}_s \cdot \boldsymbol{r}_p T} dk_{sx}dk_{sy}, \tag{6.4-15a}$$

where we have used the transpose of $\boldsymbol{r}_s^t$ in Eq. (6.4-12b) to establish

$$\boldsymbol{r}_s = \boldsymbol{r}_p T. \tag{6.4-15b}$$

Similarly, a propagation vector in the source and parallel local coordinates can also be transformed like position vectors as follows:

$$\boldsymbol{k}_p^{\,t} = T\boldsymbol{k}_s^{\,t}, \tag{6.4-16a}$$

and

$$\boldsymbol{k}_s^{\,t} = T^{-1}\boldsymbol{k}_p^{\,t}, \tag{6.4-16b}$$

where $k_p = (k_{px}, k_{py}, k_{pz})$. From Eq. (6.4-16b), we take the transpose to have

$$k_s = k_p (T^{-1})^t = k_p T. \tag{6.4-16c}$$

With this expression, Eq. (6.4-15a) becomes

$$u_p(r_p) = \frac{1}{4\pi^2} \iint_{-\infty}^{\infty} U_s(k_{sx}, k_{sy}; 0) e^{-jk_p T \cdot r_p T} dk_{sx} dk_{sy}. \tag{6.4-17}$$

Let us try to work out the dot product in the exponent. If vectors a and b are defined with row matrices, the dot product can be written as a matrix product as follows:

$$a \cdot b = ab^t.$$

Hence,

$$k_s \cdot r_s = k_p T \cdot r_p T = k_p T (r_p T)^t = k_p T (T^t r_p^t) = k_p r_p^t = k_p \cdot r_p \tag{6.4-18}$$

as $TT^t = TT^{-1} = I$ with I being an *identity matrix*. The result in Eq. (6.4-18) actually makes sense as coordinates rotation does not change the length and angles between vectors. With the result, Eq. (6.4-17) becomes

$$u_p(r_p) = \frac{1}{4\pi^2} \iint_{-\infty}^{\infty} U_s(k_{sx}, k_{sy}; 0) e^{-jk_p \cdot r_p} dk_{sx} dk_{sy}$$

$$= \frac{1}{4\pi^2} \iint_{-\infty}^{\infty} U_s(k_{sx}, k_{sy}; 0) e^{-j(k_{px}x_p + k_{py}y_p + k_{pz}z_p)} dk_{sx} dk_{sy}, \tag{6.4-19a}$$

where

$$k_{pz}(k_{px}, k_{py}) = \sqrt{k_0^2 - k_{px}^2 - k_{py}^2}. \tag{6.4-19b}$$

Note that the exponential function in Eq. (6.4-19a) represents plane wave propagating along the z_p-direction (or the z-direction) on the parallel local coordinates. To completely describe $u_p(r_p)$ by the parallel local coordinates, we need to fully convert Eq. (6.4-19a) in terms of (k_{px}, k_{py}, k_{pz}). Let us first work on $U_s(k_{sx}, k_{sy}; 0)$. Starting from Eq. (6.4-16b), we have

$$k_s^t = T^{-1} k_p^t,$$

which is equivalently equal to

$$\begin{pmatrix} k_{sx} \\ k_{sy} \\ k_{sz} \end{pmatrix} = \begin{pmatrix} a_1 & a_2 & a_3 \\ a_4 & a_5 & a_6 \\ a_7 & a_8 & a_9 \end{pmatrix} \begin{pmatrix} k_{px} \\ k_{py} \\ k_{pz} \end{pmatrix}.$$

Therefore, we have

$$k_{sx} = k_{sx}(k_{px}, k_{py}) = a_1 k_{px} + a_2 k_{py} + a_3 k_{pz}(k_{px}, k_{py}) \tag{6.4-20a}$$

and

$$k_{sy} = k_{sy}(k_{px}, k_{py}) = a_4 k_{px} + a_5 k_{py} + a_6 k_{pz}(k_{px}, k_{py}), \tag{6.4-20b}$$

which give

$$U_s\left(k_{sx}, k_{sy}; 0\right) = U_s\left(k_{sx}\left(k_{px}, k_{py}\right), k_{sy}\left(k_{px}, k_{py}\right); 0\right)$$

$$= U_s\left(a_1 k_{px} + a_2 k_{py} + a_3 k_{pz}, a_4 k_{px} + a_5 k_{py} + a_6 k_{pz}; 0\right),$$

(6.4-21)

where again $U_s\left(k_{sx}, k_{sy}; 0\right)$ is the spectrum of the given source function $u_s\left(x_s, y_s\right) = u_s\left(x_s, y_s; z_s = 0\right)$ of the polygon. Explicitly,

$$U_s\left(k_{sx}, k_{sy}; z_s = 0\right) = \mathcal{F}\left\{u_s\left(x_s, y_s; z_s = 0\right)\right\} = \int\int_{-\infty}^{\infty} u_s\left(x_s, y_s\right) e^{j\left(k_{sx}x_s + k_{sy}y_s\right)} dx_s dy_s. \quad (6.4\text{-}22)$$

Now, changing of variables from k_{sx} and k_{sy} to k_{px} and k_{py} in the differential element of Eq. (6.4-19a) is achieved by

$$dk_{sx} dk_{sy} = \left|J\left(k_{px}, k_{py}\right)\right| dk_{px} dk_{py}, \quad (6.4\text{-}23)$$

where $J\left(k_{px}, k_{py}\right)$ is the *Jacobian* of the coordinate transformation of k_{sx} and k_{sy} with respect to k_{px} and k_{py}, given by

$$J\left(k_{px}, k_{py}\right) = \frac{\partial\left(k_{sx}, k_{sy}\right)}{\partial(k_{px}, k_{py})} = \begin{vmatrix} \dfrac{\partial k_{sx}}{\partial k_{px}} & \dfrac{\partial k_{sx}}{\partial k_{py}} \\ \dfrac{\partial k_{sy}}{\partial k_{px}} & \dfrac{\partial k_{sy}}{\partial k_{py}} \end{vmatrix}. \quad (6.4\text{-}24)$$

Using Eqs. (6.4-20a), (6.4-20b), and (6.4-19b), the Jacobian can be evaluated explicitly to be

$$J\left(k_{px}, k_{py}\right) = \frac{\left(a_2 a_6 - a_3 a_5\right) k_{px}}{k_{pz}\left(k_{px}, k_{py}\right)} + \frac{\left(a_3 a_4 - a_1 a_6\right) k_{py}}{k_{pz}\left(k_{px}, k_{py}\right)} + \left(a_1 a_5 - a_2 a_4\right). \quad (6.4\text{-}25)$$

For paraxial approximations, k_{px} and k_{py} are much smaller than k_{pz}. As a result, the Jacobian becomes a constant as

$$J\left(k_{px}, k_{py}\right) \approx \left(a_1 a_5 - a_2 a_4\right). \quad (6.4\text{-}26)$$

Incorporating Eqs. (6.4-21) and (6.4-23) into Eq. (6.4-19a), the complex field propagating along the z_p direction, caused by the known polygon surface function $u_s\left(x_s, y_s; z_s = 0\right) = u_s\left(x_s, y_s\right)$, is

$$u_p\left(x_p, y_p, z_p\right)$$

$$= \frac{1}{4\pi^2} \int\int_{-\infty}^{\infty} U_s\left(k_{sx}\left(k_{px}, k_{py}\right), k_{sy}\left(k_{px}, k_{py}\right); 0\right) e^{-j\left(k_{px}x_p + k_{py}y_p + z_p\sqrt{k_0^2 - k_{px}^2 - k_{py}^2}\right)} \left|J\left(k_{px}, k_{py}\right)\right| dk_{px} dk_{py}$$

$$= \mathcal{F}^{-1}\left\{U_p\left(k_{px}, k_{py}; 0\right) e^{-jz_p\sqrt{k_0^2 - k_{px}^2 - k_{py}^2}}\right\}, \quad (6.4\text{-}27)$$

where we recognize that $U_p\left(k_{px}, k_{py}; 0\right)$ is the definition of the angular plane wave spectrum of $u_p\left(x_p, y_p, z_p\right)$ at $z_p = 0$, and, from Eq. (6.4-27), it is given by

$$U_p\left(k_{px},k_{py};0\right)=U_s\left(k_{sx}\left(k_{px},k_{py}\right),k_{sy}\left(k_{px},k_{py}\right);0\right)\left|J\left(k_{px},k_{py}\right)\right|. \quad (6.4\text{-}28)$$

Explicitly,

$$U_s\left(k_{sx}\left(k_{px},k_{py}\right),k_{px}\left(k_{px},k_{py}\right);0\right)=\mathcal{F}\left\{u_s\left(x_s,y_s;z_s=0\right)\right\}\Big|_{k_{sx}=a_1 k_{px}+a_2 k_{py}+a_3 k_{pz},\, k_{sy}=a_4 k_{px}+a_5 k_{py}+a_6 k_{pz}}$$

$$=U_s\left(a_1 k_{px}+a_2 k_{py}+a_3 k_{pz},a_4 k_{px}+a_5 k_{py}+a_6 k_{pz};0\right).$$

Equation (6.4-28) is an important result as it relates the spectrum of the polygon on the local coordinates $U_s\left(k_{sx},k_{sy};0\right)$ to that on the parallel local coordinates $U_p\left(k_{px},k_{py};0\right)$ upon coordinate transformation. In the analytical method, we will also employ this important result.

Since we have just established the polygon field due to an arbitrary polygon in the parallel local coordinates, that is, $u_p\left(x_p,y_p,z_p\right)$ in Eq. (6.4-27), we can now write the polygon field $u_i\left(x,y\right)$ (due to the ith polygon) on the hologram plane at $z_p=z_i$ with $\left(x_p,y_p\right)$ replaced by $\left(x,y\right)$ in Eq. (6.4-27):

$$u_i\left(x,y\right)=\mathcal{F}^{-1}\left\{U_{s,i}\left(a_{1,i}k_{px}+a_{2,i}k_{py}+a_{3,i}k_{pz},a_{4,i}k_{px}+a_{5,i}k_{py}+a_{6,i}k_{pz};0\right)\right.$$

$$\left.\times\left|J_i\left(k_{px},k_{py}\right)\right|e^{-jz_i\sqrt{k_0^2-k_{px}^2-k_{py}^2}}\right\}, \quad (6.4\text{-}29)$$

where $U_{s,i}\left(k_{sx,i},k_{sy,i}\right)=\mathcal{F}\left\{u_{s,i}\left(x_s,y_s;z_s=0\right)\right\}$ and $u_{s,i}\left(x_s,y_s;z_s=0\right)$ denotes the ith polygon that is z_i away from the hologram. Note that elements as from the transformation matrix T and the Jacobian are also indicated by subscript i as different polygons undergo different rotations till parallel to the hologram plane. From Eq. (6.4-29), we see that numerical calculations of each of the polygon field on the hologram require two *FFT*s and the total polygon field on the hologram is computed by Eq. (6.4-11).

Analytical Method

In the traditional method, Eq. (6.4-29) is numerically calculated using two *FFT*s. The first *FFT* is simply executed on a regular sampling grid in the local coordinate system to find the spectrum of each surface function, that is, the calculation of $\mathcal{F}\left\{u_{s,i}\left(x_s,y_s;z_s=0\right)\right\}$. However, the second *FFT* involving sampling in the parallel local coordinate system gets distorted. The reason is that coordinate system rotation from the tilted local coordinates to parallel local coordinates introduces nonlinear mapping between the two spectra [see Eq. (6.4-20)]. For this reason, interpolation is often used to ensure proper matching between the two sampling grids due to rotation. In addition, interpolation is unique for each polygon due to its geometry with respect to the hologram plane. In short, interpolation slows down the overall calculation speed of the hologram.

Ahrenberg et al. [2008] pioneered a method to analytically compute the spectrum of an arbitrary polygon, that is, $\mathcal{F}\left\{u_{s,i}\right\}$, through the use of *affine transformation*. As a result of the work, there is no need to perform *FFT* for each polygon and the subsequent interpolation needed in the traditional method. In addition, each spectrum of the polygon can be pre-calculated. The central idea of the analytical method is that

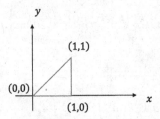

Figure 6.4-10 Unit right triangle

we can relate the Fourier transform (or the spectrum) of an arbitrary triangle to that of a *unit right triangle* (to be defined next), which is known analytically. The relation of the two spectra is established through the use of 2-D affine transformation. Once the spectrum of $u_{s,i}$ is obtained analytically, rotation of the spectrum to the parallel local coordinate system is through the use of Eq. (6.4-28), and finally diffraction between the two parallel planes is performed to have the polygon field on the hologram.

Spectrum of a Unit Right Triangle

Figure 6.4-10 shows a *unit right triangle* Δ with the coordinates of the three vertices as $(0,0)$, $(1,0)$, and $(1,1)$. The *2D unit triangle function* is then defined as

$$f_\Delta(x,y) = \begin{cases} 1, & \text{if } (x,y) \text{ lies inside } \Delta \\ 0, & \text{otherwise.} \end{cases} \tag{6.4-30}$$

The 2-D Fourier transform of $f_\Delta(x,y)$, according to Eq. (2.3-1a), is evaluated analytically and given by

$$\mathcal{F}\{f_\Delta(x,y)\} = F_\Delta(k_x,k_y) = \iint\limits_{-\infty}^{\infty} f_\Delta(x,y) e^{jk_x x + jk_y y} dxdy$$

$$= \int_0^1 \int_0^x e^{jk_x x + jk_y y}\, dydx = \begin{cases} \dfrac{1}{2}, & k_x = k_y = 0 \\[2mm] \dfrac{1-e^{jk_y}}{k_y^2} + \dfrac{j}{k_y}, & k_x = 0, k_y \neq 0 \\[2mm] \dfrac{e^{jk_x}-1}{k_x^2} - \dfrac{je^{jk_x}}{k_x}, & k_x \neq 0, k_y = 0 \\[2mm] \dfrac{1-e^{jk_y}}{k_y^2} + \dfrac{j}{k_y}, & k_x = -k_y, k_y \neq 0 \\[2mm] \dfrac{e^{jk_x}-1}{k_x k_y} + \dfrac{1-e^{j(k_x+k_y)}}{k_y(k_x+k_y)}, & \text{elsewhere.} \end{cases}$$

$$(6.4\text{-}31)$$

With the above analytical expression of the spectrum of $f_\Delta(x,y)$, we next discuss how to relate the spectrum of a general triangle to $F_\Delta(k_x,k_y)$ using an affine transformation.

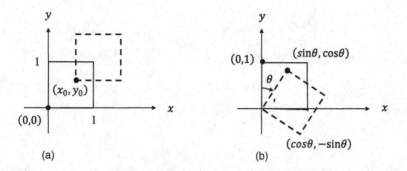

Figure 6.4-11 2-D affine operations on a unit square. (a) Translation and (b) rotation

Affine Transformation

On the basis of 2-D affine transformation, we derive an analytical frequency spectrum of a general triangle on the tilted local coordinate system. An affine transformation is a class of 2-D *geometrical transformations* that maps input coordinates (x_{in}, y_{in}) into output coordinates (x_{out}, y_{out}) according to

$$\begin{pmatrix} x_{out} \\ y_{out} \end{pmatrix} = \begin{pmatrix} a_{11} & a_{12} \\ a_{21} & a_{22} \end{pmatrix} \begin{pmatrix} x_{in} \\ y_{in} \end{pmatrix} + \begin{pmatrix} a_{13} \\ a_{23} \end{pmatrix}. \tag{6.4-32}$$

As an example, $\begin{pmatrix} a_{11} & a_{12} \\ a_{21} & a_{22} \end{pmatrix} = \begin{pmatrix} 1 & 0 \\ 0 & 1 \end{pmatrix}$ and $\begin{pmatrix} a_{13} \\ a_{23} \end{pmatrix} = \begin{pmatrix} x_0 \\ y_0 \end{pmatrix}$ represent *translation*. A translation is a geometric transformation that moves every point of an image by the same distance in a given direction. As another example, $\begin{pmatrix} a_{11} & a_{12} \\ a_{21} & a_{22} \end{pmatrix} = \begin{pmatrix} \cos\theta & \sin\theta \\ -\sin\theta & \cos\theta \end{pmatrix}$ and $\begin{pmatrix} a_{13} \\ a_{23} \end{pmatrix} = \begin{pmatrix} 0 \\ 0 \end{pmatrix}$ represent rotating the image about the origin by an angle of θ. Figure 6.4-11 illustrates the 2-D affine transformation operations on a unit square for the two examples. Other typical affine transformations are *scaling, reflection,* and *shear* [see P.6.10]. All these operations are represented in general by the affine transformation represented by Eq. (6.4-32).

Figure 6.4-12a shows an arbitrary triangle $f_T(x_s, y_s)$ on the source coordinate system with the vertex coordinates as (x_1, y_1), (x_2, y_2), and (x_3, y_3). In Figure 6.4-12b, we also show a unit right triangle on the *xy*-coordinates. Therefore, the affine transform relating the two coordinates can be written as

$$\begin{pmatrix} x_s \\ y_s \end{pmatrix} = \begin{pmatrix} a_{11} & a_{12} \\ a_{21} & a_{22} \end{pmatrix} \begin{pmatrix} x \\ y \end{pmatrix} + \begin{pmatrix} a_{13} \\ a_{23} \end{pmatrix}. \tag{6.4-33}$$

Let us find the elements a_{ij} for this particular situation.

Setting up a pairwise correspondence of the vertices between the two triangles, we have the following three matrix equations:

$$\begin{pmatrix} x_1 \\ y_1 \end{pmatrix} = \begin{pmatrix} a_{11} & a_{12} \\ a_{21} & a_{22} \end{pmatrix} \begin{pmatrix} 0 \\ 0 \end{pmatrix} + \begin{pmatrix} a_{13} \\ a_{23} \end{pmatrix},$$

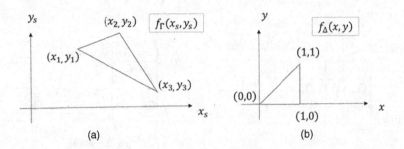

Figure 6.4-12 (a)Arbitrary triangle on the source coordinates (x_s, y_s) and (b) unit right triangle on (x, y) coordinates

$$\begin{pmatrix} x_2 \\ y_2 \end{pmatrix} = \begin{pmatrix} a_{11} & a_{12} \\ a_{21} & a_{22} \end{pmatrix} \begin{pmatrix} 1 \\ 0 \end{pmatrix} + \begin{pmatrix} a_{13} \\ a_{23} \end{pmatrix},$$

and

$$\begin{pmatrix} x_3 \\ y_3 \end{pmatrix} = \begin{pmatrix} a_{11} & a_{12} \\ a_{21} & a_{22} \end{pmatrix} \begin{pmatrix} 1 \\ 1 \end{pmatrix} + \begin{pmatrix} a_{13} \\ a_{23} \end{pmatrix}. \tag{6.4-34}$$

From the three matrix equations, we have the set of six equations allowing us to find the elements a_{ij} in terms of the vertex coordinates of the arbitrary triangle. Upon substituting the found a_{ij} into Eq. (6.4-33), we have the resulting affine operation relating the two triangles under consideration:

$$\begin{pmatrix} x_s \\ y_s \end{pmatrix} = \begin{pmatrix} x_2 - x_1 & x_3 - x_2 \\ y_2 - y_1 & y_3 - y_2 \end{pmatrix} \begin{pmatrix} x \\ y \end{pmatrix} + \begin{pmatrix} x_1 \\ y_1 \end{pmatrix}. \tag{6.4-35}$$

After establishing the affine operation for a general triangle with a unit right triangle, we shall find the spectrum of the surface function $u_s(x_s, y_s) = f_\Gamma(x_s, y_s)$. From Eq. (6.4-22), the spectrum of $f_\Gamma(x_s, y_s)$ is

$$\mathcal{F}\{f_\Gamma(x_s, y_s)\} = F_\Gamma(k_{sx}, k_{sy}) = \iint\limits_{-\infty}^{\infty} f_\Gamma(x_s, y_s) e^{j(k_{sx}x_s + k_{sy}y_s)} dx_s dy_s. \tag{6.4-36}$$

Similarly, we write the spectrum of $f_\Delta(x, y)$ as

$$\mathcal{F}\{f_\Delta(x, y)\} = F_\Delta(k_x, k_y) = \iint\limits_{-\infty}^{\infty} f_\Delta(x, y) e^{j(k_x x + k_y y)} dx dy$$

$$= \int_0^1 \int_0^x e^{j(k_x x + k_y y)} dx dy. \tag{6.4-37}$$

For an arbitrary triangle relating to the unit right triangle, we use Eq. (6.4-33) and have affine operations $x_s = a_{11}x + a_{12}y + a_{13}$ and $y_s = a_{21}x + a_{22}y + a_{23}$. With a change of variables to (x, y), Eq. (6.4-36) becomes

$$F_\Gamma(k_{sx}, k_{sy}) = \iint\limits_{-\infty}^{\infty} f_\Gamma(x_s, y_s) e^{j(k_{sx}x_s + k_{sy}y_s)} dx_s dy_s$$

$$= \int_0^1 \int_0^x f_\Delta\left(x, y\right) e^{j[k_{sx}\left(a_{11}x + a_{12}y + a_{13}\right) + k_{sy}\left(a_{21}x + a_{22}y + a_{23}\right)]} \left| J\left(x, y\right) \right| dx dy$$

$$= \int_0^1 \int_0^x e^{j[k_{sx}\left(a_{11}x + a_{12}y + a_{13}\right) + k_{sy}\left(a_{21}x + a_{22}y + a_{23}\right)]} \left| J\left(x, y\right) \right| dx dy,$$

where, according to Eq. (6.4-33), the Jacobian is

$$J\left(x, y\right) = \frac{\partial\left(x_s, y_s\right)}{\partial\left(x, y\right)} = \begin{vmatrix} \dfrac{\partial x_s}{\partial x} & \dfrac{\partial x_s}{\partial y} \\ \dfrac{\partial y_s}{\partial x} & \dfrac{\partial y_s}{\partial y} \end{vmatrix} = \begin{vmatrix} a_{11} & a_{12} \\ a_{21} & a_{22} \end{vmatrix} = a_{11}a_{22} - a_{12}a_{21}. \qquad (6.4\text{-}38)$$

Rearranging the equation $F_\Gamma\left(k_{sx}, k_{sy}\right)$ from above and using the definition of $F_\Delta\left(k_x, k_y\right)$ in Eq. (6.4-37), we have

$$F_\Gamma\left(k_{sx}, k_{sy}\right) = \left| a_{11}a_{22} - a_{12}a_{21} \right| e^{j\left(k_{sx}a_{13} + k_{sy}a_{23}\right)} \int_0^1 \int_0^x e^{j\left[k_{sx}\left(a_{11}x + a_{12}y\right) + k_{sy}\left(a_{21}x + a_{22}y\right)\right]} dx dy$$

$$= \left| a_{11}a_{22} - a_{12}a_{21} \right| e^{j\left(k_{sx}a_{13} + k_{sy}a_{23}\right)} F_\Delta\left(a_{11}k_{sx} + a_{21}k_{sy}, a_{12}k_{sx} + a_{22}k_{sy}\right). \quad (6.4\text{-}39)$$

This is an analytical expression of the spectrum of $u_s\left(x_s, y_s\right) = f_\Gamma\left(x_s, y_s\right)$, that is, $\mathcal{F}\left\{u_s\left(x_s, y_s\right)\right\} = U_s\left(k_{sx}, k_{sy}\right) = F_\Gamma\left(k_{sx}, k_{sy}\right)$, as F_Δ is given by Eq. (6.4-31) analytically. The spectrum is due to a single arbitrary triangle. Therefore, the polygon field due to the ith polygon z_i away from the hologram, according to Eq. (6.4-29) which is repeated below, is given by

$$u_i\left(x, y\right) = \mathcal{F}^{-1}\left\{ U_{s,i}\left(a_{1,i}k_{px} + a_{2,i}k_{py} + a_{3,i}k_{pz}, a_{4,i}k_{px} + a_{5,i}k_{py} + a_{6,i}k_{pz}; 0\right) \right.$$

$$\left. \times \left| J_i\left(k_{px}, k_{py}\right) \right| e^{-jz_i\sqrt{k_0^2 - k_{px}^2 - k_{py}^2}} \right\}, \qquad (6.4\text{-}40)$$

where $U_{s,i}\left(k_{sx,i}, k_{sy,i}\right)$ is now given analytically as

$$U_{s,i}\left(k_{sx,i}, k_{sy,i}\right) = \mathcal{F}\left\{u_{s,i}\left(x_s, y_s; z_s = 0\right)\right\} = F_{\Gamma,i}\left(k_{sx,i}, k_{sx,i}\right)$$

$$= \left| a_{11,i}a_{22,i} - a_{12,i}a_{21,i} \right| e^{j\left(k_{sx}a_{13,i} + k_{sy}a_{23,i}\right)} F_\Delta\left(a_{11,i}k_{sx,i} + a_{21,i}k_{sy,i}, a_{12,i}k_{sx,i} + a_{22,i}k_{sy,i}\right).$$

The total polygon field on the hologram can now be computed using Eq. (6.4-11). Hence, in the analytical method, we only need to perform a single inverse *FFT* to obtain the total polygon field, that is, the total object field, on the hologram.

As a final note to this subsection, we want to point out that while we have discussed two standard polygon-based methods, that is, the traditional method and the analytical method, there are other modern polygon-based methods which build on the idea of the two standard methods. For example, in the analytical method, we take the affine transform of the source function before rotation to align the source coordinate system to be parallel with the hologram plane. Recently, Zhang et al. [2018] have introduced a fast CGH generation method that performs polygon rotation before 2-D affine transformation. Pan et al. [2014] have developed for the first time a method using 3-D affine transformation. Avoiding the necessity for resampling, that is, the use of interpolation, of the spatial frequency grids required in the traditional method, other analytical methods have also been developed [Kim et al. (2008), Yeom and Park (2016)].

Problems

6.1 If f_s is the Nyquist rate for signal $g(x)$, find the Nyquist rate for each of the following signals:

(a) $y_a(x) = \dfrac{d}{dx} g(x)$

(b) $y_b(x) = g(x) \cos(2\pi f_0 x)$

(c) $y_c(x) = g(ax)$, where a is a real and positive constant.

(d) $y_d(x) = g^2(x)$

(e) $y_e(x) = g^3(x)$.

6.2 If B_1 and B_2 are the bandwidths of $g_1(x)$ and $g_2(x)$, respectively, find the Nyquist rate for signal $g(x) = g_1(x) g_2(x)$.

6.3 For three-step phase shifting holography, we have the following phase-shifted holograms:

$$I_0 = |\psi_0|^2 + |\psi_r|^2 + \psi_0 \psi_r^* + \psi_0^* \psi_r,$$

$$I_{\pi/2} = \left| \psi_0 + \psi_r e^{-j\pi/2} \right|^2 = |\psi_0|^2 + |\psi_r|^2 + j\psi_0 \psi_r^* - j\psi_0^* \psi_r,$$

and

$$I_\pi = \left| \psi_0 + \psi_r e^{-j\pi} \right|^2 = |\psi_0|^2 + |\psi_r|^2 - \psi_0 \psi_r^* - \psi_0^* \psi_r.$$

Show that the complex hologram of the object is given by

$$\psi_0 = \frac{(1+j)(I_0 - I_{\pi/2}) + (j-1)(I_\pi - I_{\pi/2})}{4\psi_r^*}.$$

6.4 For four-step phase shifting holography, we have the following phase-shifted holograms:

$$I_0 = |\psi_0|^2 + |\psi_r|^2 + \psi_0 \psi_r^* + \psi_0^* \psi_r,$$

$$I_{\pi/2} = |\psi_0|^2 + |\psi_r|^2 + j\psi_0 \psi_r^* - j\psi_0^* \psi_r,$$

$$I_\pi = |\psi_0|^2 + |\psi_r|^2 - \psi_0 \psi_r^* - \psi_0^* \psi_r,$$

and

$$I_{3\pi/2} = |\psi_0|^2 + |\psi_r|^2 - j\psi_0 \psi_r^* + j\psi_0^* \psi_r.$$

Show that the complex hologram of the object and it is given by

$$\psi_0 = \frac{(I_0 - I_\pi) - j(I_{\pi/2} - I_{3\pi/2})}{4\psi_r^*}.$$

6.5 Show that the coherence function of the two-pupil heterodyne scanning system shown in Figure 6.2-3 given below

$$\Gamma(x' - x'', y' - y''; z' - z'') = \int e^{j\frac{k_0}{f}[x_m(x'-x'') + y_m(y'-y'')]} e^{-j\frac{k_0(z'-z'')}{2f^2}(x_m^2 + y_m^2)} dx_m dy_m$$

is evaluated to be

$$\Gamma(x' - x'', y' - y''; z' - z'') \sim \frac{1}{z' - z''} e^{j\frac{k_0[(x'-x'')^2 + (y'-y'')^2]}{2(z'-z'')}}.$$

6.6 In code aperture imaging, the coded aperture is taken to have the following property:

$$C(x) \otimes C(x) = \int_{-\infty}^{\infty} C^*(x') C(x + x') dx' = \delta(x).$$

Show that

$$\int_{-\infty}^{\infty} C^* \left(\frac{x'' + x'}{M} \right) C \left(\frac{x + x''}{M} \right) dx'' = \delta \left(\frac{x - x'}{M} \right).$$

6.7 In COACH, $G(x, y)$ is the diffracted pattern of the coded phase mask (CPM) in the plane of the image sensor and has the following property:

$$G(x, y) \otimes G(x, y) = \delta(x, y).$$

Show that

$$G \left(x + \frac{x_i z_h}{f}, y + \frac{y_i z_h}{f} \right) \otimes G(x, y) = \delta \left(x - \frac{x_i z_h}{f}, y - \frac{y_i z_h}{f} \right).$$

6.8 In optical scanning holography, the two scanning beams are given by

$$s_1(x, y; z) = \mathcal{F} \{ p_1(x, y) \} \Big|_{k_x = \frac{k_0 x}{f}, k_y = \frac{k_0 y}{f}} * h(x, y; z)$$

and

$$s_2(x, y; z) = \mathcal{F} \{ p_2(x, y) \} \Big|_{k_x = \frac{k_0 x}{f}, k_y = \frac{k_0 y}{f}} * h(x, y; z).$$

Defining the *OTF* of the two-pupil system as

$$OTF_\Omega(k_x, k_y; z) = \mathcal{F}^* \{ s_1(x, y; z) s_2^*(x, y; z) \},$$

show that the *OTF* can be written in terms of the two pupils $p_1(x, y)$ and $p_2(x, y)$ as

$$OTF_\Omega(k_x, k_y; z) = e^{j\frac{z}{2k_0}(k_x^2 + k_y^2)} \int p_1^*(x', y') p_2 \left(x' + \frac{f}{k_0} k_x, y' + \frac{f}{k_0} k_y \right) e^{j\frac{z}{f}(x'k_x + y'k_y)} dx' dy'.$$

6.9 Show that the Jacobian of the coordinate transformation between the source coordinates and the parallel local coordinates is

$$J\left(k_{px}, k_{py}\right) = \frac{\partial\left(k_{sx}, k_{sy}\right)}{\partial\left(k_{px}, k_{py}\right)} = \frac{\left(a_2 a_6 - a_3 a_5\right)k_{px}}{k_{pz}\left(k_{px}, k_{py}\right)} + \frac{\left(a_3 a_4 - a_1 a_6\right)k_{py}}{k_{pz}\left(k_{px}, k_{py}\right)} + \left(a_1 a_5 - a_2 a_4\right)$$

in the polygon-based approach to computer-generated holography.

6.10 The general 2-D affine transformation is written as

$$\begin{pmatrix} x_{out} \\ y_{out} \end{pmatrix} = \begin{pmatrix} a_{11} & a_{12} \\ a_{21} & a_{22} \end{pmatrix} \begin{pmatrix} x_{in} \\ y_{in} \end{pmatrix} + \begin{pmatrix} a_{13} \\ a_{23} \end{pmatrix}.$$

For a unit square figure shown in P.6-10, plot the affine-transformed figure for the following cases when $\begin{pmatrix} a_{13} \\ a_{23} \end{pmatrix} = \begin{pmatrix} 0 \\ 0 \end{pmatrix}$:

(a) Shear in the x-direction:

$$\begin{pmatrix} a_{11} & a_{12} \\ a_{21} & a_{22} \end{pmatrix} = \begin{pmatrix} 1 & \tan\varphi \\ 0 & 1 \end{pmatrix}, \text{ where } \varphi < 90^0$$

(b) Shear in the y-direction:

$$\begin{pmatrix} a_{11} & a_{12} \\ a_{21} & a_{22} \end{pmatrix} = \begin{pmatrix} 1 & 0 \\ \tan\varphi & 1 \end{pmatrix}, \text{ where } \varphi < 90^0$$

(c) Scale about the origin (0,0):

$$\begin{pmatrix} a_{11} & a_{12} \\ a_{21} & a_{22} \end{pmatrix} = \begin{pmatrix} w & 0 \\ 0 & h \end{pmatrix}, \text{ where } w \text{ and } h \text{ are greater than 1.}$$

(d) Reflection about the origin (0,0):

$$\begin{pmatrix} a_{11} & a_{12} \\ a_{21} & a_{22} \end{pmatrix} = \begin{pmatrix} -1 & 0 \\ 0 & -1 \end{pmatrix}$$

(e) Reflection about the x-axis:

$$\begin{pmatrix} a_{11} & a_{12} \\ a_{21} & a_{22} \end{pmatrix} = \begin{pmatrix} 1 & 0 \\ 0 & -1 \end{pmatrix}$$

(f) Reflection about the y-axis:

$$\begin{pmatrix} a_{11} & a_{12} \\ a_{21} & a_{22} \end{pmatrix} = \begin{pmatrix} -1 & 0 \\ 0 & 1 \end{pmatrix}.$$

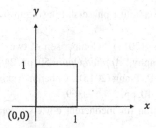

Figure P.6-10

Bibliography

Ahrenberg, L., P. Benzie, M. Magnor, and J. Watson (2008). "Computer generated holograms from three dimensional meshes using an analytic light transport model," *Applied Optics* 47, pp. 1567–1574.

Berrang, J. E. (1970). "Television transmission of holograms using a narrow-band video signal," *Bell System Technical Journal* 49, pp. 879–887.

Burckhardt, C. B. and L. H. Enloe (1969). "Television transmission of holograms with reduced resolution requirements on the camera tube," *Bell System Technical Journal* 48, pp. 1529–1535.

Cannon, T. M. and E. E. Fenimore (1980). "Coded aperture imaging:many holes make light work," *Optical Engineering* 19, pp. 283–289.

Cochran, G. (1966). "New method of making Fresnel transforms with incoherent light," *Journal of the Optical Society of America* 56, pp. 1513–1517.

Dicke, R. H. (1968). "Scatter-hole camera for X-rays and gamma rays," *The Astrophysical Journal* 153, L101.

Enloe, L. H., J. A. Murphy, and C. B. Rubinsten (1966). "Hologram transmission via television," *Bell System Technical Journal* 45, pp. 333–335.

Gabor, D. and P. Goss (1966). "Interference microscope with total wavefront reconstruction," *Journal of the Optical Society of America* 56, pp. 849–858.

Goodman, J. W. and R. W. Lawrence (1967). "Digital image formation from electronically detected holograms," *Applied Physics Letters* 11, pp. 77–79.

Guo, P. and A. J. Devane (2004), "Digital microscopy using phase-shifting digital holography with two reference waves," *Optics Letters* 29, pp. 857–859.

Indebetouw, G., T. Kim, T.-C. Poon, and B. W. Schilling (1998). "Three-dimensional location of fluorescent inhomogeneities in turbid media by scanning heterodyne holography," *Optics Letters* 23, pp. 133–137.

Indebetouw, G. and W. Zhong (2006). "Scanning holographic microscopy of three-dimensional holographic microscopy," *Journal of the Optical Society of America A* 23, pp. 2657–2661.

Kim, H., J. Hahn, and B.Lee (2008). "Mathematical modeling of triangle-mesh-modeled three-dimensional surface objects for digital holography," *Applied Optics* 47, pp. D117–D127.

Kim, T. and T.Kim (2020). "Coaxial scanning holography," *Optics Letters* 45, pp. 2046–2049.

Liu, J.-P., W.-T. Chen, H.-H. Wen and T.-C. Poon (2020). "Recording of a curved digital hologram for orthoscopic real image reconstruction," *Optics Letters* 45, pp. 4353–4356.

Liu, J.-P., T. Tahara, Y. Hayasaki, and T.-C.Poon (2018). "Incoherent digital holography: a review," *Applied Sciences* 8 (1), 143.

Liu, J.-P. and T.-C. Poon (2009). "Two-step-only quadrature phase-shifting digital holography," *Optics Letters* 34, pp. 250–252.

Liu, J. -P., T.-C. Poon, G.-S. Jhou, and P.-J. Chen (2011). "Comparison of two-, three-, and four-exposure quadrature phase-shifting holography," *Applied Optics* 50, pp. 2443–2450.

Liu, J.-P., C.-H. Guo, W.-J. Hsiao, T.-C. Poon, and P. Tsang (2015). "Coherence experiments in single-pixel digital holography," *Optics Letters* 40, pp. 2366–2369.

Lohmann, A. W. (1965). "Wavefront reconstruction for incoherent objects," *Journal of the Optical Society of America* 55, pp. 1555–1556.

Meng, X. F., L. Z. Cai, X. F. Xu, X. L. Yang, X. X. Shen, G. Y.Dong, and Y. R. Wang (2006). "Two-step phase-shifting interferometry and its application in image encryption," *Optics Letters* 31, pp. 1414–1416.

Matsushima, K. (2020). *Introduction to Computer Holography Creating Computer-Generated Holograms as the Ultimate 3D Image*, Springer, Switzerland.

Mertz, L. (1964). "Metallic beam splitters in interferometry," *Journal of the Optical Society of America*, No. 10, Advertisement vii.

Mertz, L. and N. O. Young (1962). "Fresnel transformations of images," in K. J. Habell, ed., *Proceedings of the Conference on Optical Instruments and Techniques*, pp. 305–310. Wiley and Sons, New York.

Molesini, G., D. Bertani, and M. Cetca (1982). "In-line holography with interference filters as Fourier processor," *Optica Acta: International Journal of Optics* 29, pp. 497–485.

Phan, A.-H., M.-L. Piao, S.-K. Gil, and N. Kim (2014a). "Generation speed and reconstructed image quality enhancement of a long-depth object using double wavefront recording planes and a GPU," *Applied Optics* 53, pp. 4817–4824.

Phan, A.-H., M. A. Alam, S.-H. Jeon, J.-H. Lee, N. Kim (2014b). "Fast hologram generation of long-depth object using multiple wavefront recording planes," Proc. SPIE Vol. 9006, *Practical Holography XXVIII: Materials and Applications*, 900612.

Poon, T.-C. (1985). "Scanning holography and two-dimensional image processing by acousto-optic two-pupil synthesis," *Journal of the Optical Society of America A* 2, pp. 521–527.

Poon, T.-C. (2008). "Scan-free three-dimensional imaging," *Nature Photonics* 2, pp. 131–132.

Poon, T. C. and A. Korpel (1979). "Optical transfer function of an acousto-optic heterodyning image processor," *Optics Letters* 4, pp. 317–319.

Poon, T.-C. and P. P. Banerjee (2001). *Contemporary Optical Image Processing with MATLAB®*. Elsevier, United Kingdom.

Poon, T.-C. and J.-P. Liu (2014). *Introduction to Modern Digital Holography with MATLAB*. Cambridge University Press, Cambridge, United Kingdom.

Poon, T.-C. (2007). *Optical Scanning Holography with MATLAB®*. Springer, New York.

Popescu, G. (2011). *Quantitative Phase Imaging of Cells and Tissues*. Springer, New York.

Rosen, J. and G. Brooker (2007). "Digital spatially incoherent Fresnel holography," *Optics Letters* 32, pp. 912–914.

Rosen J., V. Anand, M. R. Rai, S. Mukherjee, and A. Bulbul (2019). "Review of 3D imaging by coded aperture correlation holography (COACH)," *Applied Sciences* 9, pp. 605.

SchillingB. W. and T.-C.Poon (1995). "Real-time preprocessing of holographic information," *Optical Engineering* 34, pp. 3174–3180.

Schilling B. W., T.-C.Poon, G. Indebetouw, B. Storrie, K, Shinoda, and M. Wu (1997). "Three-dimensional holographic fluorescence microscopy," *Optics Letters* 22, pp. 1506–1508.

Schnars, U. and W. Jueptner (2005). *Digital Holography: Digital Hologram Recording, Numerical Reconstruction, and Related Techniques*. Springer, Berlin.

Shimobaba, T. and T. Tto (2019). *Computer Holography Acceleration Algorithms and Hardware Implementations*. CRC Press, Tayler & Francis Group, USA.

Shimobaba, T., N. Masuda, and T. Ito (2009). "Simple and fast calculation algorithm for computer-generated hologram with wavefront recording plane," *Optics Letters* 34, pp. 3133–3135.

Shimobaba, T., H. Nakayama, N. Masuda, and T. Ito (2010). "Rapid calculation algorithm of Fresnel computer-generated-hologram using look-up table and wavefront-recording plane methods for three-dimensional display," *Optics Express* 18, pp. 19504–19509.

Simpson, R. G., H. H. Barrett, J. A. Subach, and H. D. Fisher (1975). "Digital processing of annular coded-aperture imagery," *Optical Engineering* 14, pp. 490.

Tsai C.-M., H.-Y. Sie, T.-C. Poon, and J.-P. Liu (2021). "Optical scanning holography with a polarization directed flat lens," *Applied Optics* 60, pp. B113–B118.

Tsang, P. W. M. and T.-C. Poon (2013). "Review on theory and applications of wavefront recording plane framework in generation and processing of digital holograms," *Chinese Optics Letters* 11, 010902.

Tsang, P., W.-K. Cheung, T.-C. Poon, and C. Zhou (2011). "Holographic video at 40 frames per second for 4-million object points," *Optics Express* 19, pp. 15205–15211.

Tsang, P. W. M., T.-C. Poon, and Y. M. Wu (2018). "Review of fast methods for point-based computer-generated holography [Invited]," *Photonics Research* 6, pp. 837–846.

Vijayakumar, A., Y. Kashter, R. Kelner, and J. Rosen (2016). "Coded aperture correlation holography - a renew type of incoherent digital holograms," *Optics Express* 24, pp. 262634–262634.

Wang, Y., Y. Zhen, H. Zhang, and Y. Zhang (2004). "Study on digital holography with single phase-shifting operation," *Chinese Optics Letters* 2, pp. 141–143.

Wu, J., H. Zhang, W. Zhang, G. Jin, L. Cao, and G. Barbastathis (2020). "Single-shot lensless imaging with Fresnel zone aperture and incoherent illumination," *Light: Science & Applications* 9 (1), p. 53.

Yeom, H.-J. and J. H. Park (2016). "Calculation of reflectance distribution using angular spectrum convolution in mesh-based computer-generated hologram," *Optics Express* 24, pp. 19801–19813.

Yoneda, N., Y. Satia, and T. Nomura (2020). "Motionless optical scanning holography," *Optics Letters* 45, pp. 3184–3187.

Yoneda, N., Y. Satia, and T. Nomura (2020). "Spatially divided phase-shifting motionless optical scanning holography," *OSA Continuum* 3, pp. 3523–3555.

Yoshikawa, H., T. Yamaguchi, and R. Kitayama (2009). *In Digital Holography and Three-Dimensional Imaging, OSA Technical Digest (CD)* (Optical Society of America, 2009), paper DWC4.

Zhang, Y., T.-C. Poon, P. W. M. Tsang, R. Wang, and L. Wang (2019). "Review on feature extraction for 3-D incoherent image processing using optical scanning holography," *IEEE Transactions on Industrial informatics* 15, pp. 6146–6154.

Zhang, Y., F. Wang, T.-C. Poon, S. Fan, and W. Xu (2018). "Fast generation of full analytical polygon-based computer-generated holograms," *Optics Express* 26, pp. 19206–19224.

Zhang, Y., R. Wang, P. W. M. Tsang, and T.-C. Poon (2020), "Sectioning with edge extraction in optical incoherent imaging processing," *OSA Continuum* 3, pp. 698–708.

Zhang, Y., H. Fan, F. Wang, X. Gu, X. Qian, and T.-C. Poon (2022), "Polygon-based computer-generated holography: a review of fundamentals and recent progress [Invited]," *Applied Optics* 61, pp. B363–B374.

7 Spatial Light Modulators for Processing Optical Information

In modern optical processing and display applications, there are increased needs of a real-time device and such a device is called a spatial light modulator (SLM). Typical examples of SLMs are *acousto-optic modulators (AOMs), electro-optic modulators (EOMs),* and *liquid crystal displays.* In this chapter, we will concentrate on these types of modulators.

7.1 Information Processing with Acousto-Optic Modulators

In this section, we first discuss the acousto-optic effect. We will then cover some of its applications in laser beam modulation such as *frequency modulation* and *intensity modulation.* We conclude the section with discussions on *laser beam deflection* and *heterodyning.*

7.1.1 The Acousto-Optic Effect

In *acousto-optics*, we deal with the interaction between sound and light. As a result of the acousto-optic effect, light waves can be modulated by sound, which provides a powerful means for optical processing information.

An *acousto-optic modulator* (AOM) or commonly known as *Bragg cell* is a spatial light modulator that consists of a transparent acoustic medium (such as dense glass) to which a piezoelectric transducer is bonded. Driven by an electrical source, the transducer launches sound waves into the acoustic medium. Periodic compression and rarefaction of the index of refraction of the acoustic medium due to the applied sound waves acts as a moving phase grating with an effective spatial period equal to the wavelength Λ of the sound in the acoustic medium to diffract the incident plane wave of light into multiple diffracted orders. The situation is illustrated in Figure 7.1-1, where L denotes the length of the transducer.

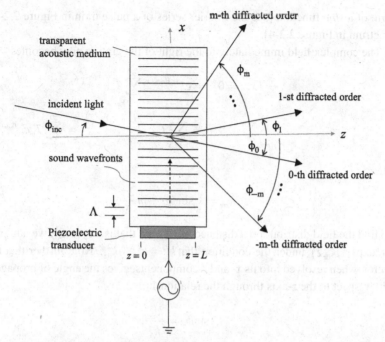

transparent
acoustic medium

incident light

ϕ_{inc}

sound wavefronts

Λ

Piezoelectric
transducer

$z = 0$ $z = L$

m-th diffracted order

ϕ_m

1-st diffracted order

ϕ_1

ϕ_0

0-th diffracted order

ϕ_{-m}

-m-th diffracted order

Figure 7.1-1 Acousto-optic modulator, illustrating diffraction by sound. Adapted from Poon and Kim (2018)

Phase Grating Approach

A simple approach is to consider the acousto-optic effect as diffraction of light by a diffraction grating. A *diffraction grating* is an optical element with a periodic structure that diffracts light into multiple beams propagating in different directions. Consider an ideal thin grating with transparency function [see Eq. (3.4-2a)] given by $t(x)$. For a unit-amplitude plane wave incident at angle ϕ_{inc}, as shown in Figure 7.1-1, we have

$$\psi_p\left(x; z = 0^+\right) = \psi_p\left(x; z = 0^-\right) t(x), \tag{7.1-1}$$

where $\psi_p\left(x; z = 0^-\right) = e^{-jk_{0x}x - k_{0z}z}\big|_{z=0^-} = e^{-jk_0 \sin\phi_{inc} x}$. The convention for angles is counterclockwise positive, measured from the z-axis as shown in Figure 7.1-1. Since the grating is a periodic structure in the x-direction, its transparency function can be described using a Fourier series [see Eq. (2.2-1)] given by

$$t(x) = \sum_{m=-\infty}^{\infty} T_m e^{-jm\frac{2\pi}{\Lambda}x}, \tag{7.1-2}$$

where T_m are the Fourier series coefficients depending on the type of grating under consideration. Note that the convention of using "$-j$" in the exponent is for spatial functions [see discussion in Section 2.3.1 for conventions]. For example, for a diffraction grating composed of a large number of parallel slits, its T_m is of the functional form of a *sinc* function [see the Fourier series of a pulse train in Figure 2.2-1 and its spectrum in Figure 2.2-4].

The complex field immediately to the right of the grating now becomes

$$\psi_p\left(x; z = 0^+\right) = e^{-jk_0\sin\phi_{inc}x} \sum_{m=-\infty}^{\infty} T_m e^{-jm\frac{2\pi}{\Lambda}x}$$

$$= \sum_{m=-\infty}^{\infty} T_m e^{-j\left(k_0\sin\phi_{inc} + m\frac{2\pi}{\Lambda}\right)x} = \sum_{m=-\infty}^{\infty} T_m e^{-jk_{xm}x},$$

where

$$k_{xm} = k_0\sin\phi_{inc} + m\frac{2\pi}{\Lambda}. \qquad (7.1\text{-}3)$$

To find the field distribution a distance of z away, that is, $\psi_p\left(x; z\right)$, we add phase factor $\exp\left(-jk_{0z}z\right)$ under the condition that $k_0^2 = k_{xm}^2 + k_{0z}^2$. Note further that the wave vector, when resolved into its x- and z-components, gives the angle of propagation ϕ_m with respect to the z-axis through the relationship

$$\sin\phi_m = \frac{k_{xm}}{k_0}, \qquad (7.1\text{-}4)$$

and $k_{0z} = k_0\cos\phi_m$. Combining Eqs. (7.1-3) and (7.1-4), we have

$$k_0\sin\phi_m = k_0\sin\phi_{inc} + m\frac{2\pi}{\Lambda},$$

and with $k_0 = 2\pi / \lambda_0$, the above equation becomes

$$\sin\phi_m = \sin\phi_{inc} + m\frac{\lambda_0}{\Lambda}, \ m = 0, \pm 1, \pm 2, \ldots \qquad (7.1\text{-}5)$$

which is known as the *grating equation*. The equation relates the incident angle to the different diffracted angles. All the angles are measured from the horizontal axis and the convention for angles is counterclockwise positive as previously mentioned.

Therefore, the complex field at z is

$$\psi_p\left(x; z\right) = \psi_p\left(x; z = 0^+\right)e^{-jk_{0z}z} = \sum_{m=-\infty}^{\infty} T_m e^{-jk_{xm}x - jk_{0z}z}$$

$$= \sum_{m=-\infty}^{\infty} T_m e^{-jk_0\sin\phi_m x - jk_0\cos\phi_m z}. \qquad (7.1\text{-}6)$$

This represents different plane waves travelling at different directions. The subscript m gives the order of the diffracted plane wave. The physical situation is shown in Figure 7.1-1, where the incident plane wave is diffracted into many plane waves with their angles governed by the grating equation. In particular, for $m = 0$, that is, the zeroth-order beam is given by

$$T_0 e^{-jk_0\sin\phi_0 x - jk_0\cos\phi_0 z},$$

where $\phi_0 = \phi_{inc}$ according to Eq. (7.1-5), and the beam is collinear with the incident plane wave.

Particle Approach

Another instinctive and more accurate approach considers the interaction of plane waves of light and sound, that is, we assume that L is sufficiently long in order to produce straight wave fronts propagating into the acoustic medium, as illustrated in Figure 7.1-1. Plane waves have well-defined momenta and particles with momentum p are associated with waves whose wavelength is $\lambda = h/p$ [see Section 1.2]. We then can consider the interaction of plane waves of sound and light as a collision of two particles such that they interact and generate a third particle. While *photons* are particles of light, *phonons* are particles of sound.

In the process of particle collision, two conservation laws have to be obeyed, namely the *conservation of energy and momentum*. If we denote the *wavevectors* (also called the *propagation vectors*) of the incident plane wave of light, diffracted plane wave of light, and sound plane wave in the acoustic medium by k_0, k_{+1}, and K, respectively, we can write the condition for *conservation of momentum* as

$$\hbar k_{+1} = \hbar k_0 + \hbar K , \qquad (7.1\text{-}7)$$

where $\hbar = h/2\pi$ and h denotes Planck's constant. Dividing Eq. (7.1-7) by $\hbar$ leads to

$$k_{+1} = k_0 + K. \qquad (7.1\text{-}8)$$

The corresponding *conversation of energy* takes the form (after division by $\hbar$)

$$\omega_{+1} = \omega_0 + \Omega, \qquad (7.1\text{-}9)$$

where ω_0, Ω, and ω_{+1} are the radian frequencies of the incident light, sound, and diffracted light, respectively. The interaction described by Eqs. (7.1-8) and (7.1-9) is called the *upshifted interaction*. Figure 7.1-2a shows the wavevector interaction diagram. Because for all practical cases $|K| \ll |k_0|$, the magnitude of k_{+1} is essentially equal to that of k_0 and therefore the wavevector momentum triangle shown is nearly isosceles. Figure 7.1-2b describes the diffracted beam being upshifted in frequency. The zeroth-order beam is the beam travelling along the same direction of the incident beam, while the +1st-order diffracted beam is the beam with frequency upshifted by the sound frequency Ω.

Suppose now that we exchange the directions of incident and diffracted light. The conservation laws can be applied again to obtain two equations similar to Eqs. (7.1-8) and (7.1-9). The two equations describing the *downshifted interaction* are

$$k_{-1} = k_0 - K \qquad (7.1\text{-}10)$$

and

$$\omega_{-1} = \omega_0 + \Omega, \qquad (7.1\text{-}11)$$

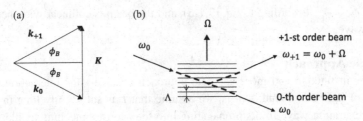

Figure 7.1-2 Upshifted diffraction: (a) wavevector diagram and (b) experimental configuration

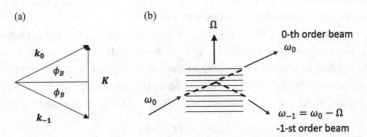

Figure 7.1-3 Downshifted diffraction: (a) wavevector diagram and (b) experimental configuration

where the subscript -1s indicate the interaction is downshifted. Figure 7.1-3a and b illustrates Eqs. (7.1-10) and (7.1-11), respectively. Note that the -1st-order diffracted beam is downshifted in frequency by Ω.

We have seen that the wavevector diagrams in Figures 7.1-2a and 7.1-3a are closed for both cases of interaction. The closed diagrams stipulate that there can only one critical incident angle (*Bragg angle*) such that plane waves of light and sound can interact. From Figures 7.1-2a or 7.1-3a, we find that the Bragg angle ϕ_B is given by

$$\sin\phi_B = \frac{|K|}{2|k_0|} = \frac{K}{2k_0} = \frac{\lambda_0}{2\Lambda},\qquad(7.1\text{-}12)$$

where λ_0 is the wavelength of light inside the acoustic medium, and Λ is the wavelength of sound. For a commercially available acousto-optic modulator [AOM-40 IntraAction Corporation] operating at sound frequency $f_s = 40$ MHz and for sound wave traveling in the glass at velocity $v_s \sim 4000$ m/s, the sound wavelength $\Lambda = v_s / f_s \sim 0.1$mm. If a He–Ne laser is used (its wavelength is about 0.6328 μm in air), the wavelength inside glass is $\lambda_0 \sim 0.6328\,\mu\text{m}\,/\,n_0 \sim 0.3743\,\mu\text{m}$ with $n_0 = 1.69$. Hence, the Bragg angle, according to Eq. (7.1-12), inside the acoustic medium is $\sim 1.9 \times 10^{-3}$ radian or about 0.1 degree; therefore, we make small angle approximations in most of the situations in the discussion of the acousto-optic effect. Figure 7.1-4 shows a typical acousto-optic modulator operating at 40 MHz. In the figure, we denote the two diffracted laser spots at a far background. The incident laser beam (not

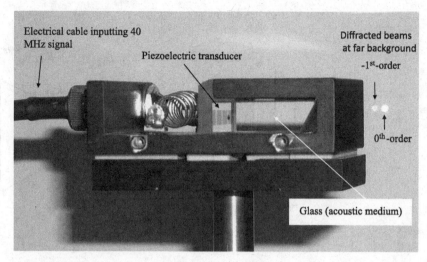

Electrical cable inputting 40 MHz signal

Piezoelectric transducer

Diffracted beams at far background

-1st-order

0th-order

Glass (acoustic medium)

Figure 7.1-4 Typical acousto-optic modulator from IntraAction Corporation, Model AOM-40. Adapted from Poon (2002)

visible as it traverses across a transparent medium of glass) is travelling along the long dimension of the transducer through the glass.

The closed diagrams in Figures 7.1-2a and 7.1-3a stipulate that there are certain critical angles of incidence ($\phi_{inc} = \pm\phi_B$) in the acoustic medium for plane waves of sound and light to interact, and also that the directions of the incident and scattered light differ in angle by $2\phi_B$. In actuality, scattering occurs even though the direction of incident light is not exactly at the Bragg angle. However, the maximum scattered intensity occurs at the Bragg angle. The reason is that for a finite length of the sound transducer, we no longer have plane wave fronts but rather the sound waves actually diffract as they propagate into the medium. As the length L of the transducer decreases, the sound field will act less and less like a plane wave and, in fact, it is now more appropriate to consider the concept of an angular plane wave spectrum of sound. Assuming a transducer of length L is along the z-direction as defined in Figure 7.1-1, the various plane waves of sound propagating over a distance x, according to Eq. (3.4-9), are given by

$$\mathcal{F}\left\{\text{rect}\left(\frac{z}{L}\right)\right\}\exp(-jK_{0z}z)\times\exp(-jK_{0x}x),$$

where $K_{0z} = \boldsymbol{K}\cdot\hat{\boldsymbol{z}} = K\sin\theta$ and $K_{0x} = \boldsymbol{K}\cdot\hat{\boldsymbol{x}} = K\cos\theta$ are the components of the sound vector $\boldsymbol{K}$ and θ is measured from the x-axis. Hence, neglecting some constant, the above equation becomes

$$\left.\text{sinc}\left(\frac{K_{0z}L}{2\pi}\right)\right|_{K_{0z}=K\sin\theta}\exp(-jK\sin\theta z - jK\cos\theta x)$$

$$\propto \text{sinc}\left(\frac{K\sin\theta L}{2\pi}\right)\exp(-jK\sin\theta z - jK\cos\theta x). \qquad (7.1\text{-}13)$$

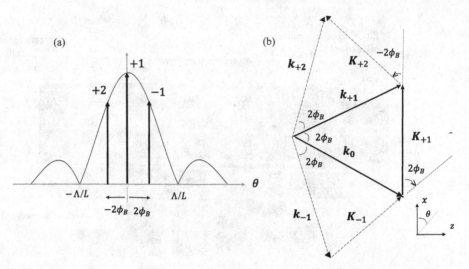

Figure 7.1-5 (a) Angular spectrum of the sound field and (b) wavevector diagrams

The *sinc* function is the angular plane wave spectrum of the sound source and the exponential term represents the direction of propagating plane waves of sound.

Figure 7.1-5a plots the absolute value of the angular spectrum as a function of θ and its first zeros are at $\theta = \pm\sin^{-1}\Lambda/L \approx \pm\Lambda/L$ for small angles. We see that the $\boldsymbol{K}$-vector of the sound field can be oriented through an angle $\pm\Lambda/L$ due to the spreading of sound. In Figure 7.1-5b, we show the wavevector diagram, where $\boldsymbol{K}_{+1}$ at $\theta = 0$ from the angular spectrum is responsible for the generation of $\boldsymbol{k}_{+1}$. In Figure 7.1-5a, we denote the amplitude of the plane wave of sound $\boldsymbol{K}_{+1}$ by an arrow labelled "+1" that is responsible for the generation of $\boldsymbol{k}_{+1}$. However, in addition to that wavevector diagram, we also show that $\boldsymbol{k}_{+2}$ (+2nd-order of the scattered light) and $\boldsymbol{k}_{-1}$ (−1st-order of the scattered light) could be generated if $\boldsymbol{K}_{+2}$ and $\boldsymbol{K}_{-1}$ are available from the angular spectrum, respectively. For example, $\boldsymbol{K}_{+2}$ is the sound plane wave travelling at $\theta = -2\phi_B$, and the amplitude of the plane wave $\boldsymbol{K}_{+2}$ is labelled "+2" by an arrow in the angular spectrum. If $\boldsymbol{K}_{+2}$ is well within the main lobe of the angular spectrum, we have $\boldsymbol{K}_{+2}$ scatters $\boldsymbol{k}_{+1}$ into $\boldsymbol{k}_{+2}$. Similarly, $\boldsymbol{K}_{-1}$ scatters $\boldsymbol{k}_0$ into $\boldsymbol{k}_{-1}$ for downshifted interaction if $\boldsymbol{K}_{-1}$ is well within the main lobe of the angular spectrum, as illustrated in Figure 7.1-5a. Note that the scattered orders are separated by $2\phi_B$, as shown in Figure 7.1-5b. For example, $\boldsymbol{k}_{+1}$ and $\boldsymbol{k}_{+2}$ are separated by $2\phi_B$.

In order to have only one diffracted order of light generated, that is, $\boldsymbol{k}_{+1}$, it is clear, from the angular spectrum shown in Figure 7.1-5a, that we have to impose the condition

$$2\phi_B = \frac{\lambda_0}{\Lambda} \gg \frac{\Lambda}{L},$$

or

$$L \gg \frac{\Lambda^2}{\lambda_0}. \tag{7.1-14}$$

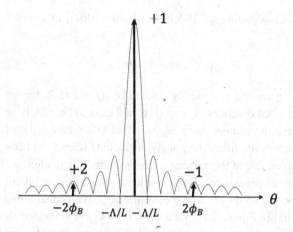

Figure 7.1-6 Narrow angular spectrum for a long transducer ($2\phi_B \gg \Lambda/L$)

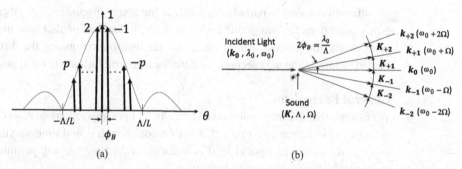

Figure 7.1-7 Raman–Nath diffraction. (a) Broad angular spectrum for a short transducer ($\Lambda/L \gg 2\phi_B$) and (b) simultaneous generation of multiple scattered orders. Adapted from Poon and Kim (2018)

The situation is shown in Figure 7.1-6, where a pertinent sound wave vector lies along K_{+2} is either not present or present in negligible amounts in the angular spectrum of the sound [see the arrow labeled "+2" in the figure, where we see that the amplitude of that sound plane wave is very small]. If L satisfies Eq. (7.1-14), the acousto-optic modulator is said to operate in the *Bragg regime* and the device is commonly known as the *Bragg cell*. However, physical reality dictates that a complete energy transfer between the two diffracted beams is impossible since there always exists more than two diffracted beams.

If we rearrange Eq. (7.1-14), we have

$$\frac{\lambda_0 L}{\Lambda^2} \gg 1,$$

or

$$2\pi L \frac{\lambda_0}{\Lambda^2} \gg 2\pi.$$

We define the *Klein–Cook parameter* [Klein and Cook (1967)] in acousto-optics as follows:

$$Q = 2\pi L \frac{\lambda_0}{\Lambda^2}. \tag{7.1-15}$$

Operation in the Bragg regime is typically defined by $Q \gg 2\pi$. In the ideal Bragg regime, only two diffracted orders exist, and Q would have to be infinity or $L \to \infty$.

If $Q \ll 2\pi$, we have the *Raman–Nath regime*. That is the case where L is sufficiently short. In Raman–Nath diffraction, many diffracted orders exist because plane waves of sound are available at the various angles required for scattering. The principle of the simultaneous generation of many orders is illustrated in Figure 7.1-7. In Figure 7.1-7a, we show a broad angular spectrum that provides many directions of sound plane waves, and in Figure 7.1-7b we show that k_{+1} is generated through the diffraction of k_0 by K_{+1}, k_{+2} is generated through the diffraction of k_{+1} by K_{+2}, and so on, where $K_{\pm p}$s ($p = 0, \pm 1, \pm 2,...$) denote the various components of the plane wave spectrum of the sound.

Although the simple particle approach in the above discussion describes the necessary conditions for Bragg diffraction to occur, it does not predict how the acousto-optic interaction process affects the amplitude distribution among the different diffracted beams. In the next section, we describe a general formalism for acousto-optics.

General Formalism

We assume that the interaction takes place in an optically inhomogeneous, nonmagnetic isotropic source-free ($\rho_v = 0$, $J = 0$) medium with μ_0 and time-varying permittivity $\tilde{\varepsilon}(R,t)$ when an optical field is incident on the time-varying permittivity. We can write the time-varying permittivity as

$$\tilde{\varepsilon}(R,t) = \varepsilon + \varepsilon'(R,t), \tag{7.1-16}$$

where $\varepsilon'(R,t) = \varepsilon C S(R,t)$ as it is proportional to the sound field amplitude $S(R,t)$ with C being the acousto-optic material constant.

When the sound field interacts with the incident light field $E_{inc}(R,t)$, the total light fields $E(R,t)$ and $H(R,t)$ in the acoustic medium must satisfy Maxwell's equations as follows:

$$\nabla \times E(R,t) = -\mu_0 \frac{\partial H(R,t)}{\partial t}, \tag{7.1-17}$$

$$\nabla \times H(R,t) = \frac{\partial\left[\tilde{\varepsilon}(R,t) E(R,t)\right]}{\partial t}, \tag{7.1-18}$$

$$\nabla \cdot \left[\tilde{\varepsilon}(R,t) E(R,t)\right] = 0, \tag{7.1-19}$$

$$\nabla \cdot H(R,t) = 0. \tag{7.1-20}$$

We shall derive the wave equation for $E(R,t)$. Taking the curl of Eq. (7.1-17) and using Eq. (7.1-18), the equation for $E(R,t)$ becomes

$$\nabla \times (\nabla \times E) = \nabla(\nabla \cdot E) - \nabla^2 E = -\mu_0 \frac{\partial^2}{\partial t^2} [\tilde{\varepsilon}(R,t) E(R,t)]. \qquad (7.1-21)$$

Now, from Eq. (7.1-19), we have

$$\nabla \cdot \tilde{\varepsilon} E = \tilde{\varepsilon} \nabla \cdot E + E \cdot \nabla \tilde{\varepsilon} = 0. \qquad (7.1-22)$$

Since the acousto-optic interaction is confined to a two-dimensional (2-D) configuration, we assume a 2-D $(x–z)$ sound field with $E(R,t)$ linearly polarized along the y-direction, that is, $E(R,t) = E(r,t)\hat{y}$, where r is the position vector in the x–z plane. According to the assumptions, we have

$$E \cdot \nabla \tilde{\varepsilon} = E(r,t)\hat{y} \cdot \left(\frac{\partial \tilde{\varepsilon}}{\partial x}\hat{x} + \frac{\partial \tilde{\varepsilon}}{\partial z}\hat{z} \right) = 0. \qquad (7.1-23)$$

As it turns out experimental results are independent of the polarization used in $E(R,t)$ during experiments. We see that the choice of $\hat{y}$ polarization for $E(R,t)$ is therefore a mathematical convenience. With Eq. (7.1-23), Eq. (7.1-22) gives us $\nabla \cdot E = 0$ as $\tilde{\varepsilon} \neq 0$. With this result, Eq. (7.1-21) reduces to the following scalar equation in $E(r,t)$:

$$\nabla^2 E(r,t) = \mu_0 \frac{\partial^2}{\partial t^2} [\tilde{\varepsilon}(r,t) E(r,t)].$$

$$= \mu_0 \left[E \frac{\partial^2 \tilde{\varepsilon}}{\partial t^2} + 2 \frac{\partial E}{\partial t} \frac{\partial \tilde{\varepsilon}}{\partial t} + \tilde{\varepsilon} \frac{\partial^2 E}{\partial t^2} \right]$$

$$\approx \mu_0 \tilde{\varepsilon}(r,t) \frac{\partial^2 E(r,t)}{\partial t^2}, \qquad (7.1-24)$$

where we have retained the last term because the time variation of $\tilde{\varepsilon}(r,t)$ is much slower than that of $E(r,t)$, that is, the sound frequency is much lower than that of the light. Now, using Eq. (7.1-16), Eq. (7.1-24) becomes

$$\nabla^2 E(r,t) - \mu_0 \varepsilon \frac{\partial^2 E(r,t)}{\partial t^2} = \mu_0 \varepsilon'(r,t) \frac{\partial^2 E(r,t)}{\partial t^2} \qquad (7.1-25)$$

This scalar wave equation is often used to investigate interaction in acousto-optics.

Conventional Interaction Configuration

Equation (7.1-25) is formidable. We shall adopt the *Korpel–Poon multiple-plane-wave theory* for further discussion on the acousto-optic effect and its applications [Poon and Kim (2018)]. Consider a conventional interaction configuration shown in Figure 7.1-8, where we assume the incident plane wave of light is

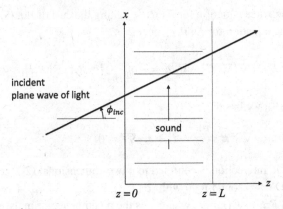

Figure 7.1-8 Conventional sharp boundary interaction configuration with plane-wave light incidence

$$E_{\mathrm{inc}}(r,t) = \mathrm{Re}\left[E_{\mathrm{inc}}(r)e^{j\omega_0 t}\right],$$

and

$$E_{\mathrm{inc}}(r) = \psi_{\mathrm{inc}} e^{-jk_0 \sin\phi_{\mathrm{inc}} x - jk_0 \cos\phi_{\mathrm{inc}} z} \qquad (7.1\text{-}26)$$

is the complex amplitude with ϕ_{inc} being the incident angle, again assumed counterclockwise positive. We also assume that a uniform sound wave of finite width L with sound propagating along the x-direction. Because of an electrical signal of the form $e(t) = \mathrm{Re}\left[|e(t)|e^{j\theta}e^{j\Omega t}\right]$ fed to the acoustic transducer, we take the time-varying permittivity in the form

$$\varepsilon'(r,t) = \varepsilon CS(r,t) = \varepsilon C\mathrm{Re}\left[S(r)e^{j\Omega t}\right],$$

where

$$S(r) = Ae^{-jKx} \qquad (7.1\text{-}27)$$

with $A = |A|e^{j\theta}$ as A, in general, can be complex. Hence, we find $\varepsilon'(r,t) = \varepsilon C|A|\cos(\Omega t - Kx + \theta)$, which represents a plane wave of sound propagating along the x-direction, as shown in Figure 7.1-8. Let us relate $C|A|$ to the refractive index variation $\Delta n(r,t)$ induced in the acousto-optic cell.

Because

$$\tilde{\varepsilon}(r,t) = \varepsilon_0 n^2(r,t) = \varepsilon_0 \left[n_0 + \Delta n(r,t)\right]^2$$

$$\approx \varepsilon_0 n_0^2 \left[1 + \frac{2\Delta n(r,t)}{n_0}\right] = \varepsilon\left[1 + \frac{2\Delta n(r,t)}{n_0}\right] = \varepsilon + \varepsilon'(r,t)$$

according to Eq. (7.1-16), we obtain

$$\varepsilon'(r,t) = \frac{2\varepsilon\Delta n(r,t)}{n_0} = \varepsilon C|A|\cos(\Omega t - Kx + \theta)$$

for sinusoidal sound wave oscillating at Ω. Therefore, we can express the change of the index of refraction within the acoustic medium explicitly as

$$\Delta n(r,t) = \frac{n_0}{2} C|A| \cos(\Omega t - Kx + \theta) = \Delta n_{\max} \cos(\Omega t - Kx + \theta).$$

It clearly shows that there is a periodic compression and rarefaction of the index of refraction of the acoustic medium as the sound wave is travelling along the x-direction.

In the theory of multiple plane-wave scattering, a parameter

$$\tilde{\alpha} = \frac{k_0 CAL}{2}$$

is defined and its magnitude is

$$|\tilde{\alpha}| = \alpha = \frac{k_0 C|A|L}{2} = k_0(\Delta n_{\max} / n_0)L, \tag{7.1-28}$$

which is proportional to $\Delta n_{\max}$. α represents the *peak phase delay* of the light through the acoustic medium of length L. We will use the peak phase delay to denote the strength of the sound amplitude in the multiple plane wave theory.

For a plane-wave incidence given by Eq. (7.1-26), we look for a solution of the form

$$E(r,t) = \sum_{m=-\infty}^{\infty} \text{Re}\left[E_m(r) e^{j(\omega_0 + m\Omega)t} \right],$$

where

$$E_m(r) = \psi_m(z,x) e^{-jk_0 \sin \phi_m x - jk_0 \cos \phi_m z} \tag{7.1-29}$$

with the choice for ϕ_m dictated by the grating equation in Eq. (7.1-5). $\psi_m(z,x)$ is the complex amplitude of the mth diffracted plane wave at frequency $\omega_0 + m\Omega$, and $k_0 \sin \phi_m$ and $k_0 \cos \phi_m$ are the x and z components of its propagation vector. In what follows, we seek to find the amplitude of the mth diffracted plane wave $\psi_m(z,x)$.

Note that the choice of the solution is motivated by previous physical discussions. First, we expect frequency shifting, that is, the term involving $\omega_0 + m\Omega$ due to sound–light interaction. Second, we also expect that the direction of the different diffracted plane waves is governed by the grating equation [see Eq.(7.1-5)].

Upon substituting Eqs. (7.1-27) and (7.1-29) into the scalar wave equation in Eq. (7.1-25), we have the following infinite coupled equations involving ψ_m, called the *Korpel–Poon equations* [Poon and Korpel (1980), Appel and Somekh (1993), Poon and Kim (2005), Zhang et al. (2022)]:

$$\frac{d\psi_m(\xi)}{d\xi} = -j\frac{\tilde{\alpha}}{2} e^{-j\frac{1}{2}Q\xi\left[\frac{\phi_{\text{inc}}}{\phi_B} + (2m-1)\right]} \psi_{m-1}(\xi) - j\frac{\tilde{\alpha}^*}{2} e^{j\frac{1}{2}Q\xi\left[\frac{\phi_{\text{inc}}}{\phi_B} + (2m+1)\right]} \psi_{m+1}(\xi) \tag{7.1-30a}$$

with

$$\psi_m(\xi) = \psi_{\text{inc}} \delta_{m0} \text{ at } \xi = z/L = 0 \tag{7.1-30b}$$

as the boundary conditions. The symbol δ_{m0} is the Kronecker delta defined by $\delta_{m0} = 1$ for $m = 0$ and $\delta_{m0} = 0$ for $m \neq 0$. So ξ is the normalized distance inside the acousto-optic cell, and $\xi = 1$ signifies the exit plane of the cell.

In deriving the above coupled equations, we take on several approximations as follows:

(1) $\phi_m \ll 1$, so that $\psi_m(z, x) \approx \psi_m(z)$,

(2) according to physical intuition, $\omega_0 \gg \Omega$,

and

(3) $\psi_m(z)$ is a slowly varying function of z in that within a wavelength of the propagation distance, the change in ψ_m is small, that is, $\Delta\psi_m \ll \psi_m$. In differential form, $\Delta\psi_m \ll \psi_m$ becomes $d\psi_m = (\partial\psi_m / \partial z)\Delta z = (\partial\psi_m / \partial z)\lambda_0 \ll \psi_m$, leading to $(\partial\psi_m / \partial z) \ll k_0\psi_m$ or equivalently to $\partial^2\psi_m(z) / \partial z^2 \ll k_0\partial\psi_m(z) / \partial z$.

As a consequence of these approximations, we have first-order coupled differential equations in Eq. (7.1-30). Detailed derivation procedures have been discussed [Poon and Kim (2018)] and we leave these procedures as a problem set at the end of the chapter [see P.7.1].

The physical interpretation of Eq. (7.1-30) is that there is a mutual coupling between neighboring orders in the interaction, that is, ψ_m is being contributed by ψ_{m-1} and ψ_{m+1}. We shall use these equations to investigate Raman–Nath and Bragg diffraction.

7.1.2   Raman–Nath and Bragg Diffraction

In this section, we will first consider two classical solutions in acousto-optics: ideal Raman–Nath and Bragg diffraction. In theory, *Raman–Nath diffraction* and *Bragg diffraction* are two extreme cases. In ideal Raman–Nath diffraction, the length of the acoustic transducer is considered to be zero, while $L \to \infty$ is for ideal Bragg diffraction. We then conclude the section with a discussion on near Bragg diffraction, where L is finite.

Ideal Raman–Nath Diffraction

In the classical case, we have normal incidence, that is, $\phi_{inc} = 0$, and $Q = 0$. When $Q = 0$, the sound column is considered to be an infinitely thin phase grating. For simplicity, we take $\tilde{\alpha}$ being real, that is, $\tilde{\alpha} = \tilde{\alpha}^* = \alpha$, Equation (7.1-30) becomes

$$\frac{d\psi_m(\xi)}{d\xi} = -j\frac{\alpha}{2}\left[\psi_{m-1}(\xi) + \psi_{m+1}(\xi)\right] \qquad (7.1\text{-}31)$$

with $\psi_m(\xi = 0) = \psi_{inc}\delta_{m0}$. Using the recursion relation for Bessel functions,

$$\frac{dJ_m(x)}{dx} = \frac{1}{2}\left[J_{m-1}(x) - J_{m+1}(x)\right], \qquad (7.1\text{-}32)$$

we can show that the solutions inside the sound column are

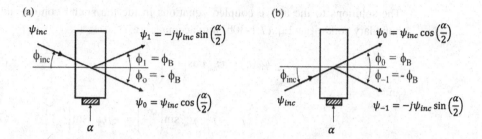

Figure 7.1-9 Bragg diffraction: (a) upshifted interaction and (b) downshifted interaction

$$\psi_m\left(\xi\right)=\left(-j\right)^{m}\psi_{\text{inc}}J_m\left(\alpha\xi\right),\tag{7.1-33}$$

which is the well-known Raman–Nath solution, and J_m is a *Bessel function of order m*. The amplitudes of the various scattering orders upon leaving the sound column, that is, at $\xi=1$, are given by

$$\psi_m\left(\xi=1\right)=\left(-j\right)^{m}\psi_{\text{inc}}J_m\left(\alpha\right),\tag{7.1-34}$$

and the solutions satisfy the conservation of energy

$$\sum_{m=-\infty}^{\infty}I_m=\sum_{m=-\infty}^{\infty}\left|\psi_m\right|^2=\left|\psi_{\text{inc}}\right|^2\sum_{m=-\infty}^{\infty}J_m^2\left(\alpha\right)=\left|\psi_{\text{inc}}\right|^2.$$

Ideal Bragg Diffraction

Ideal Bragg diffraction is characterized by the generation of only two scattered orders. This occurs when either $\phi_{\text{inc}}=\phi_B$ (downshifted Bragg diffraction) or when $\phi_{\text{inc}}=-\phi_B$ (upshifted Bragg diffraction) for $Q\to\infty$. In Figure 7.1-9, we show *upshifted and downshifted Bragg diffraction*.

For upshifted interaction, that is, $\phi_{\text{inc}}=-\phi_B$, we have diffracted orders $m=0$ and $m=1$. For small angles, the grating equation becomes

$$\phi_m=\phi_{inc}+m\frac{\lambda_0}{\Lambda}.$$

Hence, $\phi_0=\phi_{\text{inc}}=-\phi_B$, and $\phi_1=\phi_{\text{inc}}+\dfrac{\lambda_0}{\Lambda}=-\phi_B+2\phi_B=\phi_B$, as illustrated in Figure 7.1-9a. Remember that the convention for angles is counterclockwise positive. The physical situation is similar to that shown in Figure 7.1-1. Now, according to Eq. (7.1-30), we have the following coupled equations:

$$\frac{d\psi_0\left(\xi\right)}{d\xi}=-j\frac{\tilde{\alpha}^*}{2}\psi_1\left(\xi\right),\tag{7.1-35a}$$

and

$$\frac{d\psi_1\left(\xi\right)}{d\xi}=-j\frac{\tilde{\alpha}}{2}\psi_0\left(\xi\right).\tag{7.1-35b}$$

The solutions to the above coupled equations inside the sound column, taking the boundary conditions Eq. (7.1-30b) into account, are

$$\psi_0(\xi) = \psi_{inc} \cos\left(\frac{\alpha\xi}{2}\right),$$ (7.1-36a)

and

$$\psi_1(\xi) = -j\frac{\tilde{\alpha}}{\alpha}\psi_{inc} \sin\left(\frac{\alpha\xi}{2}\right) = -j\psi_{inc}\sin\left(\frac{\alpha\xi}{2}\right)$$ (7.1-36b)

for $\tilde{\alpha} = \tilde{\alpha}^* = \alpha$, where again α is the peak phase delay proportional to the strength of the sound amplitude.

Similarly, for downshifted interaction, that is, $\phi_{inc} = \phi_B$, we have diffracted orders $m = 0$ and $m = -1$. We have the following coupled equations:

$$\frac{d\psi_0(\xi)}{d\xi} = -j\frac{\tilde{\alpha}}{2}\psi_{-1}(\xi),$$ (7.1-37a)

and

$$\frac{d\psi_{-1}(\xi)}{d\xi} = -j\frac{\tilde{\alpha}^*}{2}\psi_0(\xi).$$ (7.1-37b)

The solutions are

$$\psi_0(\xi) = \psi_{inc} \cos\left(\frac{\alpha\xi}{2}\right),$$ (7.1-38a)

and

$$\psi_{-1}(\xi) = -j\frac{\tilde{\alpha}^*}{\alpha}\psi_{inc} \sin\left(\frac{\alpha\xi}{2}\right) = -j\psi_{inc}\sin\left(\frac{\alpha\xi}{2}\right)$$ (7.1-38b)

for $\tilde{\alpha} = \tilde{\alpha}^* = \alpha$. The physical situation is illustrated in Figure 7.1-9b. The intensities of the two diffracted orders leaving the sound column, that is, at $\xi = 1$, are in both cases given by

$$I_0 = |\psi_0(\xi = 1)|^2 = I_{inc}\cos^2\left(\frac{\alpha}{2}\right),$$ (7.1-39a)

and

$$I_{\pm 1} = |\psi_{\pm 1}(\xi = 1)|^2 = I_{inc}\sin^2\left(\frac{\alpha}{2}\right),$$ (7.1-39b)

where $I_{inc} = |\psi_{inc}|^2$ is the intensity of the incident light. Clearly, the solutions satisfy the conservation of energy

$$I_0 + I_{\pm 1} = I_{inc}.$$

Example: Conservation of Energy

Directly from the coupled equations, that is, Eqs. (7.1-35) or (7.1-37), we can verity the conservation of energy. Taking Eq. (7.1-35) as an example, we perform

$$\frac{d}{d\xi}\left[|\psi_0(\xi)|^2 + |\psi_1(\xi)|^2\right] = \frac{d}{d\xi}\left[\psi_0\psi_0^* + \psi_1\psi_1^*\right]$$

$$= \frac{d\psi_0(\xi)}{d\xi}\psi_0^*(\xi) + \psi_0(\xi)\frac{d\psi_0^*(\xi)}{d\xi} + \frac{d\psi_1(\xi)}{d\xi}\psi_1^*(\xi) + \psi_1(\xi)\frac{d\psi_1^*(\xi)}{d\xi}$$

$$= -j\frac{\tilde{\alpha}^*}{2}\psi_1(\xi)\psi_0^*(\xi) + \psi_0(\xi)j\frac{\tilde{\alpha}}{2}\psi_1^*(\xi) - j\frac{\tilde{\alpha}}{2}\psi_0(\xi)\psi_1^*(\xi) + \psi_1(\xi)j\frac{\tilde{\alpha}^*}{2}\psi_0^*(\xi) = 0,$$

where we have used Eq. (7.1-35) in the above equation to obtain the final result. The result implies that $|\psi_0(\xi)|^2 + |\psi_1(\xi)|^2 = \text{constant}$, an alternative statement for the *conservation of energy*.

Example: Solutions to the Coupled Equations

Equations (7.1-35a) and (7.1-35b) are coupled differential equations and these equations can be combined into a second-order differential equation. Differentiating Eq. (7.1-35a) with respect to ξ and combining with Eq.(7.1-35b), we have

$$\frac{d^2\psi_0(\xi)}{d\xi^2} = -j\frac{\tilde{\alpha}^*}{2}\frac{d\psi_1(\xi)}{d\xi} = -j\frac{\tilde{\alpha}^*}{2}\left(-j\frac{\tilde{\alpha}}{2}\right)\psi_0(\xi) = -\left(\frac{\alpha}{2}\right)^2\psi_0(\xi). \quad (7.1\text{-}40)$$

The *general solution* of the differential equation is

$$\psi_0(\xi) = Ae^{j\frac{\alpha}{2}\xi} + Be^{-j\frac{\alpha}{2}\xi} \quad (7.1\text{-}41)$$

as we can verify by direct substitution into the second-order differential equation. A and B can be found by knowing initial conditions $\psi_0(\xi)|_{\xi=0}$ and $\left.\frac{d\psi_0(\xi)}{d\xi}\right|_{\xi=0}$. According to Eq. (7.1-30b), we have $\psi_0(\xi)|_{\xi=0} = \psi_{\text{inc}}$, which leads to

$$A + B = \psi_{\text{inc}} \quad (7.1\text{-}42a)$$

from Eq. (7.1-41). We also find, from Eq. (7.1-35a), the second initial condition:

$$\left.\frac{d\psi_0(\xi)}{d\xi}\right|_{\xi=0} = -j\frac{\tilde{\alpha}^*}{2}\psi_1(\xi)|_{\xi=0} = 0.$$

This condition together with Eq. (7.1-41) leads to

$$A = B. \quad (7.1\text{-}42b)$$

From Eqs. (7.1-42a) and (7.1-42b), we get $A = B = \psi_{\text{inc}}/2$, and finally we obtain the solution to the zeroth order as

$$\psi_0(\xi) = \left(\frac{\psi_{\text{inc}}}{2}\right)e^{j\frac{\alpha}{2}\xi} + \left(\frac{\psi_{\text{inc}}}{2}\right)e^{-j\frac{\alpha}{2}\xi} = \psi_{\text{inc}}\cos\left(\frac{\alpha}{2}\xi\right). \quad (7.1\text{-}43)$$

The 1st order can then be found by using Eq. (7.1-35a):

$$\psi_1(\xi) = \frac{2}{-j\tilde{\alpha}^*}\frac{d\psi_0(\xi)}{d\xi} = -j\frac{\tilde{\alpha}}{\alpha}\psi_{\text{inc}}\sin\left(\frac{\alpha}{2}\xi\right), \quad (7.1\text{-}44)$$

which is Eq. (7.1-36b).

Near Bragg Diffraction

Near Bragg diffraction is when the length of the transducer L is finite, or equivalently Q is finite. It is usually characterized by the generation of multiple diffracted orders. In general, Eq. (7.1-30) is used to analyze the situation. We will discuss two cases as examples.

Case 1: Incident Plane Wave off the Exact Bragg Angle and Limiting to Two Diffracted Orders

We consider upshifted diffraction. We let $\phi_{inc} = -(1+\delta)\phi_B$, where δ represents the deviation of the incident plane wave away from the Bragg angle. With reference to Eq. (7.1-30), we again assume $\tilde{\alpha} = \tilde{\alpha}^* = \alpha$ for simplicity. We have

$$\frac{d\psi_0(\xi)}{d\xi} = -j\frac{\alpha}{2}e^{-j\delta Q\xi/2}\psi_1(\xi), \tag{7.1-45a}$$

and

$$\frac{d\psi_1(\xi)}{d\xi} = -j\frac{\alpha}{2}e^{j\delta Q\xi/2}\psi_0(\xi) \tag{7.1-45b}$$

with the initial conditions $\psi_0(\xi=0) = \psi_{inc}$, and $\psi_1(\xi=0) = 0$. Equation (7.1-45) has analytical solutions given by the well-known *Phariseau formula* [see Poon and Kim (2018)]:

$$\psi_0(\xi) = \psi_{inc}e^{-j\delta Q\xi/4}\left\{\cos\left[(\delta Q/4)^2 + (\alpha/2)^2\right]^{1/2}\xi + \frac{j\delta Q}{4}\frac{\sin\left[(\delta Q/4)^2 + (\alpha/2)^2\right]^{1/2}\xi}{\left[(\delta Q/4)^2 + (\alpha/2)^2\right]^{1/2}}\right\},$$

$$\tag{7.1-46a}$$

and

$$\psi_1(\xi) = \psi_{inc}e^{j\delta Q\xi/4}\left\{-j\frac{\alpha}{2}\frac{\sin\left[(\delta Q/4)^2 + (\alpha/2)^2\right]^{1/2}\xi}{\left[(\delta Q/4)^2 + (\alpha/2)^2\right]^{1/2}}\right\}. \tag{7.1-46b}$$

For $\delta = 0$, these solutions reduce to those given for *ideal Bragg diffraction* [see Eq. (7.1-36)], and there is a complete power exchange between the two orders at $\alpha = n\pi$ ($n = 1, 2, 3...$). The situation is shown in Figure 7.1-10a, where the normalized intensities $I_0/I_{inc} = |\psi_0(\xi=1)|^2/|\psi_{inc}|^2$ and $I_1/I_{inc} = |\psi_1(\xi=1)|^2/|\psi_{inc}|^2$ are plotted against α at the exit of the sound column, $\xi = 1$. For the results shown in Figure 7.1-10b, we have used $\delta = 0.25$, that is, a deviation from the Bragg angle of $0.25\,\phi_B$, and for $Q = 14$ [Typical value for model AOM-40 from IntraAction Corporation]. δ is "d" in the m-file. We note that the power transfer from I_0 to I_{inc} and vice versa is quasi-periodic and initial incomplete when α is relatively small around 3.

The above plots are generated using the m-file shown below.

(a)

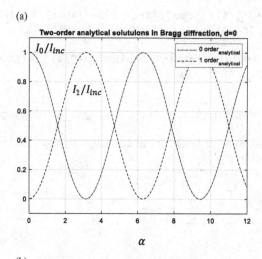

(b)

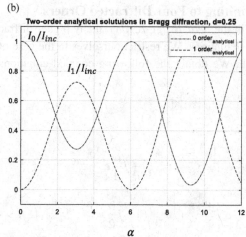

Figure 7.1-10 (a) Power exchange between orders in pure Bragg diffraction ($\delta = 0$) and (b) partial power exchange in near Bragg diffraction ($\delta = 0.25, Q = 14$).

```
%AO_Bragg_2order.m, plotting of Eq. (7.1-46)
clear

d=input('delta = ?')
Q=input('Q = ?')
al_e=input('End point of alpha = ?')

n=0;
    for al=0:0.01*pi:al_e
    n=n+1;
    AL(n)=al;
```

```
    ps0_a(n)=exp(-j*d*Q/4)*( cos(((d*Q/4)^2+ (AL(n)/2)^2)^0.5) +
j*d*Q/4*sin(((d*Q/4)^2+(AL(n)/2) ^2)^0.5)/(((d*Q/4)^2+
(AL(n)/2)^2)^0.5) );
    ps1_a(n)=exp(j*d*Q/4)*-j*AL(n)/2*sin(((d*Q/4)^2+(AL(n)/2)
^2)^0.5)/(((d*Q/4)^2+ (AL(n)/2)^2)^0.5) ;

end
    plot(AL, ps0_a.*conj(ps0_a), '-.', AL, ps1_a.*conj(ps1_a),
      '--')
    title('Two-order analytical solutions in Bragg
      diffraction, d=0.25')
    %xlabel(,alpha')
    axis([0 al_e -0.1 1.1])
    legend('0 order_a_n_a_l_y_t_i_c_a_l', '1
      order_a_n_a_l_y_t_i_c_a_l')
    grid on
```

Case 2: Finite Q and Limiting to Four Diffracted Orders

For a more accurate result in the *near Bragg regime*, many higher diffracted orders should be included in the calculations. We restrict ourselves to the case of only four orders, that is, $2 \geq m \geq -1$. We obtain the following coupled equations from Eq. (7.1-30) for $\tilde{\alpha} = \tilde{\alpha}^* = \alpha$:

$$\frac{d\psi_2(\xi)}{d\xi} = -j\frac{\alpha}{2}e^{-j\frac{1}{2}Q\xi\left[\frac{\phi_{inc}}{\phi_B}+3\right]}\psi_1(\xi),$$

$$\frac{d\psi_1(\xi)}{d\xi} = -j\frac{\alpha}{2}e^{-j\frac{1}{2}Q\xi\left[\frac{\phi_{inc}}{\phi_B}+1\right]}\psi_0(\xi) - j\frac{\alpha}{2}e^{j\frac{1}{2}Q\xi\left[\frac{\phi_{inc}}{\phi_B}+3\right]}\psi_2(\xi),$$

$$\frac{d\psi_0(\xi)}{d\xi} = -j\frac{\alpha}{2}e^{-j\frac{1}{2}Q\xi\left[\frac{\phi_{inc}}{\phi_B}-1\right]}\psi_{-1}(\xi) - j\frac{\alpha}{2}e^{j\frac{1}{2}Q\xi\left[\frac{\phi_{inc}}{\phi_B}+1\right]}\psi_1(\xi),$$

$$\frac{d\psi_{-1}(\xi)}{d\xi} = -j\frac{\alpha}{2}e^{j\frac{1}{2}Q\xi\left[\frac{\phi_{inc}}{\phi_B}-1\right]}\psi_0(\xi). \tag{7.1-47}$$

We demonstrate some numerical results. For $\phi_{inc} = -(1+\delta)\phi_B$ and $Q = 14$, we run Bragg_regime_4.m for $\delta = 0$ and 0.25 with results shown in Figure 7.1-11a and b, respectively. The results in Figure 7.1-11a and b can be directly compared with those given in Figure 7.1-10a and b, respectively. Specifically, we want to point out that as α gets higher toward around 10, the other two orders, that is, I_2 and I_{-1}, begin to appear in Figure 7.1-11a. I_2 and I_{-1} begin to appear around much earlier at $\alpha \approx 4$ for *off-Bragg angle diffraction* when $\delta = 0.25$, as shown in Figure 7.1-11b. In Figure 7.1-11c, we show the sum of all normalized intensities to be unity, that is, $\sum_{m=-1}^{m=2} I_m / I_{inc} = 1$. This gives us some assurance of the numerical calculations.

```
% Bragg_regime_4.m
% Adapted from "Engineering Optics with MATLAB, 2nd ed."
```

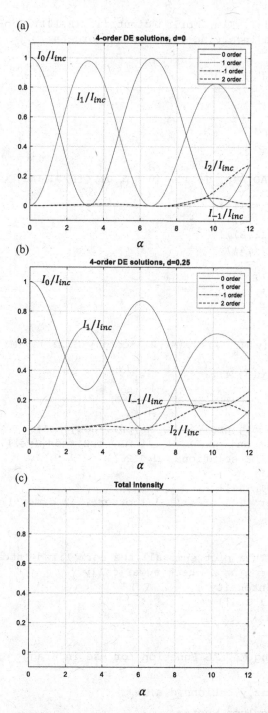

Figure 7.1-11 (a) Power exchange between four orders in near Bragg diffraction $(\delta = 0, Q = 14)$, (b) power exchange between four orders in near Bragg diffraction $(\delta = 0.25, Q = 14)$, and (c) sum of normalized intensities

```
% by T.-C. Poon and T. Kim, World Scientific (2018), Table 4.1.
% Near Bragg regime involving 4 diffracted orders
clear

d=input('delta =')
Q=input('Q = ')
n=0;

for al=0:0.01*pi:14
    n=n+1;
    AL(n)=al;
    [nz,y]=ode45('AO_B4', [0 1], [0 0 1 0], [], d, al, Q) ;

    [M1 N1]=size(y(:,1));
    [M2 N2]=size(y(:,2));
    [M3 N3]=size(y(:,3));
    [M4 N4]=size(y(:,4));

    psn1(n)=y(M4,4);
    ps0(n)=y(M3,3);
    ps1(n)=y(M2,2);
    ps2(n)=y(M1,1);

I(n)=y(M1,1).*conj(y(M1,1))+ y(M2,2).*conj(y(M2,2))+y(M3,3)*co
  nj(y(M3,3)) ...
    +y(M4,4)*conj(y(M4,4));
end

    figure(1)
    plot(AL, ps0.*conj(ps0), '-', AL,ps1.*conj(ps1), ':', ...
        AL, psn1.*conj(psn1), '-.', AL, ps2.*conj(ps2), '--')
    title('4-order DE solutions, d=0.25')
    %xlabel('alpha')
    axis([0 12 -0.1 1.1])
    legend('0 order', '1 order', '-1 order', '2 order')
    grid on

    figure(2)
    plot(AL, I) % This plot sums all the normalized intensities
    % and should be 1 as a check numerically.
    title('Total Intensity')
    axis([0 12 -0.1 1.1])
    %xlabel('alpha')
    grid on

%AO_B4.m    %creating MATLAB Function for use in Bragg_
regime_4.m

function dy=AO_B4(nz,y,options,d,a,Q)

dy=zeros(4,1); %a column vector

% -1<= m <=2
% d=delta
% nz=normalized z
```

```
%m=2   -> y(1)
%m=1   -> y(2)
%m=0   -> y(3)
%m=-1  -> y(4)

dy(1)=-j*a/2*y(2)*exp(-j*Q/2*nz*(-(1+d)+3))     +     0          ;

dy(2)=-j*a/2*y(3)*exp(-j*Q/2*nz*(-(1+d)+1))     +     -j*a/2*y(1)*-
exp(j*Q/2*nz*(-(1+d)+3));

dy(3)=-j*a/2*y(4)*exp(-j*Q/2*nz*(-(1+d)-1))     +     -j*a/2*y(2)*-
exp(j*Q/2*nz*(-(1+d)+1));

dy(4)=0     +     -j*a/2*y(3)*exp(j*Q/2*nz*(-(1+d)-1));

return
```

7.1.3   Typical Applications of the Acousto-Optic Effect

Laser Intensity Modulation

Optical modulation is the process of changing the amplitude, intensity, frequency, phase, or polarization of an optical wave to convey an information signal (such as an audio signal). The reverse process, that is, extracting the information signal from the optical wave, is called *de-modulation*. In this section, we discuss laser intensity modulation by use of the acousto-optic effect. By changing the amplitude of the sound, that is, though α, we can achieve *intensity modulation* of the diffracted laser beams. This fact is clearly shown by inspecting Eqs. (7.1-39a) and (7.1-39b). In intensity modulation, the intensity of the laser beam is varied in proportion to the information signal $m(t)$ (known as the *modulating signal* in communication theory). The resulting modulated intensity $I(t)$ is known as the *modulated signal*. For a given $m(t)$, we first need to generate the so-called *AM (amplitude modulation)* signal through an AM modulator. In amplitude modulation, the amplitude of a sinusoidal signal, whose frequency and phase are fixed, is varied in proportion to $m(t)$.

A basic AM modulator consists of an electronic multiplier and an adder, as shown in Figure 7.1-12 schematically. Accordingly, the output of the AM modulator is an amplitude-modulated signal $y_{AM}(t) = [A_c + m(t)]\cos(\Omega t)$, where the sinusoidal $\cos(\Omega t)$ is known as the *carrier* of the AM signal, and $A_c > |m(t)|_{max}$ in standard AM. This modulated signal is then used to drive the acoustic transducer of the acousto-optic modulator. Hence, $\alpha(t) \propto [A_c + m(t)]$ as $\alpha \propto |A|$ [see Eq. (7.1-28)]. For convenience, we take $\alpha = [A_c + m(t)]$ and plot it for $m(t)$ as a sinusoidal variation with A_c being a constant bias on point P of the sine squared intensity curve. The situation is shown in Figure 7.1-13. The acousto-optic modulator should be biased to the point which results in a 50% intensity transmission for linear operation.

The variation of α produces a time-varying intensity on the intensity of the first-order beam $I(t)$, as we see, for example, that points a and b on the $m(t)$ curve give intensities a' and b' as outputs on the $I(t)$ curve, respectively. To demodulate $I(t)$, we simply direct the modulated laser beam toward a photodetector. According

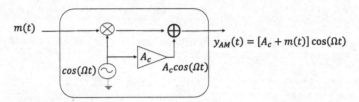

Figure 7.1-12 AM modulator: The triangle represents a circuit symbol of an amplifier of gain A_c

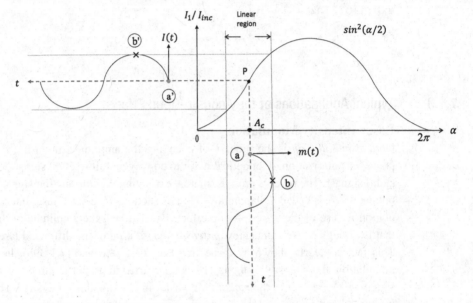

Figure 7.1-13 Acousto-optic intensity modulation principle: relationship between the modulating signal, $m(t)$, and the modulated intensity, $I(t)$, of the first-order beam

to Eq. (4.4-3), the current output $i(t)$ from the photodetector is proportional to the incident intensity; therefore, $i(t) \propto I(t) \propto m(t)$. Figure 7.1-14a shows an experimental setup used to demonstrate intensity modulation. $m(t)$ is an audio signal. Figure 7.14b shows the output of the two photodetectors PD1 and PD0. Note that there is a negative sign between the two electrical signals. The reason is that the two diffracted orders process information in the linear regions with opposite slopes as the zeroth-order diffracted order works with the cosine squared intensity curve, that is, $\cos^2\left(\dfrac{\alpha}{2}\right)$.

Acousto-Optic Laser Beam Deflector

In intensity modulation, we change the amplitude of the sound signal. For applications in *laser deflection*, we change the frequency of the sound. The situation is shown in Figure 7.1-15.

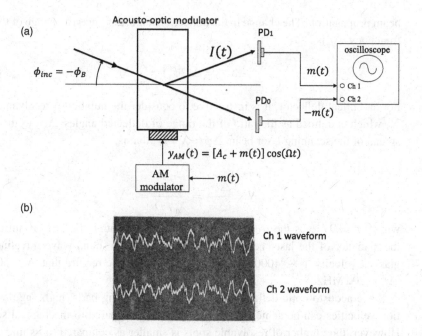

Figure 7.1-14 (a) Experimental setup for AM modulation and demodulation and (b) demodulated signals. Reprinted from Poon et al., "Modern optical signal processing experiments demonstrating intensity and pulse-width modulation using an acousto-optic modulator," *American Journal of Physics* 65 917–925 (1997), with permission

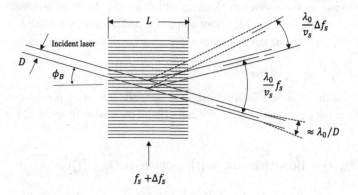

Figure 7.1-15 Acousto-optic laser beam deflector

We define the angle between the first-order beam and the zeroth-order beam as the *defection angle ϕ_d*:

$$\phi_d = 2\phi_B \approx \frac{\lambda_0}{\Lambda} = \frac{\lambda_0 f_s}{v_s}$$

as the Bragg angle is in general small. By changing the frequency of the sound wave f_s, we can change the deflection angle, thereby controlling the direction of the light beam propagation. The change in the deflection angle $\Delta\phi_d$ upon a change of the sound frequency Δf_s is

$$\Delta\phi_d = \Delta(2\phi_B) = \frac{\lambda_0}{v_s}\Delta f_s. \tag{7.1-48}$$

For an optical deflector, it is instructive to consider the number of resolvable angles N, which is defined as the ratio of the range of deflected angles $\Delta\phi_d$ to the angular spread of the scanning laser beam $\Delta\phi \approx \lambda_0 / D$, that is,

$$N = \frac{\Delta\phi_d}{\Delta\phi} = \frac{\frac{\lambda_0}{v_s}\Delta f_s}{\lambda_0 / D} = \tau\Delta f_s, \tag{7.1-49}$$

where $\tau = D / v_s$ is the *transit time* of the sound through the laser beam as D is the diameter of the laser beam. For $D = 1\,\text{cm}$ and for sound wave traveling in the glass at velocity $v_s \sim 4000\,\text{m/s}$, then $\tau = 2.5\,\mu\text{s}$. If we require that $N = 500$, then $\Delta f_s = 200\,\text{MHz}$.

Since acousto-optic deflectors do not include moving parts, high angular deflection velocities can be achieved accurately, when compared to mechanical scanners. However, the number of resolvable spots is smaller as compared to its mirror-based scanner counterpart.

Phase Modulation

While we have discussed intensity modulation using the acousto-optic effect, let us see how to use acousto-optics to phase modulate a laser. In *phase modulation (PM)*, the phase of a sinusoidal signal, whose amplitude and frequency are fixed, is varied in proportion to $m(t)$. The conventional expression for phase modulation is

$$y_{PM}(t) = A\cos\left(\Omega t + k_{pm}m(t)\right), \tag{7.1-50}$$

where k_{pm} is the *phase-deviation constant* in radian/volt if $m(t)$ is measured in volts. Again, Ω is the carrier frequency. We can expand the cosine function in Eq. (7.1-50) to have

$$y_{PM}(t) = A\left[\cos(\Omega t)\cos\left(k_{pm}m(t)\right) - \sin(\Omega t)\sin\left(k_{pm}m(t)\right)\right].$$

When $\left|k_{pm}m(t)\right| \ll 1$, $\cos\theta \approx 1$ and $\sin\theta \approx \theta$. In that case, we have narrowband phase modulation (NBPM) and the above equation becomes

$$y_{NBPM}(t) = A\left[\cos(\Omega t) - k_{pm}m(t)\sin(\Omega t)\right]. \tag{7.1-51}$$

Figure 7.1-16a shows a block diagram of a narrowband phase modulator. In the diagram, the block with $90°$ denotes a phase shifter that phase shifts the carrier signal by $90°$. In order to generate an expression given by Eq. (7.1-50), k_{pm} must be made large. NBPM can be converted to *wideband phase modulation (WBPM)* by using frequency multipliers. Figure 7.1-17b illustrates the block diagram of generating a

(a)

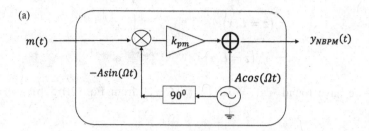

(b)

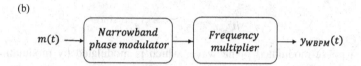

Figure 7.1-16 (a) Narrowband phase modulator and (b) wideband phase modulator

wideband phase modulation (WBPM) signal, $y_{\text{WBPM}}(t)$. A *frequency multiplier* is a nonlinear device. For example, a *square-law device* can multiply the frequency of a factor of 2. For a square-law device, the input $x(t)$ and the output $y(t)$ are related by

$$y(t) = x^2(t).$$

For $x(t) = \cos(\Omega t + k_{\text{pm}} m(t))$, $y(t) = \cos^2(\Omega t + k_{\text{pm}} m(t)) = \frac{1}{2} + \frac{1}{2}\cos(2\Omega t + 2k_{\text{pm}} m(t))$. The DC term is filtered out to give the output with carrier frequency and phase deviation constant doubled. Repeated use of *multipliers* gives us a large phase-deviation constant.

For phase modulation of a laser by $m(t)$ using the acousto-optic effect, we feed a phase modulated electrical signal of the form given by Eq. (7.1-50) to the acoustic transducer. Under the situation, the sound field, according to Eq. (7.1-27), is modelled by $A = |A|e^{j\theta} = e^{jk_{\text{pm}}m(t)}$ for pure phase modulation, giving $\tilde{\alpha} \propto A = e^{jk_{\text{pm}}m(t)}$ [see Eq. (7.1-28)].

Let us now consider upshifted Bragg diffraction. Assuming a modest sound pressure is used, that is, $\alpha \ll 1$, the diffracted orders at the exit of the sound column $(\xi = 1)$ from Eqs. (7.1-36a) and (7.1-36b) are approximated by

$$\psi_0(\xi = 1) = \psi_{\text{inc}} \cos\left(\frac{\alpha}{2}\right) \approx \psi_{\text{inc}}, \tag{7.1-52a}$$

and

$$\psi_1(\xi = 1) = -j\frac{\tilde{\alpha}}{\alpha}\psi_{\text{inc}} \sin\left(\frac{\alpha}{2}\right) \approx -j\psi_{\text{inc}}\frac{\tilde{\alpha}}{2} \propto -j\psi_{\text{inc}}e^{jk_{\text{pm}}m(t)}. \tag{7.1-52b}$$

Note that the first-order light carries the phase modulation. According to Eq. (7.1-29), the full expression of the first-order diffracted plane wave at the exit of the sound column is ($\phi_1 = \phi_B$ for upshifted interaction)

$$\psi_m\left(z=L,x\right)e^{-jk_0\sin\phi_m x-jk_0\cos\phi_m L}\Big|_{m=1}e^{j(\omega_0+\Omega)t}$$

$$=\psi_1\left(z=L,x\right)e^{-jk_0\sin\phi_B x-jk_0\cos\phi_B L}e^{j(\omega_0+\Omega)t}.$$

Since we have found $\psi_1\left(z=L,x\right)\approx\psi_1\left(\xi=1\right)$ from Eq. (7.1-52b), the equation becomes

$$\psi_1\left(\xi=1\right)e^{-jk_0\sin\phi_B x-jk_0\cos\phi_B L}e^{j(\omega_0+\Omega)t}.$$

$$\propto-j\psi_{inc}e^{jk_{pm}m(t)}e^{-jk_0\sin\phi_B x}e^{j(\omega_0+\Omega)t}. \tag{7.1-53}$$

This is a phase-modulated plane wave, which is modulated by modulating signal $m(t)$.

Heterodyning

Heterodyning, also known as *frequency mixing*, is a frequency translation process. Heterodyning has its root in radio engineering. The principle of heterodyning was discovered in the late 1910s by radio engineers experimenting with radio vacuum tubes. For a simple case when signals of two different frequencies are heterodyned or mixed, the resulting signal produces two new frequencies, the sum and difference of the two original frequencies. Heterodyning is performed by applying the two sinusoidal signals to a nonlinear device. The two new frequencies are called *heterodyne frequencies*. Typically, only one of the heterodyne frequencies is required and the other signal is filtered out of the output.

Figure 7.1-17a illustrates how signals of two frequencies are mixed or heterodyned to produce two new frequencies, $\omega_1+\omega_2$ and $\omega_1-\omega_2$, by simply electronic multiplying the two signals $\cos(\omega_1 t+\theta_1)$ and $\cos(\omega_2 t)$, where θ_1 is the phase angle between the two signals. In this particular case, the multiplier is the nonlinear device to achieve the heterodyne operation. Note that when the frequencies of the two signals to be heterodyned are the same, that is, $\omega_1=\omega_2$, the phase information of $\cos(\omega_1 t+\theta_1)$ can be extracted to obtain $\cos\left(\theta_1\right)$ if we use an electronic low-pass filter (LPF) to filter out the term $\cos(2\omega_1 t+\theta_1)$. This is shown in Figure 7.1-17b. Heterodyning is referred to as *homodyning* for the mixing of two signals of the same frequency. Homodyning allows the extraction of the phase information of the signal and this is the basic operation of a *lock-in amplifier* to retrieve the phase information from an electrical signal [see Figure 6.2-4]. Indeed, we have used the principle to extract the holographic information in optical scanning holography [see Section 6.2.2].

Let us now discuss how the principle of heterodyning is used in optics. We consider heterodyning of two plane waves with different temporal frequencies on the surface of a *photodetector*. The situation is shown in Figure 7.1-18, where ϕ is the angle between the two plane waves. According to Eq. (3.3-9), a plane wave of frequency $\omega_0+\Omega$ in general is expressed as

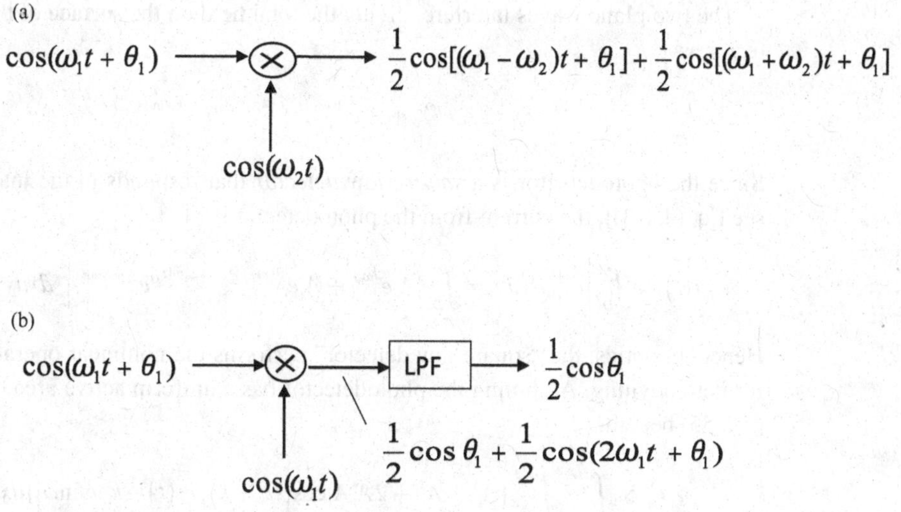

(a)

$$\cos(\omega_1 t + \theta_1) \longrightarrow \otimes \longrightarrow \frac{1}{2}\cos[(\omega_1 - \omega_2)t + \theta_1] + \frac{1}{2}\cos[(\omega_1 + \omega_2)t + \theta_1]$$

$$\uparrow$$

$$\cos(\omega_2 t)$$

(b)

$$\cos(\omega_1 t + \theta_1) \longrightarrow \otimes \longrightarrow \boxed{\text{LPF}} \longrightarrow \frac{1}{2}\cos\theta_1$$

$$\uparrow \qquad\qquad \frac{1}{2}\cos\theta_1 + \frac{1}{2}\cos(2\omega_1 t + \theta_1)$$

$$\cos(\omega_1 t)$$

Figure 7.1-17 (a) Heterodyning and (b) homodyning and phase retrieval

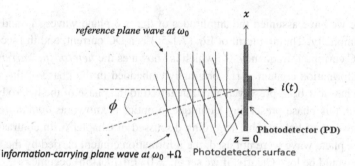

Figure 7.1-18 Optical heterodyning

$$e^{-j(k_{0x}x + k_{0y}y + k_{0z}z)}e^{j(\omega_0 + \Omega)t}.$$

For plane wave propagating at angle ϕ, as shown in Figure 7.1-18, and with the photodetector located at $z = 0$, we have $k_{0x} = k_0\hat{k}\cdot\hat{x} = \cos(90° - \phi) = k_0\sin\phi$ and $k_{0y} = 0$ [also see Eq. (5.1-13)]. Hence, the *information-carrying plane wave* (also called *signal plane wave*) on the photodetector surface can be expressed in general as, according to Eq. (7.1-53),

$$A_s e^{j[(\omega_0 + \Omega)t + k_{pm}m(t)]}e^{-jk_0\sin\phi x}. \tag{7.1-54}$$

where $m(t)$ is the information embedded on the phase of the plane wave. The other plane wave of frequency ω_0 is incident normally to the photodetector and is given by $A_r e^{j\omega_0 t}$, which is called a *reference plane wave* in contrast to the signal plane wave.

The two plane waves interfere, giving the total field on the surface of the photodetector as

$$\psi_t = A_r e^{j\omega_0 t} + A_s e^{j[(\omega_0 + \Omega)t + k_{pm}m(t)]} e^{-jk_0 \sin\phi x}.$$

Since the photodetector is a *square-law detector* that responds to the intensity [also see Eq. (4.4-3)], the current from the photodetector is

$$i(t) \propto \int_D |\psi_t|^2 dxdy = \int_D \left| A_r e^{j\omega_0 t} + A_s e^{j[(\omega_0 + \Omega)t + k_{pm}m(t)]} e^{-jk_0 \sin\phi x} \right|^2 dxdy. \quad (7.1\text{-}55)$$

Hence, in optics, the "square-law detector" performs the nonlinear operation needed for heterodyning. Assuming the photodetector has a uniform active area of $l \times l$, Eq. (7.1-55) becomes

$$i(t) \propto \int_{-l/2}^{l/2} \int_{-l/2}^{l/2} \{A_r^2 + A_s^2 + 2A_r A_s \cos[\Omega t + k_{pm} m(t) - k_0 x \sin\phi]\} dxdy$$

$$\propto l^2(A_r^2 + A_s^2) + l \int_{-l/2}^{l/2} 2A_r A_s \cos[\Omega t + k_{pm} m(t) - k_0 x \sin\phi] dx, \quad (7.1\text{-}56)$$

where we have assumed the amplitudes of the two plane waves A_r and A_s are real for simplicity. The first term of Eq. (7.1-56) is a DC current, and the second team is an AC current at frequency Ω, which is known as the *heterodyne current*. Note that the information content $m(t)$ originally embedded in the phase of the signal plane wave has now been preserved and transferred to the phase of the heterodyne current. Indeed, this phase-preserving technique in optics is known as *optical heterodyning*, reminiscent of holographic recording discussed in Chapter 5. In contrast, if the reference plane wave were not used, the information content carried by the signal plane wave would be lost. Clearly, if we set $A_r = 0$, Eq. (7.1-56) becomes $i(t) \propto A_s^2$, a DC current proportional to the intensity of the signal plane wave.

Let us now consider the AC part of the current further, that is, the heterodyne current given by Eq. (7.1-56). The heterodyne current of frequency Ω is given by

$$i_\Omega(t) \propto \int_{-l/2}^{l/2} A_r A_s \cos[\Omega t + k_{pm} m(t) - k_0 \sin\phi x] dx$$

$$\propto A_r A_s \int_{-l/2}^{l/2} \text{Re}\left[e^{j[\Omega t + k_{pm} m(t) - k_0 \sin\phi x]} \right] dx$$

$$= A_r A_s \text{Re}\left[e^{j[\Omega t + k_{pm} m(t)]} \int_{-l/2}^{l/2} e^{-jk_0 \sin\phi x} dx \right]$$

$$= A_r A_s \text{Re}\left[e^{j[\Omega t + k_{pm} m(t)]} \frac{e^{-jk_0 \sin\phi x}}{-jk_0 \sin\phi} \bigg|_{x=-l/2}^{x=l/2} \right]$$

$$= A_r A_s \text{Re}\left[e^{j[\Omega t + k_{pm} m(t)]} \frac{2\sin\left(k_0 \frac{l}{2} \sin\phi \right)}{k_0 \sin\phi} \right].$$

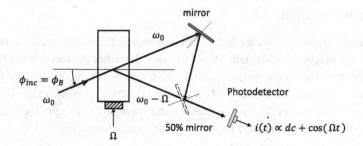

Figure 7.1-19 Acousto-optic heterodyne detection

Using the definition of the *sinc* function [see Section 2.2.1] and completing the real operation, the above equation becomes

$$i_\Omega(t) \propto lA_r A_s \, \text{sinc}\left(\frac{k_0 l}{2\pi}\sin\phi\right)\cos\left[\Omega t + k_{pm}m(t)\right]. \qquad (7.1\text{-}57)$$

The result indicates that the amplitude of the heterodyne current is controlled by a *sinc* function whose argument is a function of $\frac{k_0 l}{2\pi}\sin\phi \approx l\phi/\lambda_0$ for small angles. When $\phi = 0$, that is, the two plane waves are completely aligned, the heterodyne current is at maximum, and falls off as $\text{sinc}(k_0 l\phi/2\pi)$. We have no heterodyne current when the argument of the *sinc* function is 1, that is,

$$\frac{k_0 l\phi}{2\pi} = 1,$$

or

$$\phi = \frac{\lambda_0}{l}. \qquad (7.1\text{-}58)$$

For the size of the photodetector $l = 1$cm and red laser of $\lambda_0 = 0.6\,\mu\text{m}$, ϕ is calculated to be around 2.3×10^{-3} degrees. Therefore, in order to have a sizable heterodyne current, the angle between the two plane waves must be much less than 2.3×10^{-3} degrees. We, therefore, need to have high-precision opto-mechanical mounts for angular rotation to minimize the angular separation between the two plane waves for efficient heterodyning. In Figure 7.1-19, we show a simple experimental configuration to illustrate acousto-optic heterodyning, which recovers the sound frequency Ω as we take $\phi = 0$ in Eq. (7.1-56) with $m(t) = 0$.

7.2   Information Processing with Electro-Optic Modulators

We will first introduce the concept of polarization as electro-optic devices typically manipulate polarization of an incoming optical wave.

7.2.1　　Polarization of Light

Polarization occurs for vector fields. In electromagnetic waves, the vectors are the electric and magnetic fields. Indeed, polarization describes the locus of the tip of the E vector at a given point in space as time advances. Since the direction of the magnetic field H is related to that of E, a separate description of the magnetic field is, therefore, not necessary. Let us consider a plane wave propagating in the $+z$ direction and the electric field is oriented in the x–y plane. According to Eq. (3.3-19a), we have

$$E = E_{0x}e^{j(\omega_0 t - k_0 z)}\hat{x} + E_{0y}e^{j(\omega_0 t - k_0 z)}\hat{y}. \tag{7.2-1a}$$

E_{0x} and E_{0y} are complex in general and can be written as

$$E_{0x} = |E_{0x}|e^{-j\phi_x} \text{ and } E_{0y} = |E_{0y}|e^{-j\phi_y},$$

or

$$E_{0x} = |E_{0x}| \text{ and } E_{0y} = |E_{0y}|e^{-j\phi_0}, \tag{7.2-1b}$$

where $\phi_0 = \phi_y - \phi_x$ is the relative phase shift between the x- and y-components of the electric field. The physical quantity of the electric field is

$$\text{Re}[E] = \text{Re}\left[|E_{0x}|e^{j(\omega_0 t - k_0 z)}\hat{x} + |E_{0y}|e^{-j\phi_0}e^{j(\omega_0 t - k_0 z)}\hat{y}\right]. \tag{7.2-2}$$

Linear Polarization

Let us consider the case where $\phi_0 = 0$ or $\phi_0 = \pm\pi$. For $\phi_0 = 0$, according to Eq. (7.2-2), we have

$$\mathcal{E}(t) = \text{Re}[E] = [|E_{0x}|\hat{x} + |E_{0y}|\hat{y}]\cos(\omega_0 t - k_0 z). \tag{7.2-3}$$

The x- and y-components are in phase and $\mathcal{E}$ is fixed in the x–y plane as the wave is propagating along the $+z$ direction. Since the values of the cosine function go from 1 to -1 as time advances, $\mathcal{E}$ is either on the first quadrant or the third quadrant as illustrated in Figure 7.2-1a but it is always fixed on a plane that is referred to as the *plane of polarization*. The plane of polarization contains the direction of propagation and the electric field vector, and on this plane the electric field oscillates as time advances. The plane wave given by Eq. (7.2-3) is said to be *linearly polarized*, in particular, if $|E_{0y}| = 0$, the plane wave becomes $|E_{0x}|\hat{x} \cos(\omega_0 t - k_0 z)$. It is linearly polarized along the x-direction, and often called x-polarized. Similarly, for $\phi_0 = \pm\pi$, according to Eq. (7.2-2), we have

$$\mathcal{E}(t) = \text{Re}[E] = \text{Re}\left[|E_{0x}|e^{j(\omega_0 t - k_0 z)}\hat{x} + |E_{0y}|e^{j(\omega_0 t - k_0 z)}e^{\mp j\pi}\hat{y}\right]$$

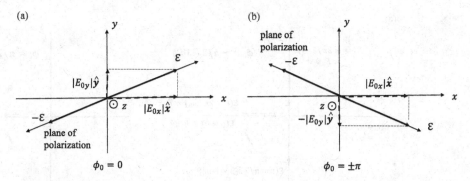

Figure 7.2-1 Linear polarization

$$= \left[|E_{0x}|\hat{x} - |E_{0y}|\hat{y}\right] \cos(\omega_0 t - k_0 z). \tag{7.2-4}$$

For this case, the situation is shown in Figure 7.2-1b. The plane of polarization now lies on the second and third quadrants.

Circular Polarization

Circular polarization is obtained when $\phi_0 = \pm\pi/2$, and $|E_{0y}| = |E_{0x}| = E_0$. For $\phi_0 = -\pi/2$, according to Eq. (7.2-2), we have

$$\mathcal{E}(t) = \text{Re}[E] = \text{Re}\left[|E_{0x}|e^{j(\omega_0 t - k_0 z)}\hat{x} + |E_{0y}|e^{j\pi/2}e^{j(\omega_0 t - k_0 z)}\hat{y}\right]$$

$$= E_0\hat{x}\cos(\omega_0 t - k_0 z) - E_0\hat{y}\sin(\omega_0 t - k_0 z). \tag{7.2-5}$$

Let us monitor the direction of the electric field at a certain position $z = z_0 = 0$ as time advances. Eq. (7.2-5) then becomes

$$\mathcal{E}(t) = E_0\hat{x}\cos(\omega_0 t) - E_0\hat{y}\sin(\omega_0 t).$$

Let us find $\mathcal{E}(t)$ for different times as follows:

At $t = 0$, $\mathcal{E}(t = 0) = E_0\hat{x}$,

$$t = \pi/2\omega_0, \ \mathcal{E}(t = \pi/2\omega_0) = -E_0\hat{y},$$

$$t = \pi/\omega_0, \ \mathcal{E}(t = \pi/\omega_0) = -E_0\hat{x},$$

$$t = 3\pi/2\omega_0, \ \mathcal{E}(t = 3\pi/2\omega_0) = E_0\hat{y},$$

and finally

$$t = 2\pi/\omega_0, \ \mathcal{E}(t = 2\pi/\omega_0) = E_0\hat{x}.$$

We see that when $t = 0$, the wave is x-polarized, and when $t = \pi/2\omega_0$, the wave is negative y-polarized, and so on until when $t = 2\pi/\omega_0$, the wave returns to x-polarization,

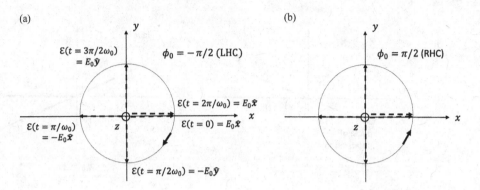

Figure 7.2-2 (a) LHC polarization and (b) RHC polarization

completing a cycle. Let us take red laser as an example where $\lambda_0 = 0.6\,\mu m$, and the frequency for the laser is $f = c/\lambda_0 \approx \dfrac{3\times10^8\,\text{m}}{0.6\times10^{-6}\,\text{m}} = 5\times10^{14}$ Hz. It, therefore,

merely takes $t = \dfrac{2\pi}{\omega_0} = \dfrac{1}{f} = 2\times10^{-15}$ s to complete a rotation as the wave is propa-

gating. In Figure 7.2-2a, we plot $\mathcal{E}(t)$ according to the different times. We see that as time advances, that is, when the wave is propagating along the z-direction, the tip of $\mathcal{E}(t)$ traces out a clockwise circle in the x–y plane, denoted by an arrow in the figure, as seen head-on at $z = z_0$. We have the so-called *clockwise circularly polarized wave*. Such a wave is also called *left-hand circularly (LHC) polarized*. The reason is that when the thumb of the left hand points along the direction of propagation, that is, the $+z$ direction, the other four fingers point in the direction of rotation of $\mathcal{E}(t)$. By the same token, for $\phi_0 = \pi/2$ we can show that the tip of $\mathcal{E}(t)$ traces out a counter-clockwise circle in the x–y plane. We have the *counter-clockwise circularly polarized wave* or *right-hand circularly (RHC) polarized wave*. The situation is shown in Figure 7.2.2b.

For plane waves that are not linearly or circularly polarized, we have *elliptical polarization*. The two electric field components might have different amplitudes, that is, $|E_{0x}| \neq |E_{0y}|$, and also might contain an arbitrary ϕ_0 between the two components.

Common Polarization Devices

Linear Polarizer

A linear polarizer is an optical element that transmits only the component of the electrical field in the direction of its *transmission axis*. This selective process can be done, for example, by selective or anisotropic absorption using certain anisotropic materials such as the well-known Polaroid H-sheet. This anisotropy in absorption is called *dichroism*. The sheet is made by stretching in one direction a thin film of poly-vinyl alcohol (PVA) stained with iodine and becomes highly dichroic. Figure 7.2-3 shows a linear polarizer with its transmission axis (denoted with arrows) along $\hat{x}$. With the

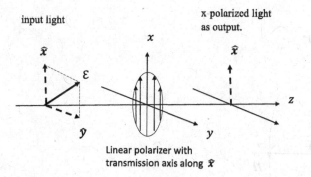

Figure 7.2-3 Linear polarizer with transmission axial along the x-direction

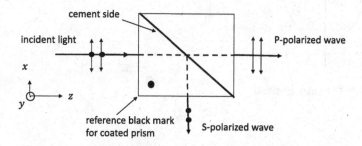

Figure 7.2-4 Polarizing beam splitter

input linear polarized wave decomposed into two orthogonal directions $\hat{x}$ and $\hat{y}$, the output wave from the polarizer has become x-polarized. *Natural light* (such as sunlight and light from bulbs) consists of randomly polarized waves and transmission of natural light through a linear polarizer gives a polarized light as output.

Let us formulate the action of the polarizer mathematically. According to Eq. (7.2-1), a linearly polarized plane wave propagating along the $+z$ direction can be written as

$$\left[\left|E_{0x}\right|\hat{x}+\left|E_{0y}\right|\hat{y}\right]e^{j(\omega_0 t-k_0 z)}.$$

Hence, at the location of the linear polarizer ($z=0$), the incident plane wave can be written as

$$E=\left[\left|E_{0x}\right|\hat{x}+\left|E_{0y}\right|\hat{y}\right]e^{j\omega_0 t}.$$

Since the transmission axis of the polarizer is along the x-direction, as shown in Figure 7.2-3, at the exit of the polarizer, the x-component of the incident electric field is [see the use of the dot product in Chapter 3]

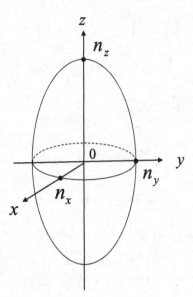

Figure 7.2-5 The index ellipsoid

$$(\boldsymbol{E}\cdot\hat{\boldsymbol{x}})\hat{\boldsymbol{x}} = \left(\left[|E_{0x}|\hat{\boldsymbol{x}}+|E_{0y}|\hat{\boldsymbol{y}}\right]\cdot\hat{\boldsymbol{x}}\right)\hat{\boldsymbol{x}}\,e^{j\omega_0 t} = |E_{0x}|\hat{\boldsymbol{x}}e^{j\omega_0 t}.$$

Polarizing Beam Splitter (PBS)

Polarizing beam splitters are typically constructed using two right-angle prisms. The hypotenuse surface of one prism is coated with a polarization sensitive thin film designed to separate an incident, unpolarized beam into two orthogonal polarization components. The two prisms are then cemented together to form a cubic shape. Figure 7.2-4 shows the situation. The incident light shows two orthogonal polarization directions (one polarization axis is along the x-axis and the other along the y-axis). The transmitted polarized light with its electric field parallel along the plane of incidence is commonly known as the *p-polarized wave*, while the reflected light whose electrical field is normal to the plane of incidence is called the *s-polarized wave*. Manufactures usually denote a reference black mark to show the coated prism and it is recommended that the light be transmitted into the coated prism in order not to damage the cement.

7.2.2   Index Ellipsoid and Birefringent Wave Plates

Crystals in general are anisotropic. The medium is anisotropic if its properties are different in different directions from any given point. In general, there are two methods for analyzing light wave propagating through crystals: the method that employs the rigorous electromagnetic theory and that employs the *refractive index ellipsoid*. The use of the index ellipsoid is intuitive and convenient and is often used in practice. We can express the optical properties of crystals in terms of the refractive index ellipsoid. The equation of the index ellipsoid for a crystal is expressed as

$$\frac{x^2}{n_x^2} + \frac{y^2}{n_y^2} + \frac{z^2}{n_z^2} = 1,$$

(7.2-6)

where $n_x^2 = \varepsilon_x / \varepsilon_0$, $n_y^2 = \varepsilon_y / \varepsilon_0$, and $n_z^2 = \varepsilon_z / \varepsilon_0$ with ε_x, ε_y, and ε_z denoting the principal dielectric constants of the crystal. As we will see, the index of ellipsoid is used conveniently to describe light wave propagation in the crystal. Figure 7.2-5 shows the index ellipsoid in the coordinate system of the *principal axes*, where x, y, and z represent the directions of the principal axis. n_x, n_y, and n_z are the refractive indices of the index ellipsoid. Three crystal classes can be identified in terms of these refractive indices.

In crystals where two of the refractive indices are equal, that is, $n_x = n_y \neq n_z$, the crystals are called *uniaxial crystals*. The two identical refractive indices are usually called the *ordinary refractive index* n_0 , that is, $n_x = n_y = n_0$. The distinct refractive index is called the *extraordinary refractive index* n_e , that is, $n_z = n_e$. Furthermore, the crystal is said to be *positive uniaxial crystal* if $n_e > n_0$ (such as quartz) and *negative uniaxial crystal* if $n_e < n_0$ (such as calcite). The z-axis of a uniaxial crystal is called the *optic axis* as the other axes have the same index of refraction. When the three indices are equal, that is, $n_x = n_y = n_z$, the crystal is optical isotropic (such as diamond). Crystals for which the three indices are different, that is, $n_x \neq n_y \neq n_z$, are called *biaxial* (such as mica). We will only discuss the applications of uniaxial crystals.

Birefringent Wave Plates

A *wave plate* introduces a phase shift between two orthogonal polarization components of the incident light wave. It is clear from Figure 7.2-5 that when the light propagates along the x-direction in uniaxial crystals, linearly polarized light can be decomposed along the y- and z-directions. Light polarized along the y-direction sees an index of $n_y = n_0$, while light polarized along the z-direction sees an index of $n_z = n_e$. The effect is to introduce a phase shift between the two polarized waves. The crystal is usually cut into a plate, with the orientation of the cut chosen so that the optic axis is parallel to the surface of the plate. Furthermore, if the plate surface is perpendicular to the principal x-axis, the crystal plate is called *x-cut*. Figure 7.2-6 shows an *x-cut wave plate* of thickness d. Let a linearly polarized optical field incident on the plate at $x = 0$ be

$$E_{inc} = \left(|E_{0y}| \hat{y} + |E_{0z}| \hat{z} \right) e^{j\omega_0 t}.$$

(7.2-7)

In Figure 7.2-6, $\theta = \tan^{-1} \left(|E_{0y}| / |E_{0z}| \right)$. At the exit of the wave plate, the field is represented by

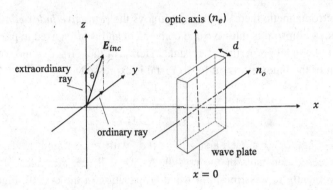

Figure 7.2-6 x-cut wave plate

$$E_{out} = \left(|E_{0y}| e^{-jk_{n0}d} \, \hat{y} + |E_{0z}| e^{-jk_{ne}d} \, \hat{z} \right) e^{j\omega_0 t}, \tag{7.2-8}$$

where $k_{n0} = (2\pi/\lambda_v) n_0$, and $k_{ne} = (2\pi/\lambda_v) n_e$ with λ_v being the wavelength in vacuum. The plane wave polarized along $\hat{y}$ is called the *ordinary wave* (*o-ray*), and the plane wave polarized along $\hat{z}$ is called the *extraordinary wave* (*e-ray*). If $n_e > n_0$, the e-ray travels slower than the o-ray. The z-principal axis is, therefore, often referred to as the *slow axis*, and the y-principal axis is the *fast axis*. If $n_0 > n_e$, the z-principal axis becomes the fast axis and the y-axis is the slow axis. Hence, the two rays of different polarizations acquire different phase shifts as they propagate through the crystal, and this phenomenon is called *birefringence*. So, we see that the optic axis of a uniaxial crystal is a direction in which a ray does not suffer birefringence. Indeed, birefringence is the optical property of a crystal having a refractive index that depends on the polarization (e-ray or o-ray) and the direction of propagation of light. Keeping track of the phase difference between the two rays and rewriting Eq. (7.2-8), we have

$$E_{out} \propto \left(|E_{0y}| \hat{y} + |E_{0z}| e^{-j(k_{ne}-k_{n0})d} \, \hat{z} \right) e^{j\omega_0 t} = \left(|E_{0y}| \hat{y} + |E_{0z}| e^{-j\Delta\phi} \, \hat{z} \right) e^{j\omega_0 t}, \tag{7.2-9}$$

where $\Delta\phi = (k_{ne} - k_{n0})d = \left(\dfrac{2\pi}{\lambda_v}\right)(n_e - n_0)d$ is the relative phase shift between the

two rays. Comparing this equation with Eq. (7.2-2), we can recognize that $\Delta\phi$ is ϕ_0. By controlling $\Delta\phi$ through the right thickness d of the wave plate, we have different types of wave plate.

Quarter-Wave Plate (QWP)

Quarter-wave plates are used to change linearly polarized light into circularly polarized light and vice versa. When $\Delta\phi = \left(\dfrac{2\pi}{\lambda_v}\right)(n_e - n_0)d = \pi/2$ by designing the correct thickness, we have a quarter-wave plate. Since $\Delta\phi = \phi_0 = \pi/2$, and with $|E_{0y}| = |E_{0z}|$ in Eq. (7.2-9), we have right-hand circularly (RHC) wave as output from the wave plate, according to Figure 7.2-2b. The situation is shown in Figure 7.2-7.

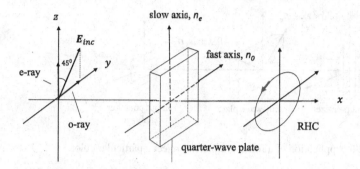

Figure 7.2-7 Quarter-wave plate

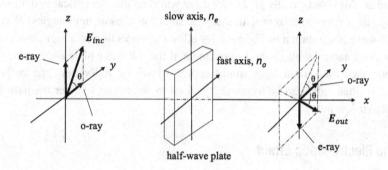

Figure 7.2-8 Half-wave plate

Half-Wave Plate (HWP)

Half-wave plates are used to rotate the plane of linearly polarized light. When $\Delta\phi = \left(\dfrac{2\pi}{\lambda_v}\right)(n_e - n_0)d = \pi$ by designing the correct thickness, we have a half-wave plate. From Eq. (7.2-9), for $\Delta\phi = \pi$, we have

$$E_{\text{out}} \propto \left(|E_{0y}|\,\hat{\boldsymbol{y}} + |E_{0z}|\,e^{-j\pi}\,\hat{\boldsymbol{z}}\right)e^{j\omega_0 t} = \left(|E_{0y}|\,\hat{\boldsymbol{y}} - |E_{0z}|\,\hat{\boldsymbol{z}}\right)e^{j\omega_0 t}.$$

The situation is shown in Figure 7.2-8. We see that effectively the half-wave plate rotates the input electrical field vector through an angle 2θ across the y-axis to give E_{out} at the exit of the half-wave plate.

Full-Wave Plates (FWP)

When $\Delta\phi = \left(\dfrac{2\pi}{\lambda_v}\right)(n_e - n_0)d = 2\pi$ by designing the correct thickness, we have a *full-wave plate*. A full-wave plate introduces exactly an optical path of exactly one wavelength between the o-ray and the e-ray. Since a full-wave plate is typically designed for a particular wavelength as $e^{-j\Delta\phi} = e^{-j2\pi} = 1$, linearly polarized wave with that wavelength exiting the wave plate remains the same polarization. Linearly polarized

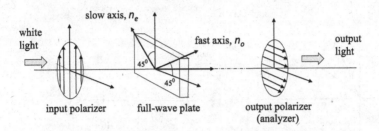

Figure 7.2-9 Application of a full-wave plate to reject a párticular color

light with other wavelengths will, therefore, exit the wave plate elliptically polarized as $\Delta\phi$ is not exactly at 2π for these wavelengths. An optical system with *cross polarizers* (two polarizers with their transmission axes at right angles) along with a full-wave plate shown in Figure 7.2-9 allows the rejection of a particular color from the input natural white light. For example, if the full-wave plate is designed for green light, then the output light would appear somewhat red/violet. The output polarizer is often called an *analyzer* as it is used to determine whether the light has been polarized or not.

7.2.3   The Electro-Optic Effect

The *electro-optic effect* is the change of the refractive index of crystals under the presence of external applied electric field. We shall use the index ellipsoid method to discuss the electro-optic effect, which is widely used to realize control over the light wave propagating through the crystals (such as phase modulation, intensity modulation, and polarization of the light wave).

We will only discuss the *linear (or Pockels-type) electro-optic effect*, meaning the case where the change in n_e and n_0 is linearly proportional to the applied electric field. Mathematically, the electro-optic effect can be best modelled as a deformation of the index ellipsoid due to an external electric field. Instead of specifying the changes in the refractive indices, it is more convenient to describe the changes in $1/n^2$, that is, $\Delta(1/n^2)$, due to the external electric field. Restricting our analysis to the Pockels effects and uniaxial crystals, the expression for the deformed ellipsoid becomes

$$\left[\frac{1}{n_o^2}+\Delta\left(\frac{1}{n^2}\right)_1\right]x^2 + \left[\frac{1}{n_o^2}+\Delta\left(\frac{1}{n^2}\right)_2\right]y^2 + \left[\frac{1}{n_e^2}+\Delta\left(\frac{1}{n^2}\right)_3\right]z^2$$

$$+2\Delta\left(\frac{1}{n^2}\right)_4 yz + 2\Delta\left(\frac{1}{n^2}\right)_5 xz + 2\Delta\left(\frac{1}{n^2}\right)_6 xy = 1, \qquad (7.2\text{-}10)$$

where

$$\Delta\left(\frac{1}{n^2}\right)_i = \sum_{j=1}^{3} r_{ij}E_j, i = 1,\dots,6,$$

and r_{ij} are called *linear electro-optic (or Pockels) coefficients*. The E_js are the components of the externally applied electric field in the direction of the principal axes x, y, and z for $j = 1, 2$, and 3, respectively. Note that the appearance of the crossed terms, that is, yz, xz, and xy, in Eq. (7.2-10) signifies the equation of a rotated ellipsoid centered at the origin. When the applied electric field is zero, that is, $E_j = 0$, Eq. (7.2-10) reduces to Eq. (7.2-6), where $n_x = n_0 = n_y$ and $n_z = n_e$. We can express $\Delta\left(\dfrac{1}{n^2}\right)_i$ in matrix form as

$$
\begin{pmatrix}
\Delta\left(\dfrac{1}{n^2}\right)_1 \\[2mm]
\Delta\left(\dfrac{1}{n^2}\right)_2 \\[2mm]
\Delta\left(\dfrac{1}{n^2}\right)_3 \\[2mm]
\Delta\left(\dfrac{1}{n^2}\right)_4 \\[2mm]
\Delta\left(\dfrac{1}{n^2}\right)_5 \\[2mm]
\Delta\left(\dfrac{1}{n^2}\right)_6
\end{pmatrix}
=
\begin{pmatrix}
r_{11} & r_{12} & r_{13} \\
r_{21} & r_{22} & r_{23} \\
r_{31} & r_{32} & r_{33} \\
r_{41} & r_{42} & r_{43} \\
r_{51} & r_{52} & r_{53} \\
r_{61} & r_{62} & r_{63}
\end{pmatrix}
\begin{pmatrix}
E_1 \\ E_2 \\ E_3
\end{pmatrix}.
\tag{7.2-11}
$$

The 6×3 matrix is often written as $[r_{ij}]$ and known as the *linear electro-optic tensor*. The tensor contains 18 elements and they are necessary in the most general case, when no symmetry is present in the crystal. But many of the 18 elements are zero and some of the nonzero elements have the same value, depending on the symmetry of the crystal. Table 7.1 lists the nonzero values of some of the representative crystals. Using Eq. (7.2-10) and Table 7.1, we can find the equation of the index ellipsoid in the presence of an external applied field.

Example: Index Ellipsoid of KDP Under the Presence of Applied Electric Field

In the presence of an external electric field $\mathbf{E} = E_x\,\hat{x} + E_y\,\hat{y} + E_z\,\hat{z}$, let us make an analysis with the frequently used KDP crystal. According to Table 7.1, the electro-optic tensor of such a crystal is

$$
[r_{ij}] =
\begin{pmatrix}
0 & 0 & 0 \\
0 & 0 & 0 \\
0 & 0 & 0 \\
r_{41} & 0 & 0 \\
0 & r_{41} & 0 \\
0 & 0 & r_{63}
\end{pmatrix}.
\tag{7.2-12}
$$

Table 7.1 Pockels coefficients

Material	r_{ij} (10^{-12} m/V)	λ_r (μm)	Refractive index
LiNbO$_3$ (Lithium Niobate)	$r_{13} = r_{23} = 8.6$ $r_{33} = 30.8$	0.63	$n_0 = 2.2967$
	$r_{22} = -r_{61} = -r_{12} = 3.4$		$n_e = 2.2082$
	$r_{51} = r_{41} = 28$		
KDP (Potassium dihydrogen phosphate)	$r_{41} = r_{52} = 8.6$ $r_{63} = 10.6$	0.55	$n_0 = 1.50737$ $n_e = 1.46685$
ADP (Ammonium dihydrogen phosphate)	$r_{41} = r_{52} = 2.8$ $r_{63} = 8.5$	0.55	$n_0 = 1.52$ $n_e = 1.48$

Note that $r_{52} = r_{41}$. Hence, these crystals only have two independent electro-optic coefficients, that is, r_{41} and r_{63}. Substituting Eq. (7.2-12) into Eq. (7.2-11), we have

$$\Delta\left(\frac{1}{n^2}\right)_1 = \Delta\left(\frac{1}{n^2}\right)_2 = \Delta\left(\frac{1}{n^2}\right)_3 = 0,$$

and

$$\Delta\left(\frac{1}{n^2}\right)_4 = r_{41}\mathrm{E}_x, \Delta\left(\frac{1}{n^2}\right)_5 = r_{41}\mathrm{E}_y, \Delta\left(\frac{1}{n^2}\right)_6 = r_{63}\mathrm{E}_z.$$

With these values, Eq. (7.2-10) becomes

$$\frac{x^2}{n_o^2} + \frac{y^2}{n_o^2} + \frac{z^2}{n_e^2} + 2r_{41}\mathrm{E}_x yz + 2r_{41}\mathrm{E}_y xz + 2r_{63}\mathrm{E}_z xy = 1. \qquad (7.2\text{-}13)$$

Example: Rotation of the Index Ellipsoid of KDP Under E_z Only

When there is only an electrical field applied along the principal z-axis, Eq. (7.2-13) becomes

$$\frac{x^2}{n_o^2} + \frac{y^2}{n_o^2} + \frac{z^2}{n_e^2} + 2r_{63}\mathrm{E}_z xy = 1. \qquad (7.2\text{-}14)$$

In Eq. (7.2-14), we see that there is a crossed term xy present, suggesting the original ellipsoid has been rotated in the xy plane. We can find a new coordinate system so that Eq. (7.2-14) will become the "principal axes" in the coordinate system, allowing us to find the effect of the electrical field on light propagation. From analytic

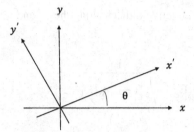

Figure 7.2-10 An xy-coordinate system rotated through an angle θ to an $x'y'$ coordinate system.

geometry, the following rotation matrix gives the relation between the original coordinates (x, y) to the rotated coordinates (x', y'), where θ is the rotation angle:

$$\begin{pmatrix} x \\ y \end{pmatrix} = \begin{pmatrix} \cos\theta & -\sin\theta \\ \sin\theta & \cos\theta \end{pmatrix} \begin{pmatrix} x' \\ y' \end{pmatrix}. \tag{7.2-15}$$

The relationship between the original and the rotated coordinate systems is shown in Figure 7.2-10. The direction of rotation is according to the right-hand rule with rotation about the z-axis.

Substituting Eq. (7.1-15) into Eq. (7.1-14), we find

$$\left[\frac{1}{n_0^2} + r_{63}E_z\sin2\theta \right] x'^2 + \left[\frac{1}{n_0^2} - r_{63}E_z\sin2\theta \right] y'^2 + \frac{1}{n_e^2}z^2 + 2r_{63}E_z\cos2\theta x'y' = 1. \tag{7.2-16}$$

To eliminate the crossed term, we let $\cos2\theta = 0$, which means $\theta = \pm45°$. We choose to take $\theta = -45°$ for our further mathematical analysis, and then Eq. (7.2-16) becomes

$$\left(\frac{1}{n_0^2} - r_{63}E_z \right) x'^2 + \left(\frac{1}{n_0^2} + r_{63}E_z \right) y'^2 + \frac{1}{n_e^2}z^2 = 1. \tag{7.2-17}$$

This is the ellipsoid after an electric field is applied on the KDP crystal along the z-axis. Hence, the index ellipsoid equation is in the coordinate system aligned with the new principal axes, x', y' and z:

$$\frac{x'^2}{n_x'^2} + \frac{y'^2}{n_y'^2} + \frac{z^2}{n_e^2} = 1,$$

where

$$\frac{1}{n_x'^2} = \frac{1}{n_0'^2} - r_{63}E_z \quad \text{and} \quad \frac{1}{n_y'^2} = \frac{1}{n_0'^2} + r_{63}E_z. \tag{7.2-18}$$

Let us now estimate the value of the refractive indices along the x' and y' principal axes. From Eq. (7.2-18),

$$n'^2_x = \frac{1}{\frac{1}{n_0^2} - r_{63}E_z} = \frac{1}{\frac{1}{n_0^2}\left[1 - n_0^2 r_{63}E_z\right]}.$$

Hence,

$$n'_x = \frac{n_0}{\sqrt{1 - n_0^2 r_{63}E_z}} \approx \frac{n_0}{1 - \frac{1}{2}n_0^2 r_{63}E_z} \tag{7.2-19}$$

as r_{63} is very small (of the order of 10^{-12} m/V, see Table 7.1) and we have used approximation $\sqrt{1-\epsilon} \approx 1 - \epsilon/2$ for ϵ small. Furthermore, we use $1/(1-\epsilon) \approx 1+\epsilon$ to rewrite Eq. (7.2-19) to finally obtain

$$n'_x \approx n_0 + \frac{1}{2}n_0^3 r_{63}E_z. \tag{7.2-20a}$$

Similarly, for n'_y, we have

$$n'_y \approx n_0 - \frac{1}{2}n_0^3 r_{63}E_z. \tag{7.2-20b}$$

We see that when an electric field is applied on the KDP crystal along the z-direction, the crystal is turned from a uniaxial crystal [see. Eq. (7.2-14) for $E_z = 0$] into a biaxial crystal, i.e., $n'_x \ne n'_y$ as n'_x and n'_y are the principal refractive indices in the new coordinate system. The new principal axes of the ellipsoid is rotated around the z-axis an angle of $45°$. The rotational angle does not depend on the magnitude of the applied electric field, but the change of the index of refraction is proportional to the electric field along the z-axis, that is, E_z.

7.2.4   Electro-Optic Intensity Modulation and Phase Modulation

Intensity Modulation

Let us investigate how to use the KDP crystal for electro-optic intensity modulation. The modulation system is shown in Figure 7.2-11. It consists of the KDP crystal placed between two crossed polarized whose transmission axes are perpendicular to each other. With the electric field (E_z) applied along the principal z-axis direction of the crystal through an external voltage $V = E_z L$, where L is the thickness of the crystal, the index ellipsoid has been rotated on the $\hat{x} - \hat{y}$ plane by $45°$ with the new principal axes $\hat{x}'$ and $\hat{y}'$, as indicated in the figure. This rotation is consistent to that shown in Figure 7.2-10 for $\theta = -45°$.

The input $\hat{x}$-linearly polarized light, $E_{inc} = E_0 \hat{x} e^{j\omega_0 t}$, propagates along the optic axis direction and can be resolved into two orthogonal components polarized along $\hat{x}'$ and $\hat{y}'$ at $z = 0$ as

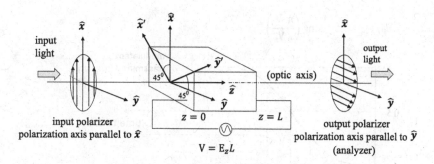

Figure 7.2-11 Electro-optic intensity modulation system

$$E_{inc} = \frac{E_0}{\sqrt{2}} (\hat{x}' + \hat{y}')e^{j\omega_0 t}$$

At the exit of the crystal, that is, $z = L$, the field is represented by

$$E_{exit} = \frac{E_0}{\sqrt{2}}\left(e^{-jk_{n'_x} L}\hat{x}' + e^{-jk_{n'_y} L}\hat{y}'\right)e^{j\omega_0 t}, \qquad (7.2\text{-}21)$$

where $k_{n'_x} = 2\pi n'_x / \lambda_v$ and $k_{n'_y} = 2\pi n'_y / \lambda_v$. The phase difference between the two components is called the *retardation* Φ and is given by

$$\Phi = (k_{n'_x} - k_{n'_y})L = \frac{2\pi}{\lambda_v}(n'_x - n'_y)L = \frac{2\pi}{\lambda_v}n_0^3 r_{63}E_z L = \frac{2\pi}{\lambda_v}n_0^3 r_{63}V, \qquad (7.2\text{-}22)$$

where we have used the values of n'_x and n'_y from Eq. (7.2-20). At this point, let us introduce an important parameter called the *half-wave voltage* V_π. The half-wave voltage is the voltage required for inducing a phase change of π, that is, when $\Phi = \pi$. Therefore, from Eq. (7.2-22), we have

$$V_\pi = \frac{\lambda_v}{2n_0^3 r_{63}}. \qquad (7.2\text{-}23)$$

From Table 7.1 and at $\lambda_v = 0.55\ \mu m$, $V_\pi \approx 7.58KV$ for KDP crystals. When the half-wave voltage is low, the power needed for the modulator will be low. Reducing the half-wave voltage of a modulator is an ongoing research.

Returning to Figure 7.2-11, let us find the electrical field passed by the analyzer or the output polarizer. This is done by finding the components of E_{exit} along the $\hat{y}$ direction. We can take advantage of the dot product, and the output electric field from the analyzer is

$$E_{out} = (E_{exit} \cdot \hat{y})\hat{y} = \left[\frac{E_0}{\sqrt{2}}\left(e^{-jk_{n'_x} L}\hat{x}' + e^{-jk_{n'_y} L}\hat{y}'\right)e^{j\omega_0 t} \cdot \hat{y}\right]\hat{y}.$$

Since $\hat{x}' \cdot \hat{y} = \cos(90° + 45°) = -1/\sqrt{2}$ and $\hat{y}' \cdot \hat{y} = \cos(45°) = 1/\sqrt{2}$, the above equation becomes

$$E_{out} = \frac{E_0}{2}\left[e^{-jk_{n'_y} L} - e^{-jk_{n'_x} L}\right]e^{j\omega_0 t}\hat{y} = \frac{E_0}{2}\left[e^{-j\frac{2\pi}{\lambda}n'_y L} - e^{-j\frac{2\pi}{\lambda}n'_x L}\right]e^{j\omega_0 t}\hat{y}.$$

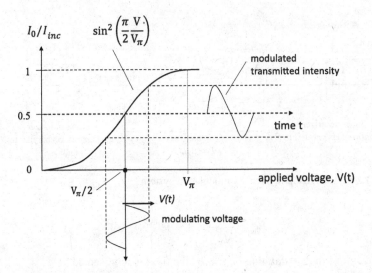

Figure 7.2-12 Relationship between the modulating voltage $V(t)$ and the modulated intensity

This is an $\hat{y}$-linearly polarized wave and the ratio of the output intensity, that is, $I_0 = |E_{\text{out}}|^2$, and the input intensity, that is, $I_{\text{inc}} = |E_{\text{inc}}|^2$, is shown to be

$$\frac{I_0}{I_{\text{inc}}} = \sin^2\left[\frac{\pi}{\lambda_v}(n_x' - n_y')L\right] = \sin^2\left(\frac{\Phi}{2}\right) = \sin^2\left(\frac{\pi}{2}\frac{V}{V_\pi}\right). \qquad (7.2\text{-}24)$$

This sine-squared characteristic is reminiscent of the intensity modulation method used in acousto-optics [see Figure 7.1-13]. Figure 7.2-12 illustrates the relationship between the modulating signal and the modulated transmitted intensity. Again, in order to obtain linear modulation, that is, modulated intensity being proportional to the modulating signal, we need to bias the modulator at $I_0 / I_{\text{inc}} = 0.5$. The bias can be achieved by adding an additional bias voltage $V_\pi / 2$ from the modulating signal already applied on the crystal.

Phase Modulation

Figure 7.2-13 shows a diagram of the principle of electro-optic phase modulation, which consists of an input polarizer and a KDP crystal. The transmission axis of the polarizer can be along one of the "rotated" axes of the crystal, that is, either $\hat{x}'$ or $\hat{y}'$. The crystal is under the influence of the external applied electric field E_z along its optical axis. In the figure, we show that the input polarizer is along the $\hat{x}'$- axis. The applied electric field along the z-direction does not change the state of polarization but only the phase. At $z = 0$, we let $E_{\text{inc}} = E_0 \hat{x}' e^{j\omega_0 t}$. At the exit of the crystal, we have

$$E_{\text{exit}} = E_0 e^{-jk_{n_x'}L} \hat{x}' e^{j\omega_0 t} = E_0 e^{-j\frac{2\pi}{\lambda_v}n_x' L} \hat{x}' e^{j\omega_0 t}. \qquad (7.2\text{-}25)$$

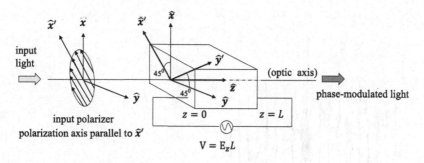

Figure 7.2-13 Principle of electro-optic phase modulation

According to Eq. (7.2-20a), we have $n'_x \approx n_0 + \frac{1}{2} n_0^3 r_{63} E_z$. Hence, the above signal is a phase modulated signal:

$$E_{\text{exit}} = E_0 e^{-j\frac{2\pi}{\lambda_v}\left(n_0 + \frac{1}{2}n_0^3 r_{63}E_z\right)L} \hat{x}' e^{j\omega_0 t} = E_0 e^{-j\frac{2\pi}{\lambda_v}n_0 L} e^{-j\frac{\pi}{\lambda_v}n_0^3 r_{63} k_p m(t)} \hat{x}' e^{j\omega_0 t} \qquad (7.2\text{-}26)$$

as $V = E_z L = k_p m(t)$ and $m(t)$ is the modulating signal with k_p being the phase-deviation constant.

7.3 Information Processing and Display with Liquid Crystal Cells

7.3.1 Liquid Crystals

There is a definite melting point for an ordinary crystal. At the melting point, the crystal makes a phase transition from the solid state into the liquid state. Some crystals, however, do not directly change from solid into liquid but process the property of fluidity of liquids as well as the characteristics of a crystal. Materials with this type of transitional phase are known as *liquid crystals*. Hence, liquid crystals (LCs) have fluidlike properties such as fluidity and elasticity and yet the arrangement of molecules within them exhibit structural orders such as anisotropicity. Liquid crystal molecules can be visualized as ellipsoids or have rodlike shapes and have a length-to-breath ratio of about 10 with a length of several nanometers. In the so-called nematic phase of the crystals, the randomly located molecules tend to align themselves with their long axes parallel. The situation is shown in Figure 7.3-1, where the vector $\hat{n}$ is called a *director*, which represents the macroscopic direction of the aligned nematic liquid crystal molecules.

When liquid crystal molecules are contained without any alignment, their appearance is often milky as they scatter light due to the randomness of liquid crystal clusters. In order to realize useful electro-optic effects, liquid crystals need to be aligned though an external applied electrical field.

The nematic liquid-crystal cell is composed of two glass plates (or glass substrates) containing a thin liquid crystal layer (typically about 5–10μm thick) inside. The two

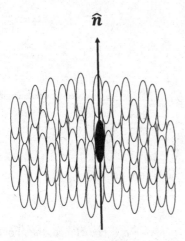

Figure 7.3-1 Molecular ordering of nematic phase of liquid crystal

glass plates impose boundary conditions on the alignment of nematic liquid crystal molecules. Each glass plate is coated with a thin electrically conductive but optically transparent metallic film (such as indium-tin-oxide, ITO) called the alignment layer and then the layer is rubbed with a fine cotton cloth in one direction. Fine grooves about several nanometers wide are formed by rubbing and thus causing the liquid crystal molecules to lie parallel to the direction of the grooves. This rubbing method has been widely used for fabricating large panel liquid crystal devices. High-quality alignment can be made by vacuum deposition of a fine silicon monoxide (SiO) layer to create micro-grooves onto the surface of the glass for aligning LC molecules.

There are two popular alignments. If each alignment layer is polished with different directions, the molecular orientation rotates helically about an axis normal to plates such as the situation shown in Figure 7.3-2a, where we see that the left side of the LC cell is polished along the y-direction and the right side along the x-direction. This situation is known as the twist alignment as the back glass plate is twisted with an angle with respect to the front plate. If the alignment directions between the two plates are $90°$, we have the so-called perpendicular alignment. If the alignment directions are parallel, the LC molecules are parallelly aligned, and we have parallel alignment. The parallel alignment is shown in Figure 7.3-2b.

7.3.2   Phase Modulation and Intensity Modulation Using Liquid Crystal Cells

We consider the so-called *parallel-aligned liquid crystal (PAL) cell* and discuss its basic principle of operation for modulation applications. When there is no applied electric field, the LC molecules is aligned along the y-direction. The LC is an organic compound that processes a very strong electric dipole moment. Upon an external applied electric field **E** along the z-direction, a dipole moment **p** of the liquid crystal is induced and the dipole experiences a torque τ given by

(a)

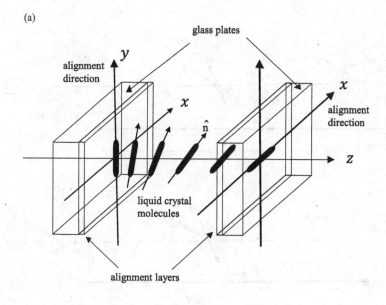

(b)

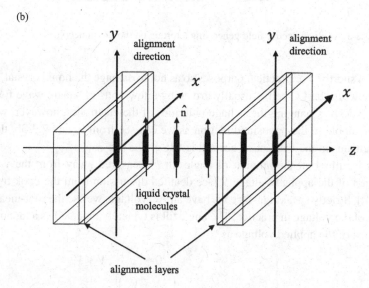

Figure 7.3-2 (a) Twist alignment and (b) parallel alignment

$$\tau = p \times E \tag{7.3-1}$$

and the *torque* is along the x-direction, trying to align the dipole with the electric field and turning clockwise in the y–z plane by the force **F**. The situation is illustrated in Figure 7.3-3a. Remember that the cross product is given with direction given by the right-hand rule [see the cross product example in Chapter 3]. The rotation of the dipole of the LC molecules changes the optical properties of the LC cell as we will

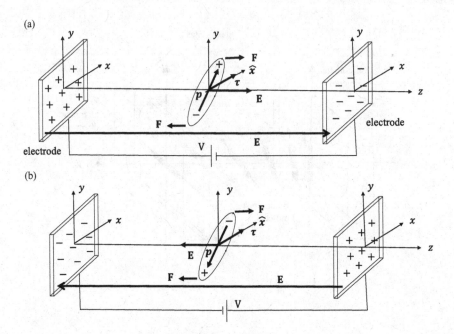

Figure 7.3-3 Applied electric field generating a torque on the LC molecule

explain shortly. For practical purposes so as not to damage the liquid crystal, the voltage between the LC cell is usually driven by a few volts of square wave function of about 1 KHz. Changing the voltage polarities of the electrodes, however, would not tilt the dipole in the opposite direction as we can see from Figure 7.3-3b, the torque still remains in the x-direction according to Eq. (7.3-1).

As it turns out, the amount of angle tilt θ measured away from the y-axis is a function of the applied voltage V [see detailed discussion from the book by Gennes (1974)]. Strictly speaking, since we have an AC voltage, we use the root-mean-square value of the voltage in practice. The angle tilt is typically modeled as a monotonically function of the applied voltage as

$$\theta = \begin{cases} 0 & V \leq V_{th} \\ \dfrac{\pi}{2} - 2\tan^{-1}\left[\exp\left(-\dfrac{V-V_{th}}{V_0}\right)\right] & V \geq V_{th}, \end{cases} \tag{7.3-2}$$

where V_{th} is a *threshold voltage* about 1–2 V and below which no tilting of the molecules occurs and V_0 is a constant voltage around 1–2 V. For $V > V_{th}$, the angle θ keeps increasing with V until reaching a saturation value of $\pi/2$. Figure 7.3-4 plots the angle θ in degrees as a function of $(V - V_{th})/V_0$ for $V \geq V_{th}$. When the voltage is turned off, the molecules return to their original positions.

Let us find the index of refraction along the y-axis after the electric dipole is tilted due to the external applied field. Figure 7.3-5a shows a cross section (in the y–z plane) of the index of ellipsoid with the *tilt angle* θ. The circle with a cross at the center

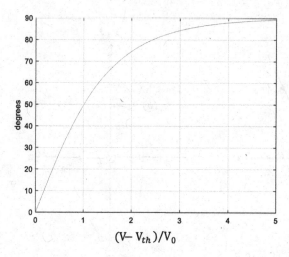

Figure 7.3-4 The tilt angle θ as a function of $(V - V_{th})/V_0$

signifies that the direction of the x-axis is into the paper, following the right-handed coordinate system convention. Our objective is to find $n_y(\theta)$, which is a tilt-angle-dependent index of refraction. For ease of calculation, we rotate the cross section of the index of ellipsoid back to the original location, as shown in Figure 7.3-5b. Now, the equation of the ellipse on the y–z plane is

$$\frac{y^2}{n_e^2} + \frac{z^2}{n_o^2} = 1. \tag{7.3-3}$$

By inspecting the geometry in Figure 7.3-5b, we find the relations

$$n_y^2(\theta) = y^2 + z^2, \tag{7.3-4a}$$

and

$$\sin\theta = z/n_y(\theta). \tag{7.3-4b}$$

Combining Eqs. (7.3-3) and (7.3-4), we find

$$\frac{1}{n_y^2(\theta)} = \frac{\cos^2\theta}{n_e^2} + \frac{\sin^2\theta}{n_o^2}. \tag{7.3-5}$$

Modulation Schemes

Let us consider a wave traveling along the z-direction with polarizations in the x- and y-directions. From Eq. (7.3-5) along with Figure 7.3-5b, we see that when there is no tilt, that is, $\theta = 0$, the waves polarized along in the x- and y-directions have refractive indices n_o and n_e, respectively and the retardation is maximum at $\Phi = (k_{n_y} - k_{n_x})d = 2\pi[n_e - n_o]d/\lambda_v$, where d is length of the LC cell. After the tilt, the index refraction along the y-axis is changed to $n_y(\theta)$ with the index of refraction along the x-axis remaining unchanged. The retardation becomes $2\pi[n_y(\theta) - n_o]d/\lambda_v$.

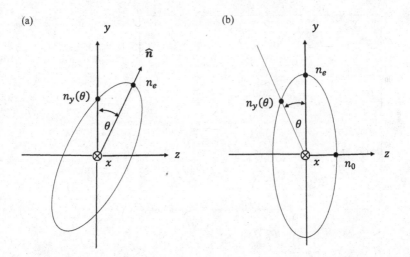

Figure 7.3-5 Cross section of index ellipsoid : (a) showing tilted by θ due to applied electric field and (b) rotating back to the original position for ease of calculation of $n_y(\theta)$

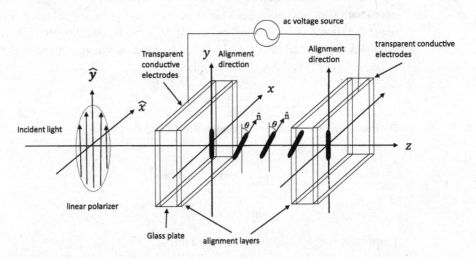

Figure 7.3-6 Phase modulation scheme using a PAL cell

Figure 7.3-6 shows a phase modulation scheme using a PAL cell. When a linearly polarized light is incident on a PAL cell with the polarization parallel to the direction of the alignment direction of the LC cell and in this case, the output y-linearly polarized wave is phase modulated with the term $2\pi n_y(\theta)d/\lambda_v$. In other words, if the input to the LC cell is $E_{inc} = \hat{y}E_0 e^{j\omega_0 t}$, at the exit of the cell, we have

$$E_{exit} = \hat{y}E_0 e^{-j2\pi n_y(\theta)d/\lambda_v} e^{j\omega_0 t}.$$

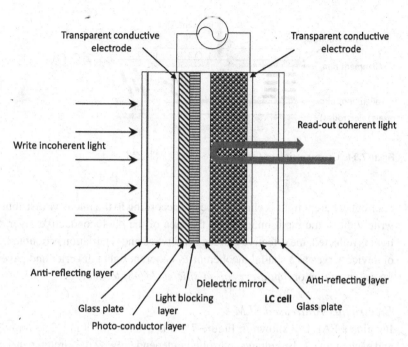

Figure 7.3-7 Structure of the optically addressed SLM

Therefore, the output light experiences a pure phase change and remains the same polarization at the exit of the LC crystal. The applied AC voltage tilts the dipoles governed by Eq. (7.3-2), and the tilt of the dipole changes $n_y(\theta)$ according to Eq. 7.3-5, thereby modulating the phase of the linearly polarized wave along $\hat{y}$. We have voltage-controlled phase modulation.

If the linear polarizer is rotated, the input linearly polarized wave can be decomposed into two orthogonal components, that is, along the x-direction and the y-direction, and the LC cell serve as a voltage-controlled wave plate, similar to the situation shown in Figure 7.2-6. If the linear polarizer is rotated 45° away from the y-direction and a crossed polarizer is placed at the output of the LC cell, we achieve voltage-controlled intensity modulation, reminiscent of the electro-optic intensity modulation system shown in Figure 7.2-11.

Liquid Crystal-Based Spatial Light Modulators (SLMs)
Optically Addressed SLM

Figure 7.3-7 shows the structure of the SLM. The AC voltage between the electrodes is to provide a bias electric field for the LC layer. The field across the electrodes is then varied according to the change of the impedance of the photoconductive layer upon illumination by the write incoherent light. In areas where the layer is dark, its impedance is high, and a small voltage is dropped across the LC layer. However, when the layer is illuminated, its conductivity increases and hence, the impedance decreases and the voltage dropped across the liquid increases. The phase of the

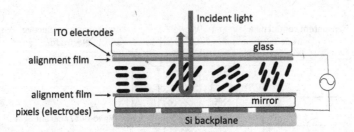

Figure 7.3-8 Cross section of the structure of a LCoS SLM

read-out coherent light is changed locally according to the intensity distribution of the write light or the input image upon the area of the photoconductive layer. Since the light is reflected, the effective cell length is twice and retardation is doubled. This type of device serves as a spatial incoherent-to-coherent light converter and paves the way for the invention of *liquid-crystal-on-silicon (LCoS) technology*.

Electrically Addressed SLM

Imagine a PAL cell shown in Figure 7.3-6 is a single pixel in a 2-D array of pixels, and each pixel can be voltage controlled independently by the circuitry in a computer. We have a 2-D spatial light modulator. A popular modern liquid crystal SLM known as liquid crystal on silicon or LCoS SLM is shown in Figure 7.3-8. Each liquid crystal pixel is built on a silicon backplane. The backplane is simply a printed circuit board containing circuitry connections to pixelated electrodes and allows for communication between all connected computer boards.

Problems

7.1 In this problem, we outline the procedures to obtain the Korpel–Poon equations from the scalar-wave equation in Eq. (7.1-25).

(a) Show that by substituting Eqs. (7.1-27) and (7.1-29) into the scalar equation in Eq. (7.1-25), we have

$$\nabla^2 E_m(\mathbf{r}) + k_0^2 E_m(\mathbf{r}) + \frac{1}{2} k_0^2 CS(\mathbf{r}) E_{m-1}(\mathbf{r}) + \frac{1}{2} k_0^2 CS^*(\mathbf{r}) E_{m+1}(\mathbf{r}) = 0,$$

where $k_0 = \omega_0 \sqrt{\mu_0 \varepsilon}$ is the propagation constant of the light in the medium. $E_m(\mathbf{r})$ is the phasor amplitude of the mth order light at frequency $\omega_0 + m\Omega$. In deriving the above equation, we assume $\omega_0 \gg m\Omega$, and hence we use $\omega_0 + m\Omega \approx \omega_0$ in the derivation process.

(b) Letting $S(\mathbf{r}) = Ae^{-jKx}$ [see Eq. (7.1-27)] and $E_m(\mathbf{r}) = \psi_m(z,x)e^{-jk_0\sin\phi_m x - jk_0\cos\phi_m z}$ [see Eq. (7.1-29)] along with the grating equation in Eq. (7.1-5), show that the result of part a) is given by

$$\frac{\partial^2 \psi_m}{\partial x^2} + \frac{\partial^2 \psi_m}{\partial z^2} - 2jk_0 \sin\phi_m \frac{\partial \psi_m}{\partial x} - 2jk_0 \cos\phi_m \frac{\partial \psi_m}{\partial z}$$

$$+ \frac{1}{2}k_0^2 CA^* \psi_{m+1} e^{-jk_0 z(\cos\phi_{m+1} - \cos\phi_m)} + \frac{1}{2}k_0^2 CA \psi_{m-1} e^{-jk_0 z(\cos\phi_{m-1} - \cos\phi_m)} = 0.$$

(c) Assuming the incident angle is small, that is, $\phi_{inc} \ll 1$, $\psi_m(x,z) \approx \psi_m(z)$, and $\psi_m(z)$ is a slowly varying function of z, that is,

$$\frac{\partial^2 \psi_m}{\partial z^2} \ll k_0 \frac{\partial \psi_m}{\partial z},$$

show that the result of part b) becomes a first-order differential equation in ψ_m, and we have

$$\frac{d\psi_m(z)}{dz} = -j\frac{k_0 CA}{4\cos\phi_m} \psi_{m-1}(z) e^{-jk_0 z(\cos\phi_{m-1} - \cos\phi_m)}$$

$$- j\frac{k_0 CA^*}{4\cos\phi_m} \psi_{m+1}(z) e^{-jk_0 z(\cos\phi_{m+1} - \cos\phi_m)}.$$

This equation is one variant of the well-known *Raman–Nath equations* [Raman and Nath (1935)].

(d) Using the grating equation, that is, Eq. (7.1-5), show that

$$\cos\phi_{m-1} - \cos\phi_m \approx \frac{K}{k_0}\sin\phi_{inc} + \left(m - \frac{1}{2}\right)\left(\frac{K}{k_0}\right)^2 + \cdots,$$

and

$$\cos\phi_{m+1} - \cos\phi_m \approx -\frac{K}{k_0}\sin\phi_{inc} - \left(m + \frac{1}{2}\right)\left(\frac{K}{k_0}\right)^2 + \cdots$$

(e) For small angles, that is, $\sin\phi_{inc} \approx \phi_{inc}$, and $\cos\phi_m \approx 1$, show that the result in part c) along with the approximations from part d), where we only keep the first term of the approximation expansion, becomes

$$\frac{d\psi_m(\xi)}{d\xi} = -j\frac{\tilde{\alpha}}{2} e^{-j\frac{1}{2}Q\xi\left[\frac{\phi_{inc}}{\phi_B} + (2m-1)\right]} \psi_{m-1}(\xi) - j\frac{\tilde{\alpha}^*}{2} e^{j\frac{1}{2}Q\xi\left[\frac{\phi_{inc}}{\phi_B} + (2m+1)\right]} \psi_{m+1}(\xi),$$

where $\tilde{\alpha} = k_0 CA/2$, $\xi = z/L$, and $Q = L\frac{K^2}{k_0}$. This is Eq. (7.1-30).

7.2 A laser beam of temporal frequency ω_0 is incident at the Bragg angle to the system of two Bragg acousto-optic modulators, as shown in Figure P.7.2. Two sinusoidal signals at frequencies Ω_1 and Ω_2 drive the two acoustic-optic modulators. Find the temporal frequency of laser beam 1, beam 2, and beam 3.

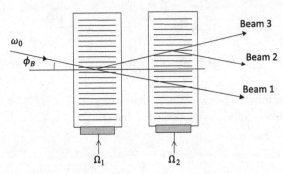

Figure P.7.2

7.3 Suppose that a red laser beam of wavelength 0.6326 μm is to be deflected by a dense-glass Bragg acousto-optic modulator. The highest acoustic frequency is 1 GHz. How much deflection (in degrees) is possible? Assume that the speed of sound in the glass is 7.4 km/s. In your calculation, do not assume the small-angle approximations.

7.4 Show that

$$\psi_m(\xi) = (-j)^m \psi_{inc} J_m(\alpha\xi)$$

are the solutions to

$$\frac{d\psi_m(\xi)}{d\xi} = -j\frac{\alpha}{2}[\psi_{m-1}(\xi) + \psi_{m+1}(\xi)]$$

subject to boundary conditions $\psi_m = \psi_{inc}\delta_{m0}$ at $\xi = z/L = 0$.

7.5 For near Bragg diffraction, Q is finite. Considering the incident angle is off the exact Bragg angle, that is, $\phi_{inc} = -(1+\delta)\phi_B$, where δ represents the deviation of the incident plane wave away from the Bragg angle, we have the following coupled equations:

$$\frac{d\psi_0(\xi)}{d\xi} = -j\frac{\alpha}{2}e^{-j\delta Q\xi/2}\psi_1(\xi),$$

and

$$\frac{d\psi_1(\xi)}{d\xi} = -j\frac{\alpha}{2}e^{j\delta Q\xi/2}\psi_0(\xi).$$

Show that the conservation of energy is satisfied by showing

$$\frac{d}{d\xi}\left[|\psi_1(\xi)|^2 + |\psi_2(\xi)|^2\right] = 0.$$

7.6 Given a beam of light that might be unpolarized or circularly polarized, how might you determine its actual state of polarization? You may use a waveplate and a linear polarizer.

7.7 A clockwise circularly polarized light is incident on an x-cut quartz ($n_0 = 1.544$, $n_e = 1.553$) of thickness d.

(a) Give an expression of the incident light field, assuming the x-cut crystal is located at $x = 0$.

(b) Design the thickness of the crystal d such that the relative phase between the z- and y-components of the electric field $\phi_{zy} = \phi_z - \phi_y = -3\pi$.

(c) Give the expression of the optical field after the crystal. What is the polarization state of the output field.

7.8 From Figure 7.2-7, we rotate the quarter-wave plate by $90°$ to achieve the situation shown in Figure P.7.8.

(a) Describe the wave field at the exit of the quarter-wave plate. Justify your answer with some mathematical deliberation.

(b) Describe what happens when the incident electric field is not at $45°$ but at $30°$.

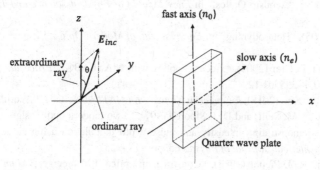

Figure P.7.8

7.9 (a) Find the electro-optic tensor of Lithium Niobate ($LiNbO_3$) and express it similar to Eq. (7.2-12).

(b) Find the equation of the index of ellipsoid in the presence of an external applied electric field $\mathbf{E} = E_z \hat{z}$.

(c) Assuming $n_0 r_{ij} E_z \ll 1$ and $n_e r_{ij} E_z \ll 1$, show that

$$n_x = n_y \approx n_0 - \frac{1}{2}n_0^3 r_{13} E_z \text{ and } n_z \approx n_0 - \frac{1}{2}n_e^3 r_{33} E_z.$$

This shows that when the applied electric field is along the z-direction, the ellipsoid does not undergo any rotation but only the lengths of the axes of the ellipsoid change.

Bibliography

Appel, R. and M. G. Somekh (1993). "Series solution for two-frequency Bragg interaction using the Korpel-Poon multiple-scattering model," *Journal of the Optical Society of America A* 10, pp. 466–476.

Banerjee, P. P. and T.-C. Poon (1991). *Principles of Applied Optics*. Irwin, Illinois.

Brooks, P. and C. D. Reeve (1995). "Limitations in acousto-optic FM demodulators," *IEE Proceedings - Optoelectronics* 142, pp. 149–156.

de Gennes, P. G. (1974). *The Physics of Liquid Crystals*. Clarendon Press, Oxford.

IntraAction Corp., 3719 Warren Avenue, Bellwood, IL 60104. https://intraaction.com/

Klein, W.R. and B. D. Cook (1967). "Unified approach to ultrasonic light diffraction," *IEEE Transactions on Sonics Ultrasonic* SU-14, pp.123–134.

Korpel, A. (1972). "Acousto-optics", in *Applied Solid State Science*, Vol. 3 (R.Wolfe, ed.) Academic, New York.

Korpel, A. and T.-C. Poon (1980). "Explicit formalism for acousto-optics multiple plane scattering," *Journal of the Optical Society of America* 70, pp. 817–820.

Phariseau, P. (1956). "On the diffraction of light by progressive supersonic waves," *P. Proc. Indian Acad. Sci.* 44, p.165.

Pieper, R.J. and T.-C. Poon (1985). "An acousto-optic FM receiver demonstrating some principles of modern signal processing," *IEEE Transactions on Education* E-27, No. 3, pp. 11–17.

Poon, T.-C. (2002). "Acousto-Optics," in *Encyclopedia of Physical Science and Technology*, Academic Press.

Poon, T.-C. (2005). "Heterodyning," in *Encyclopedia of Modern Optics*, Elsevier Physics, pp. 201–206.

Poon, T.-C. and T. Kim (2005). "Acousto-optics with MATLAB®," *Proceedings of SPIE*, 5953, 59530J-1–59530J-12.

Poon, T.-C. and T. Kim (2018). *Engineering Optics with MATLAB®*, 2nd ed.,World Scientific.

Poon, T.-C., M. D. McNeill, and D. J. Moore (1997). "Two modern optical signal processing experiments demonstrating intensity and pulse-width modulation using an acousto-optic modulator," *American Journal of Physics* 65, pp. 917–925.

Poon, T.-C. and R. J. Pieper (1983). "Construct an optical FM receiver," *Ham Radio*, pp. 53–56.

Raman, C. V. and N. S. N. Nath (1935). "The diffraction of light by high frequency sound waves: Part I," *Proceedings of the Indian Academy of Sciences* 2, pp. 406–412.

Saleh, B. E. A. and M. C. Teich (1991). *Fundamentals of Photonics*. John Wiley & Sons, Inc.

VanderLugt, A. (1992). *Optical Signal Processing*. John Wiley & Sons, Inc., New York.

Zhang, Y., H. Fan, and T.-C. Poon (2022). "Optical image processing using acousto-optic modulators as programmable volume holograms: a review [Invited]," *Chinese Optics Letters* 20, pp. 021101-1.

Index